A River No More

PHILIP L. FRADKIN

A River No More

The Colorado River and the West

Photographs by the Author

THE UNIVERSITY OF ARIZONA PRESS
Tucson, Arizona

About the Author

PHILIP L. FRADKIN, a resident of California since 1960, has reported for various newspapers in that state, including the *Los Angeles Times*, where he shared a Pulitzer Prize for coverage of the Watts riots. He served as assistant secretary of the California Resources Agency during the first administration of Governor Edmund G. Brown, Jr. Fradkin has taught courses in writing at Stanford University and been western editor of *Audubon* magazine. He is the author of *California, The Golden Coast* (1974).

THE UNIVERSITY OF ARIZONA PRESS
First printing 1984
Manufactured in the U.S.A.

91 90 89 88 87 86
7 6 5 4 3 2

Published by special arrangement with Alfred A. Knopf, Inc.
Certain passages in this book first appeared in somewhat different form in *Audubon* magazine and the *Los Angeles Times*.

Library of Congress Cataloging in Publication Data
Fradkin, Philip L.
   A river no more.

   Originally published: 1st ed. New York : Knopf, 1981.
   Bibliography: p.
   Includes index.
   1. Colorado River (Colo.-Mexico)—Description and
travel.   2. Colorado River (Colo.-Mexico)—History.
3. Colorado River Valley (Colo.-Mexico)—Description
and travel.   4. Colorado River Valley (Colo.-Mexico)—
History.   5. Fradkin, Philip L.   I. Title.
F788.F75   1984   979.1'3   83-18033

ISBN 0-8165-0823-2

For my father,

LEON HENRY FRADKIN,

who first brought me West

CONTENTS

MAP OF THE

COLORADO RIVER BASIN

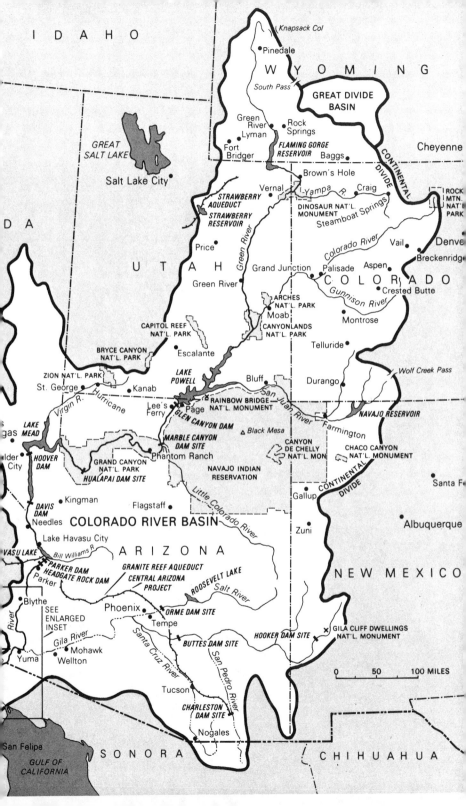

ACKNOWLEDGMENTS

Many people made it possible for me to write this book, which contains material gathered intermittently over the last ten years. During that time, particularly in the last four years, I either interviewed formally or chatted with some 250 persons about Colorado River matters, from a former Attorney General of the United States in New York City to a ditch tender in the Imperial Valley of California. Almost all were generous with their time and some welcomed me, a weary traveler on occasion, into their homes.

A few deserve special mention. Myron B. Holburt, chief engineer of the Colorado River Board of California and an acquaintance of ten years' standing, read the entire manuscript in draft form, as did Gary D. Weatherford, director of the John Muir Institute for Environmental Studies. Weatherford, a lawyer and former law professor, served as a special assistant to the solicitor for the Department of the Interior, a consultant to the Attorney General, deputy secretary of the California Resources Agency, and a senior investigator for the Lake Powell Research Project, funded by the National Science Foundation—all of which positions touched upon Colorado River matters. Gary, a good friend and co-worker while I was with the resources agency, encouraged and advised me along the way and traveled with me on foot and in a kayak to the end of the river. Another friend, Michael E. Bry, a photographer who printed the pictures for this book, accompanied me on two trips to the river. Holburt, an outstanding technical authority on the river and a staunch defender of the status quo, at least as far as California's interests are concerned, differed with certain conclusions and, like Weatherford, corrected inaccuracies. Needless to say, neither is responsible for the final product.

I would like to thank the members of one group too numerous to name individually: all those librarians from Los Angeles and Berkeley to Denver and on to New York City and Washington, D.C., and from Yuma, Arizona, to Pinedale, Wyoming, who invariably conducted me with care and interest through their individual realms. Those who follow should know that the library and files maintained by the Colorado River Board of California in Los Angeles is the single best depository of material on the subject, having been put together with thought and care. But by no means is that library's collection all-inclusive. Probably no other single natural feature in this country has attracted so many written words, mostly on the technical and legal levels. The abundance of material is staggering, and my main problem was deciding what not to use.

There were others. With the cooperation of former Interior Secretary Stewart L. Udall, I was the first researcher to peruse his valuable collection of papers at the University of Arizona. His brother, Morris, an Arizona congressman and chairman of the House Interior Committee, let me read his raw files pertaining to President Jimmy Carter's attempt to cut back on water projects, especially the Central Arizona Project, in 1977. The papers of Wayne N. Aspinall, the former chairman of that same House committee, at the University of Denver and those of the late Senator Carl T. Hayden at Arizona State University were also quite helpful.

A series of black-bound notebooks kept by the Imperial Irrigation District contained a wealth of material on the 160-acre-limitation issue. I obtained documents that dealt with the salinity negotiations with Mexico in the early 1970s from the personal files of Herbert Brownell, who negotiated the agreement for the United States. The former ambassador and Attorney General was generous with his time. I had to file a formal Freedom of Information Act request with the Department of State before receiving an avalanche of documents on the same subject. From the Department of the Interior and its various subagencies I also obtained heretofore unpublished material, some of it by informally invoking the provisions of the act. I was helped with these requests by various departmental information officers and others.

There is a group of nameless bureaucrats who never get credit for what they write. They are the authors of the many technical reports and environmental-impact statements that are the single most important source of information on the contemporary West. I want them to know that their efforts are gratefully appreciated.

Lastly, I owe a great debt to the three vehicles, two of which were equipped with four-wheel drive, that transported me over 75,000 miles of the West without breaking down. In some places, had that happened, it would have been days before I could have reached civilization. My travels were also on foot, skis, horseback, kayak, wooden dory, rubber raft, helicopter, small plane, and commercial airliner. Martin Litton twice arranged for me to run the Colorado River within the Grand Canyon in his dories. I hesitated to write about a place until I had seen it and spent some time there. But I did not travel every mile of the river system. Nor did I gather the material on any one trip taking in the whole Colorado River basin. The climatic extremes and distances are too great, and the issues too complex, for such a single sweep. The material from many different trips was combined in the text, though, in an attempt to give a sense of narrative continuity.

I know of no one else who has made the journey—at times it seemed like a pilgrimage—to the headwaters of the four main tributaries and the ends of the Colorado. But I strongly suspect someone has done it before, and simply chosen not to write about it. I would hope that this person, or persons, was a user of Colorado River water who one day was struck with curiosity about this unifying source of life in the West and set off, as I did, to learn a little more about it.

P.L.F.

Why another book on the Colorado River? people sometimes asked. It would invariably turn out the questioner was thinking about the Grand Canyon. There have, indeed, been many books written about the most spectacular natural feature the river passes through, in a relatively short portion of its total length, but there have been very few books on the entire river and none on it as an organic whole that succors most of the West.

Four previous books deserve mention. *The Grand Colorado,* edited by T. H. Watkins, concentrates on the history of the river and the land surrounding it. *Water and the West* and *Dividing the Waters,* both by Norris Hundley Jr., deal respectively with the 1922 Colorado River Compact and the 1944 treaty with Mexico, the two key legal documents that apportion the water. Because of the excellence of these three books, I have chosen not to go into great depth in these areas. I recommend them for those persons seeking a fuller understanding of the subject. On the other hand, *The Colorado* by Frank Waters is a disappointment. I quote Francis P. Farquhar, who wrote in his *The Books of the Colorado River and the Grand Canyon,* after referring to Waters's chapters on the Imperial Valley and the delta: "The rest of the book is largely irrelevant or deals with subjects treated better by others." Because of Waters's cursory treatment and dated material—the book was written thirty-five years ago—I decided there was room for another book on the river. Undoubtedly someone after me will decide the same, because the subject is inexhaustible and has a certain continuing drama.

It was my aim to sketch the background, give a sense of place and people, define the issues, set forth the problems, and offer a few thoughts about the continued availability of Colorado River water and the viability of the West—all the time emphasizing the politics of natural resources. To me the river, in its present state, is primarily a product of the political process, whether conducted in Salt Lake City or Washington, D.C., rather than a natural phenomenon. The policies and laws that determine where water goes mean life itself in this dry region—not only life but death, as the river has been depleted to serve the lands and people surrounding it. This oasis civilization will ultimately face that same process of withering when shortages occur in western water supplies in coming years.

I embarked on this project first out of a sense of curiosity and then growing alarm, heightened by the recent drought in the western states. The book offers no specific solutions, but it is my fervent hope that it will help to increase public awareness of the importance of water to the West, and of how its distribution can be used as the single most effective tool in this arid region to control and direct the rampant growth. If the debate is widened beyond the western water bloc, the conservationists, and even the few writers who have participated so far, then the effort will have been worthwhile. For too long, a narrow group of men has distributed the ultimate source of wealth in the West with diminishing wisdom. It is my hope that from such a widened debate fair and rational solutions will emerge.

Philip L. Fradkin
Inverness, California
June, 1980

A River No More

CHAPTER ONE # Watershed:
A Separate Totality

1

On February 21, 1977, President Jimmy Carter sent a message to Congress declaring his Administration's intention to review the necessity for nineteen water projects already authorized but awaiting funding for the next fiscal year. So began what was briefly referred to as the "War on the West" and what will long be remembered in that section of the country as the first serious challenge by a president to the West's primal shibboleth—its essential aridity, and the need for dams and ditches to assure a dependable living and some measure of prosperity in a dry land.

In a relatively short time President Carter and his environmentally-minded, cost-conscious aides would leave their policy-making positions; but the nearly one-hundred-year-old western institution they had tinkered with—reverently referred to as Reclamation and written with a capital R—would remain. True, it would never be the same again, but this would be less a function of what Carter had accomplished than of the forces he had unleashed. For the last quarter-century, what had once seemed an unassailable concept—the idea of structural solutions for the region's aridity—had increasingly been challenged on environmental, economic, and social grounds. Dams destroyed beautiful places. Their costs could exceed benefits. Indians were not getting their fair share of water. But the West's water establishment, with strings running from the smallest water district to those congressional committees that act almost invisibly on natural-resource matters, had successfully circumvented these growing challenges. Dams and aqueducts were necessary for growth. Dams offered protection against floods, had recreational advantages, and, along with the reservoirs they formed, possessed a certain beauty.

Jimmy Carter's contribution was to bruise this water establishment. That this happened in the midst of one of the West's periodic droughts proved it doubly vulnerable. Partly because of the durability of the long-cherished institution and the Carter Administration's heavy-handedness, there was no clearcut victory; but for the first time such an effort, which can be politically costly with few tangible gains, had actually been made by a president. Historically, Congress has ruled on water projects with only an occasional dissent from the Chief Executive's budget watchers. Conceivably, President Carter's example could make it easier for subsequent administrations to act on water matters, and thus they would have a greater hand in directing the future of the West as such national priorities as increasing domestic energy production became more important. The arid West has always feared executive dominance, and the sparsely settled interior states have successfully worked with their sympathizers in Congress to thwart it. Certainly the region, as symbolized by the Colorado River and the lands it succors, was approaching a new era in which the ultimate limits of what had always been considered a limitless frontier were in sight. Within a few more years, perhaps twenty or so, there was not going to be enough water to fulfill everybody's desires. The river was running dry.

The furor that arose in the West and the halls of Congress following the President's announcement had all the angry and fearful characteristics of a disturbed bull, an image that was frequently applied to the Colorado River in its pre-dam state. One early river historian referred to it as "a blooded bull," and a later writer described the river as "a wild bull of destruction." To a Bureau of Reclamation official, it was "an angry bull —slashing its way through canyons and across deserts to the mother sea. . . ." In a similar manner the western states and their elected representatives reacted to what became known as the "hit list," a term with gangster connotations. The Western States Water Council, which represents western governors on water matters, stated with careful restraint, "Officials in the West appeared shocked by the proposed budget announcement and expressed plans to fight the decision." Representative Morris K. Udall of Arizona, chairman of the House committee with jurisdiction over such water projects and contender for the 1976 Democratic presidential nomination, declared that he was "shocked and angered" at the inclusion of the Central Arizona Project on the list. Udall, along with seventy-three other congressmen, had sent a letter one week earlier to the President stating "our support for your efforts to reform the water resource programs of the Army Corps of Engineers and the Bureau of Reclamation," the two principal federal dam-building agencies. They never imagined such a massive, potentially disruptive response from the Administration. Old animosities were instantly reawakened. Arizonans muttered about a California plot to steal their water, while California agencies attempted to reassure the neighboring state about their continued solidarity on water matters by passing resolutions asking for completion of the Central Arizona Project. A whole interlocking structure of common allegiances, carefully constructed over the last few years, was threatening to unravel.

Later the President would admit he had made a mistake by not discussing the proposed budget deletions with the affected lawmakers. Administration aides looked back at the hastily compiled list, and the manner in which it was released, as a public-relations disaster. Barely a month after taking office, the Carter Administration had precipitated its first major confrontation with Congress over nineteen obscure water projects (four of which were in the Colorado River basin) and put its entire domestic program into jeopardy. Some powerful members had been offended. It was time to compromise.

If President Carter had misread the West's tenacious resolve to hold onto its prime legislative interest—the disposal of water as it wished—so had the region and its elected representatives neglected to read the signs of what was obviously going to happen. In regard to water projects, Carter was an unusual chief executive in office at an unusual time. When the list was announced, there was no overwhelming foreign or domestic crisis. The President could afford to spend some time on water projects. Other presidents who served at crucial times in the development of the West and were knowledgeable about water policies were diverted from influencing them—Herbert Hoover, who had helped negotiate the Colorado River Compact of 1922, by the Great Depression, and Lyndon B. Johnson, who had hauled some irrigation pipe in his day, by the Vietnam War. Jimmy Carter, although he was from a state with an average annual rainfall of forty-eight inches (seven times the amount of Arizona), had an interest in water projects and the luxury of having no crisis to impede his doing something about it. He was, to say the least, skeptical of them, as were other key members of his Administration. Interior Secretary Cecil D. Andrus, while governor of Idaho, witnessed the immediate aftermath of the Teton Dam disaster. Charles Warren, who headed the President's Council on Environmental Quality, had strong environmental credentials, as did Kathy Fletcher, a member of the transition team and later of the White House domestic staff.

Environmental concerns rated high with President Carter, at least at the start of his term, and he possessed the rational, analytical mind of a civil-engineering graduate of the Naval Academy. He had not served in Congress, thus missing the camaraderie involved in passing on public-works projects, more commonly and somewhat mistakenly known as pork-barrel politics. (Some of the projects are definite necessities.) Besides, the President had vivid memories of a dispute with the dam builders while governor of Georgia.

For a long time there had been plans to construct a dam at Spewrell Bluff on the Flint River. Those plans by the Army Corps of Engineers solidified in the early 1970s, as did the opposition to the dam. Since the corps had a policy of not building a dam if the governor of a state did not want it, Governor Carter's decision was crucial. He took two canoe trips and a helicopter ride down the river and personally studied all the reports the project had generated, including a report by the General Accounting Office, the investigatory arm of Congress. This report seriously questioned the methods the corps had used to compute the dam's cost-benefit ratio. On October 1, 1973, Carter issued a fifteen-page hand-written statement opposing the dam. The determination of feasibility, Carter declared, was "based on incorrect data and unwarranted assumptions." He urged that Congress and responsible agencies investigate the bias of the dam builders. "The construction of unwarranted dams and other projects at public expense should be prevented," he stated.

The corps tried an end run around the governor, something no politician likes, and sought approval for the dam from the Georgia legislature. In this way the Engineers sought to pressure Carter into changing his mind. But despite some intense lobbying by the corps and other development interests, the state senate voted 27 to 23 against the dam. One year later Carter, who by then was considering the presidential race and was a hero to antidam interests, spoke at a Los Angeles rally against a California dam project. "Similar distortions exist in the New Melones project," he stated. While campaigning for the presidency during a time of concern about inflation and high taxes, Carter said there was no need to spend "tens of millions of dollars" on such projects. He foresaw the end of the dam-building era, pointing out that "most beneficial projects have been built." On New Year's Day of 1977 the Associated Press carried a story about sixty-one water projects being picked by the incoming Administration for reevaluation. It was these pronouncements that had inspired the letter of support from the congressmen. Clearly, they had had no idea the list would, at least for some, strike so close to home.

So when the two top officers in the Army Corps of Engineers later testified on the hit list before a House appropriations subcommittee, they simply read a statement prepared by the White House and, smiling, sat back as one congressman after another attacked the President's position.

"Before the President pulled his new economic criteria out of the sky to make these projects look bad, they were all economically justified, weren't they?" asked one congressman sympathetic to the water projects.

"Yes, sir," replied the major general.

During the first few months of 1977 the West experienced the second of two extremely dry years. A map of the eleven western states, issued on March 1 by the Soil Conservation Service, forecast river runoff between o and 60 percent of normal for most of the area. The warmer tones of red and pink were used almost solidly over most of the region to depict river flows far below average; there were only a few splotches of cooler purple and light blue indicating 60 to 90 percent of average flows. For a traveler in the West, the land felt the same way: a few oases in the midst of crackling dryness. It is the flow of streams from the annual spring snowmelt that counts in the interior West, not rainfall in the deserts. Thus, southern California and portions of Arizona would have above-normal rainfall; but that would not come near to equalling the much greater amounts of precipitation that normally fell in the mountains along the Continental Divide in the form of snow, to melt in spring and tumble down, to be trapped behind numerous dams for release and use in the dry summer and early fall months. From the snowmelt, the West gets 70 percent of its water supply. Farmers cultivating irrigated fields, who account for between 85 and 90 percent of western water use, do not want rain, particularly in the hot desert areas. It is too unpredictable. Irrigation water can be delivered and applied when needed.

So it was the snowpack that mattered in the first few months of 1977, and as the winter began to end there was precious little of it in the mountains; a high-pressure system off the California coast continued to deflect the "normal" winter storms. What many westerners had forgotten —had perhaps never been aware of, since most were not around during the Dust Bowl years of the mid-1930s—was that dry years were part of the normal weather cycle. For the users of Colorado River water, this would be a crucial factor, since the river's waters had been divided in 1922 on the basis of records kept over a series of relatively wet years. Although some would use the drought as an argument for more dams and ditches and others would use it to claim that the great system of water-works that links the West saved the region from disaster, the prime lesson of the drought was that it clearly demonstrated the region's vulnerability to the lack of water, especially should the dry cycle continue or intensify and the demands for more water keep escalating. Novelist and essayist Joan Didion, that deft chronicler of life on the West Coast, wrote of its quintessential dryness, "The apparent ease of California life is an illusion, and those who believe the illusion real live here in only the most temporary way." *Time* magazine said of the drought, "It raised once again basic questions of how the nation should use one of its most vital resources, just how much population growth the available water can sustain." In the arid and semiarid West, water is *the* most vital resource. In the early months of 1977, as the drought worsened and the Carter administration attempted to cut back on water projects, people began to realize this basic fact. The sense of a civilization's vulnerability was extremely disquieting at the time.

Most of the attention concerning the drought focused on California. It had the largest population, the greatest amount of agricultural products and industry, the worst dryness, and the reputation for the most profligacy among western water users. It also came up with the most imaginative solutions—water-rationing programs that cut consumption more than 50 percent in some areas, and a massive water trade, whereby critical shortages in northern California were alleviated by increased use of Colorado River water in southern California. The trade allowed northern California water that would normally have flowed south through the State Water Project to be diverted to the north. Thus, Marin County, just north of San Francisco, which came to epitomize the extremes of the urban drought situation, was rescued by the figurative use of Colorado River water—it could not physically flow uphill through a nonexistent aqueduct to northern California, but it could be traded on paper. Southern California then took more than its normal share of Colorado River water. Meanwhile, the runoff in the Colorado River basin hit a new record low, 40 percent of the average flow for the past fifty-six years as measured at Lee's Ferry. But, at least for the first and, as it turned out, the last year of the water trade, this did not mean much, because the river's reservoirs were nearly full in anticipation of years to come, when the demands on the Colorado would dramatically rise.

What the drought proved was that water could be consumed in amounts significantly less than normal; and there would be no great alteration in the western way of life. When the ultimate pinch came, the drought being just a precursor, separate administrative entities could cooperate for their own common welfare. Priorities could be established, although they would involve hard choices and some sacrifices. Acrimony could be expected, along with some selfish use of water, symbolized in that drought year by the filling of an artificial lake in a southern California subdivision. All the vast schemes for the importation of water into the arid West—put away during those cost-conscious and environmentally sensitive years of the early 1970s—would again surface and increase the level of contention. Towing giant icebergs from the Antarctic, shipping snow west from the inundated cities of the Northeast, and tapping the Yukon River in Alaska were some of the wilder schemes discussed that crucial year.

The mind boggled at the additional waterworks involved in such plans. Above all else, the Year of the Drought showed the West to be one big, complex plumbing system. That is the reason for its continued existence. The power and the glory, not to mention the money, center around water and the means used to convey it. Woe to any president who tries to cut back this system. It consists of ditches, flumes, penstocks, dams, canals, laterals, pipelines, aqueducts, and more ditches laid across nobody knows how many tens of thousands of miles that bisect mountains, deserts, farmlands, and cities. It represents billions of dollars of water projects and a political system to procure them that has yet to be successfully thwarted, although President Carter came closest. These basic facts tend to become lost in a normal year, particularly to urban populations, but 1977 was not a normal year in the West. It was a watershed year, a harbinger of what could come to be regarded as normal.

It was time to compromise. Besides, as the Administration's review teams worked through the early spring months of 1977 to finalize their recommendations on the water projects, there were certain interests at work within the government that were not in sympathy with what the President wanted. They were the bureaucrats in the Corps of Engineers and Bureau of Reclamation whose long careers had been molded by the structural concept of water development. They were not bad men. Disloyal to a president, yes, but not to a system, a cause, a habit named Reclamation that surely was of longer duration than one presidential administration. They were loyal, too—loyal to their constituencies: the states, the irrigation districts, the farmers, energy producers, and land developers. "Uninformed amateurs," Ellis L. Armstrong called the meddlers in the new Administration. Armstrong, who served as commissioner of Reclamation under Nixon, had retired to his home state of Utah. The amateurs had disrupted the harmony of a partnership between the federal government and western states that assured the states all they wanted in terms of water development, which meant everything in the West.

It was the harmony of this partnership that President Carter had upset and the western states wanted to restore. The partnership had been dominated in the past by the states, or, to be more precise, by those persons within the states who took a continued, active interest in water matters. At times the federal government seemed little more than an errand boy, rather than the regulator and comprehensive planner and builder it should have been. Certain members of the departmental review teams, which out of necessity included bureau personnel knowledgeable about the projects they had designed, fed information on the progress of the teams to congressmen who were sympathetic to the threatened water projects. Thus, an official in the Phoenix office of the Bureau of Reclamation passed information on the Central Arizona Project review team's findings to a state water official, who passed it on to an aide of Congressman Udall, who wrote a memo to his boss, who made sure the project was not scrapped. The review team's unfavorable conclusion on the cost-benefit ratio could not be countered "without exposing the assistance" of the bureau man, stated the memo. It was Udall who led the fight to save the billion-dollar-plus project designed to rescue Arizona from water bankruptcy. He did not rebut the Administration's economic arguments but pleaded eloquently: "Water is life in the desert. We have not always used our water wisely, but we are moving to correct our mistakes. The Central Arizona Project is a very old dream. I first heard of it from my grandfather." Officials in the Department of Interior, the parent agency of the bureau, were not unmindful of such internal espionage. They tried to exclude bureaucrats suspected of being more loyal to past concepts than the position of the present administration from the policy-making level of the review process. It was a time of great mutual distrust.

In mid-April a compromise was reached on the Central Arizona Project. It was the product of an intensive lobbying effort by Udall and others from Arizona and was reminiscent of a similar but much more prolonged lobbying effort that finally resulted in the project's authorization in 1968 —twenty-one years after the first bill for its authorization had been introduced in Congress. Such projects have long histories and deep roots in the West. Backing the continued construction of the project was the entire Arizona congressional delegation, six men who would have been committing political suicide in that water-short state had they not supported it. Senator Barry Goldwater, who traditionally held himself aloof from the nuts-and-bolts of water matters, declared, "It has to be built, and if it's not, this valley is going out of business." State and local officials, university professors, labor leaders, farmers, and miners joined the bandwagon. With the exception of Utah, Arizona has at certain times come the closest of the states in the Colorado River basin to speaking with one voice on water matters. The opponents were only a handful of environmentalists and Indians. The Indians opposed the location of one dam but wanted the benefits of increased water. The state's strategy was to stress the dire need for the project and the overall benefits for the Indians, an argument that was judged to move the sympathies of easterners. Rather than "raising Cain" with the President, it was decided to educate him.

Some also wanted a study on the importation of Columbia River water into the Colorado River basin, since a ten-year moratorium on such a study was due to expire soon. But this move was suppressed, as it had been in 1968, in order to get the support of the Northwest's congressional delegation. History was repeating itself. The compromise that was struck was to modify the project. Three of the four dams would be temporarily taken out and given more study. The state would do something about its diminishing groundwater supplies, but work would continue on the project. Nothing crucial had been tampered with, and a later administration could always turn things back to where they had been. Arizonans saw the outcome of the review process as a victory. Said Udall, "Today a President with strong environmental views has decided that CAP should be completed. I believe this will end the debate and let Arizona move ahead."

In announcing his decision April 18 on the outcome of the water project review, the President stated: "In the arid West and across the entire nation, we must begin to recognize that water is not free—it is a precious resource. As with our energy problem, the cornerstone of future water policy should be wise management and conservation." Two projects in the Colorado River basin, the Fruitland Mesa Project in Colorado and the Savery–Pot Hook Project straddling the Wyoming-Colorado border, were recommended to Congress for no funding. The Bonneville Unit of the Central Utah Project was modified, as was the Arizona project. Although the President wanted all or portions of the four projects in the basin deauthorized, the worst that would happen would be that some or all of their funds would be diminished for one fiscal year. Next year the budget process would begin all over again. Of the two-step process needed to get a water project completed—authorization and yearly funding—Congress would only go along reluctantly with the latter, dragging its feet all the way.

A few days later Senator Russell B. Long, a Louisiana Democrat who was chairman of the Senate Finance Committee, had lunch with the President. After the meeting, Long threatened: "We can amend some bills he very much wants to sign to encourage the President to do what we think he ought to do. I have suggested to him further than that, but I have no response from him. He ought to work out some compromise of this issue. Otherwise he is going to find himself at war with his own best soldiers in Congress." The stakes were getting higher. Funding was restored for two projects in Long's state and for the Central Utah Project, but was deleted in the bill signed by the President for the smaller Fruitland Mesa and Savery–Pot Hook Projects.

The next year, reacting to the continued tenacity of the West, the President would veto the public-works bill. Of the previous year's compromise he said: "There have been some cases where I have erred on the side of not vetoing a bill. I think that last year I should have vetoed the appropriations bill that authorized unnecessary water projects. If I had it to do all over again, I would have vetoed it. But that is one of the rare occasions when I think I have been too lenient in accommodating the desires of Congress. But the Congress is now trying to reimpose those water projects on me as President, and even additional ones, that are worse." Besides the two perennial contenders for the list from the Colorado River basin, Fruitland Mesa and Savery–Pot Hook, two units of the Central Utah Project and the Animas–La Plata Project in Colorado and New Mexico had not measured up to the standards of the Carter Administration. Fruitland Mesa would have benefited sixty-nine ranchers, at an investment of $1.2 million per landowner. For Savery–Pot Hook, it was a $700,000 investment per ranch. The three other projects did not have all the necessary paperwork done, nor had repayment contracts been signed. For the two units of the Central Utah Project, their final environmental-impact statements had not been filed, nor had their feasibility been determined.

Congress failed to override the veto, but in a little-noticed action the same day it passed an amendment to a Department of Interior appropriations bill negating a suit filed in federal district court by three environmental organizations. The suit, citing various well-known precedents, sought a comprehensive environmental-impact statement for the whole Colorado River basin. The Administration agreed that such a statement was necessary, had actually started work on one before Congress refused to fund it, but made no move to block the amendment. Congressmen from other river basins saw the danger in such a precedent—it could result in a further erosion of their control over water projects. As one western congressman put it: "Now, if we do allow a study on the entire river basin, who is going to control the Colorado River? Either the Department of Interior or the President's Domestic Council, one or the other—not the States. And such control will rule regardless of the interstate compacts and regardless of the laws this Congress has passed in the development of this river area." Another consequence of such a comprehensive statement could have been a definitive discussion on when shortages in the Colorado River would occur, what could be done about them, and who would suffer. Since no politician knowingly wants to be the vehicle for bringing bad news, it was decided that such a determination was better left unstated. The Colorado River system would remain fractured, with no one official document stitching it together.

Along with singling out specific projects as being unworthy of funding or in need of modification, the Carter administration set out to establish some standards for development that came to be known as water-policy reforms. Such an attempt was seen as a reform of the system that produced individual water projects, and vastly more important or dangerous —depending on one's viewpoint—than the narrower issue of what was built where and how. More public hearings were held and more fears were expressed in the western states that watershed year. Meeting in Anchorage, Alaska, in early September, the western governors vented their anger and fears on Interior Secretary Andrus, who had once been one of them—a point that was repeatedly flung in his face as he tried to explain the Administration's position. One after the other, the governors publicly berated Andrus. "When you're talking about water," Governor Scott M. Matheson of Utah lectured, "you're talking about the most important, finite resource we have, other than our human resource." As the prospect of another snowless winter approached, Andrus reflected: "I think what it is, is an underlying fear on the part of the arid West— and you have to designate arid West because it doesn't prevail in western Oregon, Washington, Alaska, or northern California—that some unforeseen hand is going to reach out and turn off the valve."

The next month, in a Denver panel discussion on water policy, President Carter attempted to reassure his audience: "I want to make clear from the very beginning that there absolutely will be no federal preemption of state or private prerogatives in the use or management of water. This is not the purpose of the policy at all." The crux of what the Administration eventually proposed was that the states pay 10 percent of any federal water project. The Administration publicly tried to sell what seemed like a reasonable proposal on the grounds that it would give the states "a more meaningful role," as President Carter put it, in the decision-making process. But privately key aides said it was an attempt to break up the "Iron Triangle"—the Bureau of Reclamation, Congress, and water users—by injecting a fourth element, a vote involving expenditure of money on the state level. Such a proposal would need the approval of Congress, and that did not seem to be immediately forthcoming.

The rains began in September of that crucial year; most areas of the West that had experienced the worst drought conditions received double their normal precipitation for that month. The forecast for October was for much of the same, but a lot of snow would have to fall in the mountains that winter to fill the reservoirs next spring, and nobody could yet be sure the drought was over. By early January of 1978 the Soil Conservation Service, which conducts snow surveys in the western mountains, expressed guarded optimism on the possibility of normal water supplies. By March, with most of the snow season past, it looked very good. There were floods in Arizona, full reservoirs in California, and above-normal snowpack along the Continental Divide in Wyoming and Colorado. On April 6, 1978, Commissioner of Reclamation R. Keith Higginson announced the two-year drought had ended for most areas in the West. One week later the Bureau of Reclamation forecast 138 percent of normal runoff for the Colorado River system above Lake Powell.

In the waning days of 1977 the Los Angeles City Council began to lift the water-rationing controls it had imposed on the largest city in the West. It had been a reluctant imposition of a 10 percent cut to begin with, since the Colorado River seemingly offered all the water that was needed. Mandatory rationing ended in Marin County in January after residents in those affluent suburbs of San Francisco had achieved a 65 percent reduction in water use. Consumption then began to rise slowly in what had been the water-short areas of the state. A poll conducted for the California Department of Water Resources indicated that 60 percent of the population thought it was "extremely important" or "very important" to continue to conserve water. There was less enthusiasm about saving water in southern California than northern California. And so ended that watershed year.

2

Down through the years the image of a river has frequently been used as a metaphor for life. Rivers have been depicted as life-giving forces that renew the fertility of a land and people, and their waters have been considered sacred while their sources were shrouded in mystery. Witness the continued search for the start of the Nile River in the nineteenth century, and the medieval belief that rivers flowed magically from the center of the earth. To float down a river is to experience its purifying aspects, as did Huckleberry Finn on the Mississippi and as do those now seeking controlled adventure on the Colorado through the Grand Canyon. To cross a river is to take a significant step, as Julius Caesar did when he came to the Rubicon and as the first Spanish explorers retreated from doing when they timidly approached the Colorado in 1540, thus delaying the exploration of the interior West for more than two hundred years. The Colorado, which came late into recorded history, was given an aura of antiquity by early references to it as "the Nile of America." E. C. La Rue, the first to make a comprehensive study of the Colorado River system, compared the two rivers in his 1916 report, *Colorado River and Its Utilization.* Both rivers carried substantial amounts of silt that piled up in large, fertile deltas. The crops grown in the two regions were similar, as were their climates. But the Nile drained a basin of 1,112,000 square miles, while the Colorado River basin was 244,000 square miles. The Nile took 4,000 miles to get to the sea, dropping 6,600 feet along the way, while the Colorado stretched only 1,700 miles but descended 14,000 feet. The average annual flow of the Nile, La Rue recorded, was four times greater than that of the Colorado. Whatever their other differences and similarities, however, the most salient single fact about these two great river systems was the vast potential of their waters for giving life to the arid lands that surrounded each of them. Rainfall was scant and irrigation a definite necessity both along the Nile Valley and in America's West.

The Aswan Dam would be built on the Nile to smooth out and store its flows. By the mid-1960s, when all the principal dams had been completed on the Colorado, it too would have a year-round tranquil flow to match the placidity of the Nile and the classical concept of a river. "Tamed" and "deficient" would be the terms now most frequently applied to the Colorado, as if the river, not the plans and works of those who used its waters, were to blame for its shortcomings. In ancient times, to stop a river or reverse its flow was regarded as the ultimate display of power, or nature gone awry.

There are many things the Colorado River is not. It is not the largest or the mightiest river. The Nile is longest and the Amazon carries the most water and has the largest watershed. The flow of the Colorado is about equal to that of the Delaware River, although the western river drains a much larger area. Another way to put it is that the Columbia River basin, seen by some as the ultimate salvation of life further south, is about the same size as the Colorado watershed, yet its *unused* flow is about twelve times greater than the virgin flow of the Colorado. If the average yearly flow of the Colorado were evenly spread over the upper Colorado River basin—essentially those parts of Utah, Wyoming, Colorado, and New Mexico drained by the river—the water would be 2.22 inches deep. The upper Columbia River basin would be covered by 15.7 inches of its own water, and the Delaware River basin by 20.9 inches. Such are the different statistics of arid and humid regions.

But the Colorado is the most used, the most dramatic, and the most highly litigated and politicized river in this country, if not the world. "The Colorado basin," the National Academy of Sciences cautiously stated in 1968, "is closer than most other basins in the United States to utilizing the last drop of available water for man's needs." This was written shortly before Congress that same year passed the Colorado River Basin Project Act that provided the structures, among them the Central Arizona Project, that would squeeze the last drops of water out of the river. Norris Hundley, Jr., a historian of Colorado River law, wrote: "As a result of the various demands placed upon the river's flow by the seven states and

Mexico, the Colorado has become one of the most litigated, regulated, and argued-about rivers in the world." The Bureau of Reclamation, the federal agency in charge of the river's development and operation, referred to it thus: "The Colorado River is not only one of the most physically developed and controlled rivers in the nation, but it is also one of the most institutionally encompassed rivers in the country. There is no other river in the Western Hemisphere that has been the subject of as many disputes of such wide scope during the last half century as the Colorado River. These controversies have permeated the political, social, economic and legal facets of seven Colorado River Basin states." The most eloquent documentation concerning the river's great use is the fact that, except for occasional local flood flows, no water has reached the Gulf of California, the river's historic outlet to the Pacific Ocean, in the last twenty years.

The Colorado also carries the most silt and has the most damaging salinity problem of any river in this country, but possesses little of those industrial and municipal wastes that are commonly associated with river pollution. It also has the greatest evaporation rate and warmest·waters. Navigation has played only a minor role in the river's history and nowadays there are few losses from floods, two characteristics which set the Colorado off from rivers further to the west and to the east. The Colorado is, in short, a different kind of river flowing across a different land that has evolved institutions peculiar to its own needs. As historian Francis P. Farquhar wrote in *The Books of the Colorado River and the Grand Canyon:*

> There is a unity about the Colorado River Country that is established by the River itself—always the River. The topographical features, although on a vast scale, have a simple relationship to the central controlling element. . . . And just as the River has formed the landscape so has it determined the course of human history within its basin.

The river's waters and the land surrounding it in the basin—the heartland of the West—are fused together in a common destiny, as are those areas outside the watershed to which Colorado River water is diverted—southern California, Salt Lake valley, Colorado's Front Range, and the Rio Grande valley in New Mexico. The quantity and quality of the river's flows are a mirror image of what is upon the land—indeed, are the prime reason for there being something built upon or scratched out of the soil in the first place. How easily this is forgotten in the urban areas of this oasis civilization. Not the Rocky Mountains nor the Pacific Ocean, but the Colorado River which flows from one toward the other, is the single most unifying geographical and political factor in the West. The river has been the most significant catalyst in the politics of the West since the turn of the century. The reason why there was a near-hysterical reaction to President Carter's "War on the West," why the war evolved into the "sagebrush rebellion," an attempt by the western states to gain control over federal lands, was not only because it involved water, but also because it encompassed all those other essential western activities that depend upon it, like livestock grazing, mining, energy production, and recreation. Going back further, it was not the six-gun or barbed wire that won this part of the West (although they helped) but the tools that evolved from the Indian's digging stick to the shovel to the giant earth-moving equipment that could be used to build the structures needed to store and move water.

It was not only in the present century that large amounts of water were diverted onto arid lands, causing problems that seem to be endemic to civilizations in dry lands that depend on storing and transporting water some distance from its source. There are precedents going back to the cradle of civilization, back to the Garden of Eden—which was located in ancient Mesopotamia and had a river running through it that was probably used to irrigate that infamous fruit tree. The Tigris-Euphrates valley is the oldest cultivated area in the world and contains a record, still visible in the ruins of ancient canals, of man diverting water for his own benefit dating back at least six thousand years. The Euphrates River, mentioned along with the Tigris in the biblical Adam and Eve story, has been the prime supplier of water for the region because, like the lower Colorado River, sediment from the uplands, transported down the river and deposited when its flow slowed, raised the riverbed so it was higher than the surrounding land, thus making it easier to divert water and irrigate lands. In such a way has the fertile Imperial Valley in southern California, part of which lies below sea level, been supplied with water from the Colorado.

All civilizations dependent on massive waterworks have needed a cohesive political and social structure to build and operate dams and canals. They are large projects needing the support of cooperative structures, such as the social and religious cohesion of the Mormons of Utah or the tight-knit irrigation districts of California with their ties to key congressmen who can obtain the authorization and funding for the large federal water projects that dot the West. Such a cohesiveness was achieved in ancient Mesopotamia during the flowering of the Sumerian civilization in the lower portion of the valley. It was a civilization dependent on the control of water. Dams were built, some of which are still in use today, and the water was diverted into canals, one being 180 miles long. The population grew with increasing prosperity. The Sumerian city of Ur contained a quarter of a million inhabitants on a land base of 1,450 acres. The Euphrates Delta was divided into a number of other city-states, each organized around an agricultural unit. Cooperation was a necessity, and gradually a centralized governing group of non-laborers evolved into a regional authority with absolute powers.

After some 1,000 to 1,500 years—it has taken only a little more than a half-century in the American West—the Sumerians began to experience a serious salinity problem. Salinity is the accumulation of dissolved minerals in water—its hardness—and is caused by natural sources, such as hot springs, or by the repeated use of water on irrigated fields. The salts are leached from the fields and deposited further downstream when the water is again used to irrigate. It is extremely harmful to plant growth, among other things. In Sumeria, sometime around 2400 B.C., population growth began to impose unrealistic demands on the ability of the land and the water-supply system to support it, and the overuse of water led to ever-increasing salinity. There were also problems with increasing sedimentation gumming up the waterworks, as well as with a rising underground water table, caused by the repeated application of surface water on fields; as a result, the soil became increasingly water-logged, which also stifles plant growth. (These, too, are problems not unknown in the modern West in recent years.) Over the next few centuries, crop yields lessened; the economic and political structure—the strong centralized authority needed to build and maintain the water system—grew progressively weaker; and around 2000 B.C. the last Sumerian Empire, the Third Dynasty of Ur, fell. Two University of Chicago professors, Thorkild Jacobsen and Robert M. Adams, have written of the decline of Sumerian civilization:

> While never completely abandoned afterwards, cultural and political leadership passed permanently out of the region with the rise of Babylon in the 18th Century B.C., and many of the great Sumerian cities dwindled to villages or were left in ruins. Probably there is no historical event of this magnitude for which a single explanation is adequate, but that growing soil salinity played an important part in the breakup of Sumerian civilization seems beyond question.

But Babylon did not remain dominant for long, and around 1740 B.C. a great silence fell over its cities. There had been a sudden end. Houses were abandoned without any sign of violence, and bowls and grinding stones were left behind. Just before the abandonment there had been a sharp increase in loans to farmers, then farmlands were quickly sold. The water supply had apparently failed.

Closer to home in the Colorado River basin, the Anasazi Indians of

Chaco Canyon, in the desert of northwestern New Mexico, and the Hoho-kam Indians, who inhabited an area near Phoenix, had similar water problems. The West has the feeling of an instant, somewhat alien culture plopped down upon the land within the last hundred years or so; and in most cases that is true. The ruins of the Anasazi Indians in the Southwest, however, give off the sense of an ancient presence, one that rivals the sophistication of other early cultures. Nowhere else did this early Indian culture flourish with such intensity as along twenty miles of Chaco Wash, most of which is now within Chaco Canyon National Monument. The Indians first arrived in the canyon around 600 A.D. At the height of occupancy in the early twelfth century, there were between five and ten thousand inhabitants living within the canyon, one-half to three-quarters of a mile wide. There were thirteen multiple-story structures, actually small towns in their own right, and perhaps fifty smaller villages scattered within shouting distance of each other. It was the most densely populated strip of land in its time in the Southwest. Just one of the large dwellings, Pueblo Bonito, was four or five stories high, with three hundred rooms on the first floor alone, eight hundred rooms altogether, and a total occupancy of twelve hundred people. It is an achievement that has rarely been equaled in modern apartment construction. To serve this popula-tion, the Indians developed a complex water-distribution system emanat-ing from side canyons and a partially paved and lined three-hundred-mile road system radiating out from the canyon, although the Anasazi had no domestic animals and had not invented the wheel. Then, starting around the mid-1100s, the canyon was gradually abandoned. It was about five hundred years before other Indians, this time Navajos, wandered back into the canyon and established homes in much diminished numbers. Because of energy developments, namely coal and uranium, that section of New Mexico is once again experiencing a population boom, and water supplies are getting short.

The causes of the abandonment of Chaco Canyon are familiar—too many people and overuse of water and other natural resources. All this happened during a time when the climate was getting drier. Chaco Wash, the main watercourse, is an ephemeral stream beginning at the 6,900-foot elevation of the Continental Divide in New Mexico and running about 150 miles to the San Juan River near Shiprock on the present-day Navajo Reservation. The drainage basin of the wash is a land of sandstone and shale, low mesas and cliffs. It is a hard, eroded, treeless land where living has always been difficult. The climate is classified as a cold desert. Only in the time of the Anasazi did the canyon hold any appreciable number of people. The wash is now cut by a sinuous arroyo, some twenty-five feet deep and one hundred to three hundred feet wide—a stretched-out U shape within the larger, flat watercourse. But the wash did not always

contain an arroyo. Drought, as is commonly perceived, was not the cause
of the abandonment of the canyon. The reasons were more subtle, more
a combination of factors historically found in dry lands.

At the start of the twelfth century there were diversion dams and
adobe-lined ditches along most of the northern side canyons to catch the
runoff and divert it onto the fields of Chaco Canyon. There were even
some reservoirs to store water for later use. Most of the canyon floor,
however, was irrigated by floodwaters from Chaco Wash before the arroyo cut into the flat, meandering streambed and lowered the surface
water flow below the level of the fields. What with the foot movements
of the extensive population, the numerous ditches and other irrigation
works, the large structures, and the fields cleared for planting, much of
the perennial cover had been stripped, making the terrain ripe for erosion. Extensive pine forests had once blanketed the region, but they can
be seen no longer. The trees were felled for construction of the large

Visitor at Anasazi ruins, Chaco Canyon National Monument.

structures, one such pueblo using 5,000 trees. It is estimated that between 75,000 and 100,000 trees were cut during that century. Such massive disruption of the vegetative cover contributed greatly to the erosion potential. Coupled with this denuding was the onslaught of a drier period. When the occasional rains did come in the mid-1100s, instead of spilling over onto the flood plain they cut a trench or arroyo down the middle of the wash, carried off the loosened topsoil, and lowered the water level. Such arroyo cutting "would have jeopardized life in Chaco Canyon towns within a few days, and in a week or two would have made them uninhabitable," according to anthropologist Edgar L. Hewett. Other researchers have found an accumulation of salts in the upper layers of the canyon's soil. A second period of arroyo cutting began in the 1860s and was blamed on overgrazing from the large influx of cattle into the West. Chaco Canyon was mostly abandoned by 1200, its inhabitants and their descendents eventually seeking refuge in the pueblos along the Rio Grande to the east. Its meteoric rise and fall comes close to approximating the time constraints of the speedier twentieth century.

The ancient Indians' contributions to the water development of the West tend to be downgraded because they have few direct, linear ties to the present tribes; and the whites, who came later, fancied themselves to be the first. Thus the Mormons of Utah are frequently given credit for initiating the diversion of water on a large scale, but this simply is not so. The Hohokam Indians of Arizona, contemporaries of the upper-basin Anasazi, had accomplished it a thousand years before the Mormons arrived in the West. Not much changed in the intervening centuries, since there are only so many ways to divert water and channel it by gravity flow to habitable areas. A map drawn in 1929 showing the prehistoric irrigation canals and present-day ditches in the Phoenix area has some lines superimposed upon or closely parallel to each other. Most of the lines are heavy, dark, and stand alone. They depict the more extensive Hohokam canal system. The map, which accompanied a report published by the Arizona state historian, was captioned: "Canal building in the Salt River Valley with a stone hoe held in the hand without a handle. These were the original engineers, the true pioneers who built, used and abandoned a canal system when London and Paris were a cluster of wild huts."

The management of water by the ancient Indians reached its apex with the Hohokam, but it started on a much more primitive scale around 200 B.C. Before that time and probably dating back to 2500 B.C. in the Southwest—then the height of the Sumerian civilization—the Indians had little use for crops, preferring to hunt and collect food. Perhaps a small trickle of water was first led from a seep or spring to a patch of cultivated ground. Later, stones were laid across drainages, forming small check dams similar to those bulldozed today by ranchers and called stock ponds. At Mesa

Verde, another Anasazi ruin north of Chaco Canyon, there were as many as a hundred of these low dams, perhaps three or four feet high and up to thirty feet long, in a single small draw atop the mesa. These small dams served as impoundments for water and soil, and intensive farming took place on the terraces. Below the cliff dwellings in the canyons were higher dams backing up pools of water that hopefully would last until the summer rains arrived. Neither Mesa Verde's water system nor those of other Indians on the Colorado Plateau matched the complexity or sophistication of those at Chaco Canyon and the Hohokam villages. Nor could the Spanish, who traveled up the Arizona river valleys from Mexico after these prehistoric Indians had disappeared, equal their engineering and structural accomplishments, which included canals thirty-five feet wide and from ten to fifteen feet deep. The Spanish, who came from an arid land and knew a thing or two about waterworks, only added a few refinements, such as reservoirs. Their permanent masonry dams frequently filled with silt, while the Indians' brush dams were designed to wash away during floods, thus allowing the flood flows, which were heavily laden with silt, to wash on downstream. Then the brush dams were replaced.

The winning of the interior West was principally a matter of moving dirt, from prehistoric times to the present. In 1979 the excavation of the 6.8-mile Buckskin Tunnel in western Arizona was completed. The tunnel is part of the Central Arizona Project, which when finished in the mid-1980s will bring water from the lower Colorado River to where the Hohokam dug their canals to divert the Salt and Gila rivers. The tunnel was dug through volcanic rock with a specially built TBM (tunnel boring machine) resembling a space laboratory. The 350-ton rotary borer had a thrust of 2.6 million pounds. Its fifteen half-inch cutting discs ate away at a record rate of 150 feet in one twenty-four-hour period. During its brief lifetime on the job, it ground out 600,000 cubic yards of rock. The TBM was a descendant, by way of the common shovel, of the Hohokam digging stick. The Indians used the sharpened stick to pry dirt loose, and then carried it away in a basket. The stick, also employed for planting, was made from a hardwood, like ironwood or mesquite. It was typically about four feet long. The chisel-like digging point was kept sharp by rubbing it with a roughened stone, much as a steel file is used on modern digging and cutting implements. Rounded stone hand tools and thin sheets of rock were also used as digging tools, and it is possible the Hohokam had primitive shovels and hoes, sharpened rocks with wooden handles attached. As Arizona governor Jack Williams said in the early 1970s, "We have been taught to believe that Samuel Colt's revolving six-shooter and the heavy-barreled Sharp's rifle conquered the western frontier. But in this Valley it was the shovel and the plow and the hands of men and women who saw a vision. . . ."

What the Hohokam built with their primitive tools was between 200 and 250 miles of canals leading off the two rivers. Some are in use to this day. The Gila River south of Phoenix is now normally dry because of upstream dam developments, but in ancient times, when not in flood, it had a depth of about six feet or so, and the flow of water was dependable, in quantity if not in quality. It was because of increased salinity and a high water table, both the results of the overuse of water, that the extensive Hohokam civilization collapsed around 1450. The present day Pima Indians later reoccupied Hohokam lands, and this much lesser tribe also discovered that the saline waters stunted the growth of crops. Their solution was the same as all desert peoples', including that of the white farmers of the Imperial Valley, and that was to pour more water onto the land to flush the salts out. The limit of such extravagant use of water is fast approaching, and there is no indication that modern technology, which has played such a prominent role in western water development, is going to do anything but delay the inevitable. After all, there are some precedents. It was atop some Hohokam ruins that a drunken Englishman named the city of Phoenix after the bird that arose from its own ashes.

One who understood the limits of the West, even before there were any dams of appreciable size on the Colorado River and its tributaries, was John Wesley Powell, the one-armed Civil War major who is generally credited with being the first to float all the way through the turbulent waters of the Grand Canyon in 1869. Powell parlayed the fame he received from that considerable achievement and a subsequent trip down the river into a bureaucratic career in Washington, D.C., and headed, among other agencies, the forerunner of what was to become the Bureau of Reclamation. Much has been written of Powell's prescience, which was formidable, but in the end he was a failure. Although a powerful and shrewd bureaucrat, he was unable to get his most important program through Congress, because it alienated too many interests. Like President Carter, he should have been more adroit in his dealings with the bloc that has been traditionally protective of what it viewed as the West's best interests. What Powell proposed, though it made rational sense, was too sweeping a reform, too methodical a process and too drastic a change for his time, and even possibly for the present. Novelist and historian Wallace Stegner wrote in the introduction to the 1962 edition of Powell's 1878 *Report on the Lands of the Arid Region of the United States*, "He wanted to start reforming the West from the grassroots, and by changing land and water laws in advance of further settlement, to change its whole institutional base." That he was opposed by the same community of interests who fought the deletion of water projects from the budget in

1977 placed him first among what was to be a growing number of victims whose rational approach toward the development of the West attracted the wrath of western congressmen. There has always been something inherently emotional about the Reclamation ethic. It was as if water projects were the West's payoff for being a colony to the rest of the nation, or a "plundered province," in the words of historian Bernard DeVoto.

Most of what Powell proposed was the result of his travels through Utah, which contained a unique communal society dependent on the diversion of water and readily able, through its religious institutions, to accomplish this purpose. Utah remains unique in the West to this day because of its cohesiveness. What evolved from the Mormon experience, Powell's writings, and the recognition elsewhere of the need for a communal organization to develop and distribute water, was the modern-day irrigation district—a western institution that is the basic governing structure in many communities. It was the Imperial Irrigation District in southern California, the largest single user of Colorado River water, that was to dictate the state's policy toward the river for many years, distribute the means toward wealth in the rich Imperial Valley, and successfully help farmers evade a key provision in the federal Reclamation Act of 1902 that comes closest in this country to an attempt at land reform. Above all else, Powell preached the uniqueness of arid lands and their need for special institutions. He used common sense and proposed that instead of the rectangular grid survey useful to the east on flat, equally watered lands, the arid West should be divided into watersheds, such as the Colorado River basin. The West has paid dearly for not following that suggestion; witness the bitter intrastate water feuds. Powell knew that the West did not have an unlimited amount of land that could be irrigated or an inexhaustible supply of water, two false impressions spread widely by various boosters. He was read out of the Reclamation movement for declaring at the National Irrigation Congress in Los Angeles in 1893, "Gentlemen, it may be unpleasant for me to give you these facts. I hesitated a good deal but finally concluded to do so. I tell you, gentlemen, you are piling up a heritage of conflict and litigation of water rights, for there is not sufficient water to supply the land."

Whoever controlled the water and how it was used, Powell pointed out time and again, determined what happened on the surrounding lands; thus its distribution should be carefully planned and carried out, preferably by an entity higher than the water users and individual states. It was Powell's downfall that he put himself, as the director of the U.S. Geological Survey, in the position of halting the development of western lands until a scientifically executed survey could be made of possible reservoir sites and irrigable lands. The timing could not have been worse, since the

West in the years of 1888 and 1889 was in the grip of an extreme drought
and there were large cattle losses. Western congressmen wanted a quick
survey to determine where the water projects should go. Powell en-
visaged something more methodical and restrictive in terms of land development, at least until the survey was completed. He had put himself
in the position of having to certify dam sites, thus deciding where they
would go and who would benefit. It was a situation intolerable to western
congressmen—so much power being in the hands of the executive branch
of government. So they set out to chop Powell down by taking away his
power and appropriations. Powell lost, and resigned shortly thereafter.
The West was not again to feel so threatened until President Carter took
office. Stegner wrote of Powell's defeat in *Beyond the Hundredth Meridian,*
"It was the West itself that beat him . . . the myth-bound West which
insisted on running into the future like a streetcar on a gravel road." But
Powell also beat himself. He had acquired too much visible power and
had confronted the western interests where they were most vulnerable.
Thomas G. Alexander, a Utah history professor, later wrote, "In some
circles, the myth of the omniscience of scientists has been carried almost
to the same point that fundamentalists have carried literal interpretations
of the Bible." He was referring to Powell.

It was appropriate that the man sharing the platform with Powell at the
1893 irrigation congress was William E. Smythe, who cried out in objec-
tion to Powell, "We have said we have homes for a million more people,
you say we have not." Smythe was the ultimate boomer. He saw room not
only for a million new homes, which actually would come about, but also
a hundred million new inhabitants between the Pacific Ocean and the
Missouri River, an impossibility given the present water supply. Smythe
foresaw irrigation spreading as far east as Boston. Aridity was seen as a
benefit: "But we have yet to mention the chief blessing of aridity. This
is the fact that it compels the use of irrigation. And irrigation is a mira-
cle," wrote Smythe in *The Conquest of Arid America,* the bible of the Reclamation movement of the 1890s. The question of whether there was
enough water in the West to supply the utopia that advocates of Reclamation foresaw, was never seriously posed, except by Powell. Smythe, a
newspaper editor who had lost his objectivity, said the idea of irrigation
was to him "a new crusade . . . a philosophy, a religion, and a programme
of practical statesmanship rolled into one." Of Smythe, Walter Prescott
Webb wrote in *The Great Plains,* the first coherent explanation of what
dryness really meant in the West, "He played upon emotion, imagination,
pecuniary desire, and patriotism, and to him irrigation became the one
means by which arid America could be conquered." The key word was
"conquer," because the literature and spoken words of those who were
to follow Smythe's exhortations during the next century continually

echoed this concept of doing battle with the twin evils of unproductive land and untamed rivers. The ultimate lesson for the West to learn in the twentieth century was the impossibility of a complete victory. The emotional element of the movement launched by Smythe would continue to the present.

The drought of the late 1880s, besides helping to spawn the Reclamation movement of the next decade, brought a halt to the first large influx of population into the interior regions of the West. Some had come because of the new railroads, the cheap land, and the booming livestock industry; others were the spillover from the California mines and other disappointed hopes on the West Coast. For whatever reason, they were a wave between 1870 and 1890 that migrated to the coast, broke, then receded back over the interior regions. They were predominantly ranchers and miners; their deserted homesteads and diggings are still visible, a reminder of the first major land-use revolution in the West, the change from wilderness to an agrarian society. The migratory pattern was to repeat itself in the 1960s and 1970s, when the cities again surrendered their population growth to the rural areas of the West, resulting in the second major change in land use. Now, as the importance of agriculture has stabilized, it is energy production and recreation, the seeking after the good life whether in retirement or making do during the peak years with a decreased income, that has brought about the urbanization of western rural areas in recent years. The West Coast and some interior cities, like Denver and Phoenix, experienced rapid spurts of growth after World War II, but the vast interior spaces did not really begin to fill up until the 1960s. The population growth of Los Angeles would stabilize in the 1970s, but dying towns like Rock Springs, Craig, Aspen, Durango, Moab, St. George, Clifton, and Prescott would become small cities and boom. The 1890 census showed the frontier was no longer a solid advancing line. The report on that census was published in 1892; one year later Frederick Jackson Turner, the historian who dignified the West by making it the goal of the primary migratory movement in this country's history, declared, "This brief official statement marks the closing of a great historic movement." But Turner, actually a man of the Middle West, did not know the arid West very well. He, too, spoke of conquering the arid lands with irrigation. Pockets of frontier existed into the 1950s and migration into and within the region continued through the 1970s.

It was the fruition—some would say the excesses—of the Reclamation ethic, based on the national policy of manifest destiny and guided by the boomer mentality, that brought about these vast changes in the Colorado River basin and the remainder of the West. The Bureau of Reclamation was one tool, albeit the most important single federal institution, to further this national policy to develop the West. As William E. Warne, a former official in both the bureau and the Department of the Interior, wrote: "The Bureau of Reclamation has identified effectively with the West. In the conduct of its business, it has been sympathetic with local leaders who have sought the economic growth and development of their communities. . . . Thus, a clientele of the most active local leaders has been developed for the Bureau in the rural areas of the West." Indeed, the clientele would come to control the bureau in time. It was the threat by the Carter Administration in 1977 to rearrange the relationship in this unequal marriage of seventy-five years that caused so much strain. Population dynamics had something to do with it, too, as outsiders moved in and established priorities were attacked on an unprecedented scale.

The Colorado River basin was the last area in the contiguous United States to be explored. It contained the last rivers and mountains to be named, some as late as 1872. A map drawn in 1857, which summarized existing knowledge to that date, had the warning "unexplored" written across the center of the basin. A chronicler of the West, Sam Bowles, wrote at the time of the Civil War, "The maps from Washington, that put down only what is absolutely, scientifically known, leave a great blank space here of three hundred to five hundred miles long and one hundred to two hundred miles broad. Is any other nation so ignorant of itself?" Besides being retarded, the interior West resembled an abused child, denied those vital fluids all the other children in the neighborhood received. Its life-giving force—water—and those products that flow from its application upon desert lands were sucked away by others. It has long been regarded as a region to quickly extract riches from, or as an obstacle to be rapidly crossed in air-conditioned comfort, with a few brief stops along the way to admire the view and the strange doings of the natives. Only recently have large numbers of people paused there for its own sake, drawn by the retirement and recreation-oriented opportunities of the western sunbelt.

From a few scattered white outposts, many of them Mormon communities, and impoverished Indian tribes, the basin's population jumped dramatically in the post–Civil War period to 260,000 in 1900. It then climbed steadily, if not so dramatically, to just under 1 million inhabitants prior to World War II. By way of comparison, the population of southern California, to which Colorado River water would eventually be diverted, grew from the same 1900 level to 3.5 million in 1940. About 70 percent of the basin's population in 1940 was classified as rural. By 1970 the predominantly white population of the basin was 2.5 million, most of it concentrated in a few urbanized areas of 50,000 inhabitants or more. Arizona, Colorado, Nevada, and Utah all had between 65 and 70 percent of their population living in such areas. California had more, while Wyoming and New Mexico had much less. Although the eleven western states would lead the nation in the post–World War II years in growth, in terms of absolute numbers there would still be comparatively little population. Thus in 1970 the basin's population was spread out at the rate of 10.5 persons per square mile, the basin states had 24.2 and the nation averaged 57.4. Again, the statistics of an arid land. As the 1960s ended the population of the basin and much of the remainder of the West was concentrated in urban areas, and a significant amount of its increase was attributable to births, rather than migration.

Then, late in that decade, something happened that took demographers and city planners by complete surprise. The national trend of the last 150 years—the greater population growth of cities at the expense of rural areas—was halted and reversed. Calvin L. Beale, a demographer for the Department of Agriculture, stated, "Thus the reversal of this flow in the early 1970s looked so unlikely that first reports of it were greeted with some skepticism. But as the decade has proceeded, every source of formal data has confirmed the new trend." Migration out of the cities into rural areas, termed the "rural renaissance," accounted for most of the growth in the interior West and began to affect established political alliances that had been forged over the years to procure western water. Rural areas of the West quickly filled up, and it was not births that accounted for this growth. Migrants—instant voters with few roots yet sunk into their new communities—arrived on the scene along with their urban outlooks and prejudices. Although the confirmatory figures from the 1980 census were not yet available, some demographers had already spotted the trend in the mid-1970s. What official figures had not yet established, anyone's extensive travels through the interior West in the last decade could confirm. It was hard to find a community that was not throbbing with activity, be it the construction of subdivisions, ski lifts, or supermarkets.

Instead of played-out gold and silver mines, it was the faltering aerospace industry, smog, the overall frenzied pace of urban life, increased personal income, and the romanticized vision of "returning to the land," an offshoot of the environmental movement of the early 1970s, that brought about California's second contribution—some would call it Californication—to the repopulation of the West's rural spaces. It was a time when population growth in southern California, the major drain on the natural resources of the West, had stabilized. In Beale's words: "The key change in this region, as I see it, is the end of the former dominance of California in Western growth. It is still by far the largest state in the region, but for the first time in over half a century, most of the population growth in the West is now occurring outside of California." Peter A. Morrison, a demographer for the Rand Corporation in Santa Monica, a

Los Angeles suburb, wrote: "California migratory experience no longer dominates the West's regional migration growth as it did prior to 1965. ... Gain in the Mountain States was especially impressive: In the first six years of the 1970s, these eight states gained 913,000 through net in-migration, compared with 307,000 during the entire decade of the 1960s." The mountain states are the heart of the interior West. Morrison wrote of the effects of this migration that "These small, once stable communities are ill-equipped to deal with sudden population growth. Like new celebrities they lack the full array of legal and institutional structures for coping with unaccustomed attention."

Massive dislocations took place in these small towns and cities, echoing past boom-bust cycles that are an integral part of western history. These changes were not only social and economic, but also political. Wayne N. Aspinall, the quintessential western congressman who had such a decisive influence on Colorado River water matters, lost what had been a safe

Recreational subdivision, near Kanab, Utah.

congressional seat in western Colorado when the urban influx changed voting patterns in 1972. But it was not a total takeover. The troops could still be mustered in sufficient numbers to deflect the main thrust of the Carter Administration's water policy.

The system of rivers that bisects the Colorado River watershed is best depicted by a splayed left hand, palm facing away and the thumb pointing down. The thick thumb is the 800-mile long main stem of the river, termed the lower Colorado, extending from just below Glen Canyon Dam to just short of the Gulf of California in Mexico. The turbulent upper 275 miles of the lower river pass through the heart of the canyon country, traversing Marble and Grand Canyons. Below Lake Mead, where the water is backed up by Hoover Dam and the end of the canyon country is drowned, the river passes serenely through the low deserts of Nevada, Arizona, California, and the Mexican states of Baja California and Sonora. Through this ultimate portion of the plumbing system, it is hemmed in and channeled by the artificial shores of reservoirs and embankments so every last drop may be wrung from the river. Seemingly wild but actually extremely controlled through the canyons, then flowing in broad, calm arcs or placid pools through the low deserts, the Colorado's water in the lower reaches of the river begins to get turgid, each drop having been used an average of at least three times and then returned somewhat more lifeless to the river. Because of such heavy use, the Gila River, originating in western New Mexico, no longer flows its entire 500-mile length to join the Colorado near Yuma, Arizona. It is the forefinger on the splayed hand, with a section missing between the joint and knuckle.

The pinkie, which should be at the one o'clock position, represents the 730-mile Green River, whose headwaters in the glaciers of the Wind River Range in Wyoming are the true start of the river system, if one is interested in a single mystical point of origin. From the perpetual, vertical ice of Knapsack Col along the Continental Divide it is 1,700 miles to the other extreme of the burning, horizontal sands of the delta in Mexico. To experience these two extremes, no matter if months apart, is to feel the ultimate in geographic dislocation bound together by a river. From where the Green River joins the upper Colorado in Canyonlands National Park of eastern Utah, it is about 200 miles through the upper portions of the canyon country to Glen Canyon Dam. Just above the confluence of the Green and Colorado, where the river begins to sink into the depths, the canyon country begins. Clockwise from the pinkie, the next finger is the upper Colorado; it was formerly called the Grand River and is mistakenly referred to as the start of the river simply because of the name change, which was more a product of local boosterism than of correct hydrology. From its beginning in Rocky Mountain National Park just below a diversion ditch that carries the first waters east across the Continental Divide, the upper Colorado flows west 420 miles to its confluence with the Green. The middle finger, in the three o'clock position, is actually the shortest of the principal tributaries. It represents the San Juan River, flowing only 360 miles from the San Juan Mountains in southwestern Colorado to Lake Powell.

From their separate, pristine mountain wildernesses these rivers plunge down onto the flat desert steppes and pasturelands of the high country. Then there is another wild descent through the confines of the canyon country before emerging onto the hot, low deserts. The end comes under the merciless sun and through the porous sand of the delta's vast, empty spaces.

Four major rivers and numerous smaller tributaries, all united in search of an end and no longer finding it because so much of their water has been drawn off on the descent from 14,000-foot heights to below sea level at the Salton Sea. The separate tributaries drain a watershed encompassing 242,000 square miles in this country—one-twelfth the land area of the contiguous states—and 2,000 miles in Mexico. Although technically not emptying into the Colorado River basin, the 7,800-square-mile Salton Sea basin in southern California and the 4,000-square-mile Great Divide Basin in southwestern Wyoming are frequently considered part of the watershed because of mutual dependence and similar characteristics. Where the water is diverted out of the basin, it serves a population exceeding twelve million. Altogether, more than half the population of the West, in all but the four most northwestern of the eleven western states, is directly dependent to some extent on Colorado River water. Additionally, the Northwest is tied into the electrical grid, part of whose power is generated by Colorado River water. So the Colorado River system as a totality and in its different forms is, in a very real sense, responsible for what a large part of the West is today.

And what is it? Very simply, in terms of the largest amounts of water consumed and land used, it is a vast feedlot for livestock. The West is cows—beef, red meat—and the grasses and feeds growing wild or cultivated for these domestic bovine animals. It is not glorious national parks and coal strip mines, although they are there too; it is predominantly cattle, and to some extent sheep, and the feed needed to fatten them for the dinner table and fast-food chains. Consider the statistics. Just over 90 percent of the water used in the four upper-basin states (Wyoming, Colorado, Utah, and New Mexico) is spread on land irrigated for crops. Of the 1.6 million irrigated acres in the upper basin, feed for livestock is raised on 88 percent of the irrigated land. There is a total of 65 million acres in the upper basin. Of this amount, 47 million acres are utilized for agricultural purposes, which means the grazing of livestock. In the lower basin (California, Arizona, and Nevada), 85 percent of the water goes toward agricultural purposes. Smaller amounts of water are expended here to raise feed for cattle; but even so, feed crops cover nearly as much land as all other types of crops, and alfalfa is the most water-consumptive of all major crops raised in the lower basin. Of the 99 million acres in the lower basin, 82 million are rangeland or pasture. Only slightly more than 0.5 million acres are classified as urban.

Cattle, the chief beneficiaries of western water.

The impression is often given of the Colorado River mainly benefiting people—meaning the urban masses in the metropolitan areas of Los Angeles, Denver, Phoenix, Las Vegas, Salt Lake City, and Albuquerque. Actually, cows are far and away the chief beneficiaries of water, even in the West's most populous region. Of the 4.7 million acre-feet of water diverted each year from the river for southern California's use in the mid-1970s (an amount equal to about one-third of the river's total annual flow) only about 800,000 acre-feet went to the urban coastal areas. (An acre-foot is a unit of measurement for large amounts of water. It is the amount of water needed to cover one acre to a depth of one foot. That equals 326,000 gallons, enough to supply the needs of a family of five for a year.) The remainder went to four remote irrigation districts, by far the largest being the Imperial Irrigation District stretching between the Mexican border and the Salton Sea. The Imperial Valley was the single largest user of Colorado River water; the most valuable single crop, taking up most of the land within the district's boundaries, was alfalfa, used for hay. The seven Colorado River basin states, including those portions of the states lying outside the basin and not receiving Colorado River water, produced only 13 percent of the total value of the nation's livestock. Probably never in history has so much money been spent, so many waterworks constructed, so many political battles fought, and so many lawsuits filed to succor a rather sluggish four-legged beast.

But if water had to be used in these vast amounts in the West, if that was what the drive toward manifest destiny dictated, then perhaps it was better that it went for such pastoral purposes rather than more urban, industrial pursuits. A ranch seems to fit the sere western landscape far more agreeably than a factory, although its consumption of water may not be that much different. But this balance was beginning to change in recent years.

That there will not always be a limitless supply of water for all desired purposes in the West is a fact just beginning to be recognized by a few, and dealt with by even fewer. The Bureau of Reclamation warned in a little-noticed 1977 publication, entitled *'75 Water Assessment,* "The average annual water supply of the Colorado River is inadequate to meet compact allocations and treaty entitlements. Deficiencies are expected to occur by year 2000. Competition for water will become increasingly severe for all uses, with many demands remaining unmet." Not only would the quantity of water shrink, but the quality would worsen. The salinity increases seventeenfold from the headwaters to Imperial Dam, just above the Mexican border, and this can be expected to rise. History has spoken clearly on the overuse of water in arid lands. The specters of the Hohokam villages, Chaco Canyon, and ancient Mesopotamia hang over the West as it approaches the end of the century.

CHAPTER TWO

Beginnings: Four Streams in Search of a River

1

It was the last night of August and the wind-driven snow was being slammed furiously against the rattling tent. The lightning would have had all the intensity and surprise value of a giant strobe set off randomly before the naked eye, had not the light been filtered by the two layers of orange fabric that formed the mountain tent. The thunder felt as though it was squatting directly over the campsite huddled beside a clump of subalpine fir permanently bowed by the wind. For a while the scene had all the ingredients of a *Walpurgisnacht;* then the fury tapered off to the steady slap of wet snow falling on the tent fly. This snow falling year after year and consolidated into glaciers formed the source of the Colorado River within two miles of the tent, pitched at 10,700 feet, between Peak Lake, the first gathering of the waters in Wyoming, and Knapsack Col.

In the morning there was a feeling of emergence. The clouds, in the aptly named Wind River Range, drifted apart and then formed again in random patterns outlined by the sun rising behind the saddlelike depression that marks the Continental Divide. It was as if with strokes of thunder and lightning and behind a curtain of swirling snow and dark the gods had created a river; like a pilgrim, I set off alone in the morning to the source. That there was a single source, a learned river historian had told me, was a nineteenth-century concept, one that generated all those searches for the beginning of the Nile.

The Green River, whose headwaters I was about to trace, has undergone a number of interesting name changes. The Indian name for the river was Seedskeedee (spelled variously), after the prairie chickens that inhabited the high desert area. Its first European name was the San Buenaventura, named after a thirteenth-century theologian by Silvestre Velez de Escalante, a Franciscan missionary who stumbled on the river while looking for a route from Sante Fe, New Mexico, to California. The Spanish who came after Escalante called it the Rio Verde. This was translated to Green River by the fur trappers in the early nineteenth century. Others simply called it the Colorado River, since what is now known as the Colorado above the confluence of the two streams was then known as the Grand River.

The appellation Green, perhaps derived from the reflection of the surrounding vegetation or from the cast of glacial silt, stuck. But it nearly became unglued sixty years ago when politics, combined with local boosterism, invaded the realm of geographical nomenclature. That politics should play a part in this process is appropriate, since so much of what has happened to this river system has had a political derivation. The Utah legislature started it off in early 1921 by introducing and then defeating a measure to change the name of the Green to the Colorado River. This proposed action, at least, had some historical and hydrologic validity. The Utah lawmakers were not to meet again for another two years, so the Colorado legislature acted quickly on the prodding of Congressman Edward F. Taylor of Glenwood Springs, who first as a state senator and then as a congressman had campaigned to change the name of the Grand to the Colorado. The Colorado legislature acted promptly, as did Congress, and on July 25, 1921, President Warren G. Harding signed the bill and gave the pen to Taylor, who could now claim that his Colorado congressional district was at the headwaters of the Colorado River. Throughout the process the Board of Geographical Names, which is supposed to rule on such matters, maintained a low profile.

Before the name change, it was commonly thought the Colorado River began at the headwaters of the Green, which was longer and drained a larger area than the Grand, whose average annual flow was greater than the Green. When the hydrologist E. C. La Rue made the first comprehensive report for the U.S. Geological Survey on the Colorado River system in 1916, he wrote, "Colorado River is formed by the junction of the Grand and the Green. Green River drains a larger area than the Grand and is considered the upper continuation of the Colorado. Including the Green, the river is about 1,700 miles long." Since the name change, clearly an exercise of the chamber-of-commerce mentality, it has commonly been assumed that the upper Colorado River in Rocky Mountain National Park is the source; and official literature distributed by the Bureau of Reclamation and National Park Service, among others, has reflected this misapprehension. The longer stem of a river, termed the master stream, is usually considered its main branch, and that clearly is the Green River. Besides, the headwaters of the Green is a more proper setting for the start of such a great river system.

I walked east up the unnamed cirque valley, the first to go that way in the undisturbed snow that now, under the rising sun, was beginning to melt. The stream, also unnamed at this point, widened a few times into what could almost be called a series of small, thin lakes: paternoster lakes, so called because of their resemblance to the beads of a rosary. A coyote howled and its cries blended with the many different sounds of running water. It was as if the mountains were weeping. There were countless glints of warm light on water and ice as I walked toward the rising sun. To the right was a waterfall splashing down in stages from the small tarn holding the meltwater from Stroud Glacier. The minute, intense reflection of the sun in the tarn almost brought my perception of the source down to a single radiant drop. But on this day there were many sources.

The climb toward the col was tedious. The rocks buried by the snow could not be seen, but were slippery underfoot. As I went higher and neared the 12,000-foot-high ridge, the snow got deeper. My struggles were mocked by a rough-legged hawk that glided effortlessly over the saddle. The only other wildlife I saw were spiders trapped in the snow. At times the snow was to my hips, and I was glad I had brought an ice ax to probe with and gaiters to keep the snow out of my boots. It was hot, yet new clouds were piling up to the west. I stopped a few hundred feet short of my goal at the base of a steep slope that looked avalanche-prone. This was far enough. I no longer heard the sound of running water.

Although I was alone this day, I did have a companion of sorts, the guidebook for the mountains, written by Finis Mitchell. I have never met Mr. Mitchell, but by all accounts he is an unusual person. He arrived at the foot of the Wind River Range in a mule-drawn wagon in 1906 and has made the mountains his abiding interest ever since. Mitchell began climbing the mountains in 1909 and began photographing them in 1920. In 1977, at the age of seventy-five, Mitchell commented to an interviewer, "I'll be climbing for another ten or fifteen years. There are about fifty-eight more peaks I want pictures from." Up to that point the indefatigable Mitchell had climbed 220 peaks and taken 101,345 pictures, all properly filed and indexed. He used to carry a seventy-five-pound pack, but has whittled that weight down to fifty pounds over the years. Mitchell's regimen is simple. He does not smoke or drink liquor, coffee, or soft drinks. In the mountains he does not eat food warmer than his body temperature and wraps himself in a piece of plastic rather than seeking shelter in a $200 tent. Mitchell likes being alone in the mountains and, although he has had some close calls, told a local undertaker, "I'm going to put daisies on top of you before I leave." He follows in the tradition of William J. Stroud, known also as Rocky Mountain Bill, who first climbed Stroud Peak above my campsite and many other mountains in the same suit he wore to transact business and visit friends in Rock Springs. Somehow these earlier travelers make us Vibram-soled, Gore-Tex-clad late arrivals seem overprotected from the elements.

That more of us are arriving in numbers that can be considered a crowd became evident two days later when I packed my gear and headed down to the parking lot at the edge of the Bridger Wilderness Area. It was the start of the long Labor Day weekend, and a steady stream of backpackers and horseback riders was heading up the trail. It would be the same at the other headwaters of the Colorado River system, all located in officially designated wilderness areas, as it would be at the end of the river and any intermediate spot I visited. The West is not an empty region. In terms of what it can support, it is already crowded; the promises being made to future arrivals should be regarded with great care, since many cannot be fulfilled.

As I descended from the col I had a feeling of accomplishment, since I had tried once before to reach the source but had been turned back far below by the deep snows of early June. That I did not quite make it to the top, as I was not going to quite make it to the end of the river, left me with the satisfaction of retaining some small sense of mystery and wonder. I felt exuberant because of the nature of the day, and as I carefully picked my way down I sang. One of the songs that spilled out unconsciously was *Cry Me a River*—a tune, it occurred to me, that fit the surroundings. The mountains were weeping this day; the river's growing salinity, like tears, threatened to stifle its life-giving force; and my overall feeling of what had happened to the world's most-used river was one of muted sadness. It was a sadness tempered by the reality of the West, an arid land where the first impulse was to make water work for man's needs. This was an accepted national policy that had gone unchallenged for a half century. That the policy had outlived its usefulness—had, indeed, gone awry at times—was to become apparent in my travels through the lands the river nurtured. On such a day, though, these thoughts did not last long. There was more a feeling of birth than of death.

I was reluctant to leave my campsite and descend into the real world. It had been a restful place, and I had spent enough time there to feel some harmony between my rhythms and those of the mountains and falling water. Peak Lake, just below the campsite, was a friendly presence. Late in the afternoon I usually sat with my back against a boulder, facing this first gathering of the waters and the warm sun. Silver streaks of wind would flash across the dark-turquoise water. The streaks and trailing wisps gave the wind a tangible substance, yet the deep water remained undisturbed after their ephemeral passage. The taste of water was also insubstantial, perhaps because it was too pure or my palate had been spoiled by coarser water available below. The lake in its rock-encrusted setting was a scene of utmost purity and clarity, to a degree that was not to be repeated as I followed the river toward the ocean. Yet I was glad, at least, to have had this experience at the start.

From Peak Lake the stream drops quickly through a narrow, boulder-strewn canyon to slender Stonehammer Lake, one more bead on the string. Stonehammer is just below the timberline, and the sparse alpine tundra gives way here to more luxuriant growth. Up higher it was all rock and water, but now a more softening effect is felt upon the landscape.

Peak Lake, the first gathering of the waters on the Green River, the start of the Colorado River system in the Wind River Range of Wyoming.

Where the water tumbles from a narrow outlet two-thirds of the way down Stonehammer into the upper reaches of the U-shaped, glacially carved Green River Valley, the vegetation becomes profuse. It is here that the topographical map first labels the stream "Green River." I camped that night at Three Forks Park, a large, gentle meadow area formed by Wells and Trail Creeks joining the Green River. In the meanders of the open, flat meadow the Green begins to acquire the grace and force of a river and lose the brashness of a tumbling stream. Here the river is perhaps fifty feet wide, and slow and demure for the first time. There were ducks on a pond ringed by waist-high grasses and sedges, tinged with the golden color of an early autumn, and at sunset some moose moved unconcernedly along the tree line.

Early the next morning a thin mist hung over the park, mixing with smoke from a campfire lit by backpackers. The walk out that day to the parking lot was hot and boring. Other campers were coming in for the long holiday weekend. At the end of the trail were all the appurtenances of a modern camping scene: a filled-to-capacity campground, large recreational vehicles, small trail bikes, children, dogs, elbow-to-elbow fishermen, trash, more parked cars, and a Forest Service sign at the entrance to the wilderness area that declared:

Are you aware overused areas create:
congestion, pollution, eroded trails, visual blight—
generates—discontentment, concern, bad experience—
instigates—enforcement, restriction (non-use),
closure (do not enter).
It's up to you, John Q. Public.
Concentration brings depredation.
Dissemination brings perpetuation.

A few days later, at the Pinedale ranger station, I talked to the Forest Service official who designed the sign. Wayne Anderson, the resource coordinator for the district, said he consulted a psychologist and landscape architect to get the right wording and design for the sign, whose main purpose was to encourage people to scatter throughout the wilderness rather than congregate in a few well-used areas. The Wind River Range, which is not as well known as other western mountain ranges, was one of the last wilderness areas to feel the impact of the new backpacking generation of the early 1970s. Anderson said he could remember when he saw only one person in the 383,300-acre wilderness area, actually then designated a primitive area. And that person was a packer looking for a stray horse. Along with the explosion in new camping equipment and the back-to-nature movement of the late 1960s and early 1970s, the recreational use of Bridger skyrocketed 800 percent from 1966 to 1973. Now 23,000 enter the wilderness area yearly for stays that average four and a half days. That may not sound like much, but when the numbers are spread over a two-month season and concentrated on a few popular trails, it can feel like too many. As has happened elsewhere, the Forest Service is considering requiring permits to enter the wilderness area, a regulatory device seemingly antithetical to the freedom inherent in such an experience; yet it is the only means such land administrators have devised to protect it from overuse.

As in the remainder of the West, which some people still mistakenly think of as a frontier, somebody—Indians, Spaniards, Mexicans, or Anglos—used the Bridger Wilderness Area, heavily at times, before the current generation of backpackers took to the trails thinking they were the first. The Indians, who summered in Three Forks Park, believed the mountains were the home of the spirits. The fur trappers slipped in and out with their catches, which were sold lower down on the Green River at the yearly rendezvous. Early explorers like Benjamin Bonneville and John C. Frémont climbed the peaks and marveled at the scenery with the inevitable nineteenth-century comparisons to the European Alps. Next came the ranchers at the end of that century to turn their stock loose in the wilderness, where they wreaked havoc on the vegetation and landscape. The ranchers also built earthen dams on the mountain streams to store and regulate irrigation flows used downstream to grow hay.

The wilderness area and adjoining multiple-purpose National Forest lands are still used extensively for grazing. There are 127,035 acres in the Upper Green River Cattle Allotment held by the Green River Cattle and Horse Growers Association and administered by the Forest Service. In the association are twenty-five permittees, either local ranchers or out-of-state corporations. During the summer months they graze 7,600 cattle and 8,000 sheep on the allotment. The association was formed in 1916, and prior to 1924 as many as 13,000 cattle and 130,000 sheep had grazed these lands during the summer months, heavily contributing to their deterioration. In 1956 and 1957 the Forest Service, thinking to improve this area for grazing purposes, sprayed 10,130 acres with the herbicide 2,4-D. They aimed for the brush but also hit willow and aspen trees with the drift. For this practice, termed "range improvement," the Forest Service was heavily criticized; it halted further spraying on the allotment after two years. A range foreman and six or seven riders now look after the cattle and sheep on the summer range; the cattle drive up to and back from the higher elevations is along one of the last remaining cattle trails in the West.

There is also some mining activity in the wilderness. There are about sixty molybdenum claims in the area and some oil and gas leases along its boundary. Within the drainage area of the Green River but west of it and outside the wilderness area, yet still within Bridger-Teton National Forest, is the Overthrust Belt—referred to by the oil industry in the late 1970s as the hottest new area for oil and gas drilling in the country. To the south are older oil fields. In the early 1970s the Atomic Energy Commission had proposed setting off a series of nuclear explosions near Pinedale to tap natural gas supplies.

Sharing the wilderness area with these human activities are black bear, elk, moose, bighorn sheep, mountain lions, coyote, bobcat, and wolverine. Predator-control programs using guns and directed at the carnivores that are thought to prey on cattle and sheep have been allowed. The endangered American peregrine falcon and threatened prairie falcon have ostensibly remained untouched here during all this activity in recent years. Along with the campers—about 15 percent of them from such populous states as California, New York, and Illinois—the competition for space at the headwaters of the Colorado River was getting fiercer.

The Green River first flows in a northerly direction, as if making toward Canada or attempting to join the Columbia River system, but then as it nears the flank of the mountain range, it hooks around and heads south toward its junction with the Colorado River 700 miles distant. The point at which it begins to change directions is just past the boundary of the wilderness area that crosses the lower end of Green River Lakes. Not only is it a political and administrative boundary, but also a marked change takes place in the character of the land. The forests fall back and the valley takes on that open, sagebrush-dotted aspect of high-country ranchlands. This is still national forest, but the crumbling buck-and-pole fences and weathered, decaying cabins speak of long use by livestock and men. This is the classic western landscape—rolling gray-green lands where cowboys like John Wayne and the Marlboro men rode against backdrops of snow-capped peaks. As the road passes onto private lands, there is a sign to see or call a man in Pinedale about ten-acre recreation ranches.

It is about this point that the Green River gets its first healthy injection of saline water. Kendall Warm Springs tumbles down a series of small travertine terraces, through a culvert under the dirt road, and into the nearby river with 9,800 tons of dissolved minerals annually. The warm water heated by the hot rocks within the earth bubbles to the surface rich in calcium carbonates and bicarbonates. These dissolved minerals build up the terraces, much as they do along Havasu Creek in the Grand Canyon, and the saline water flows into the river. The springs are the only habitat of the Kendall Warm Springs dace, two-inch long fish that were separated from the river by the rising terraces and learned to adapt to their minuscule environment.

This stretch of the river has the potential to be included in the Wild and Scenic Rivers System, but the support for such a designation under the federal act was lacking, as the energy crisis of the early 1970s spurred planners once again to consider building Kendall Dam. As far back as 1946 the Bureau of Reclamation had suggested damming the Green at the Kendall site and forming a 340,000-acre-foot reservoir. At that time it was conceived of as an irrigation project to benefit 66,050 acres of ranch lands. With the arrival of the energy crisis the bureau, here as elsewhere, switched priorities and proposed a dam and 900,000-acre-foot-reservoir. The water would be diverted from Kendall Reservoir by canal, which would back up the river twenty miles, and be pumped over the Continental Divide at South Pass, where it would be used for energy developments, mostly coal, in the North Platte River Basin. The bureau said there would be little water available from new water projects for irrigation in the Green River Basin. The ranchers were understandably incensed at the government agency that had historically considered their needs first. Between wild rivers and dams, energy and agriculture, the competition for water at the headwaters of the Colorado River was getting intense.

2

Where the upper Colorado River begins in Rocky Mountain National Park in Colorado there is a long scar along the flanks of the Never Summer Mountains. This light-colored, thin slash set against the dark forest green of spruce and fir is the 16.7-mile Grand Ditch, the first major conveyor built to transport water outside the Colorado River Basin. As such, it was to set a precedent of expropriation for later out-of-basin water transfers that serve southern California, communities along the Rio Grande in New Mexico, Denver and other eastern Colorado communities, and Great Basin communities near Salt Lake City. More water is exported from the Colorado River watershed than from any other river basin in the country. The complex of dams, reservoirs, tunnels, and canals spreading out from the Colorado River system to embrace much of the West has become the most complicated plumbing system in the world. Water can be flushed through the system in varying amounts at the flick of a switch. Once water was diverted across the Continental Divide in 1892 through the Grand Ditch, the beginning of this tributary of the Colorado River system was no longer the pristine mountain streams tumbling down from the tundra world of the high Rockies but the trickle of seepage from the ditch down desiccated streambeds. The Bureau of Reclamation and others advertise this as the start of the country's most dramatic river, but it is no proper beginning for such a river system.

As with so many other large public and private construction projects in the West, the Chinese, this time with help from the Swedes, built the Grand Ditch. (At the end of the river in Baja California in the early years of this century, other Chinese, this time aided by Japanese, constructed the levees to contain the Colorado when it flooded, and they constituted much of the work force used by a southern California syndicate to farm

The Continental Divide at La Poudre Pass, Rocky Mountain National Park, Colorado. The Grand Ditch diverts the water out of the upper Colorado River to eastern Colorado. More water is diverted out of the Colorado River basin than out of any other river basin in the nation.

these Mexican lands. There were enough Chinese and Japanese laborers in Baja California at the time to make Arizonans and Californians living near the border feel the threat of a "yellow peril." An elected official in Imperial County, just across the border in California, said of the water quality, "Who wants to drink from a stream when he knows that there are 7,000 Chinamen, Japs, and Mexicans camped on that stream a few miles above in Mexico?") Up above the 10,000-foot level in the Never Summer Mountains, the Chinese and Swedes lived in separate camps. Rice was shipped over La Poudre Pass and stored in the caves where the Chinese lived. These foreign laborers served the needs of farmers who owned lands northeast of Fort Collins, Colorado, the area celebrated by James A. Michener in his novel *Centennial.* Early on, these farmers had exhausted the supply of water available on the east slope of the Rocky Mountains. They then looked to the west slope and the Colorado River drainage. The Larimer County Ditch Company was formed in 1880, and ten years later its successor, the Water Supply and Storage Company of Fort Collins, filed for 524.6 second-feet of upper Colorado River water. The water company obtained its appropriation, since there was little interest at that time on the west slope of the Rockies in the interbasin transfer. The competition for water between the two slopes was later to get more intense when it became evident there was not enough for everyone's desires.

Tilted slightly downslope and heading northeast, the ditch trapped water flowing down the mountain streams in the summer and transported it by gravity flow over the divide at La Poudre Pass and dumped it into a creek of the same name. From Long Draw Reservoir two miles downstream the water flowed into Cache La Poudre River, thence into the South Platte River and onto 37,425 irrigated acres divided into 260 farms where corn, alfalfa, sugar beets, potatoes, vegetables, and barley were grown. In 1892 when the water company incorporated, its stated purpose was "the construction of new ditches, to take, divert and appropriate as much as possible of the unappropriated waters of the Grand. . . ." At the time of incorporation the amount of capital stock was $60,000. Seventy-five years later the value of the stock was more than $9 million. "And," Michener wrote, "any hard-working newcomer who bought irrigated land in the years from 1896 through 1910 acquired a bargain whose value would multiply with the years. This was bonanza time, when the last of the great irrigation ditches were being dug, when desert land was being made to blossom."

In 1915 Rocky Mountain National Park was created and in 1930 the Never Summer Mountains were added to the park by Congress, which allowed the ditch and an undetermined width of right-of-way to remain in the park as an incongruous inholding. The ditch, which was built piecemeal, was completed to its present length before the water company decided in the late 1960s to raise the height of the dam outside the park an additional 23.4 feet to increase the capacity of Long Draw Reservoir from 4,400 to 11,000 acre-feet. At the same time portions of the ditch were to be lined with wire mesh and concrete to cut down on seepage, which was accounting for a 30 percent loss in flows. It is this lost water that constitutes the source of the upper Colorado River.

To hike up to the ditch and spend a couple of nights at the headwaters is no simple task. What the Forest Service was considering in the Bridger wilderness had already arrived at Rocky Mountain National Park. One's movements in the wilderness are carefully controlled and rationed by a rather complex permit system. So I drove over popular Trail Ridge Road, one of the few high crossings of the Continental Divide in the Rockies, to the park headquarters just outside of Estes Park. The road bisecting the park links Estes Park on the east with Grand Lake on the west, two towns pervaded by Rocky Mountain kitsch. About 25,000 cars a year— between 500 to 700 an hour at peak periods—travel this road, which rises to 11,796-foot Fall River Pass. On this early fall day the lower slopes of the Never Summer Mountains, a fine name considering the arctic world of tundra found at higher elevations, were splashed with the vivid colors of aspen trees set against the evergreens. These mountains are more gentle, less sculptured and granitic, more flowing and lava-encrusted than the Wind River Range to the north. Because of their volcanic origins, they are darker. On this day there was a glint of bright, metallic light along the scar. I later learned that flatbed trucks owned by the water company were hauling large sections of galvanized pipe into the park along the ditch's service road. The pipe was to replace a wooden flume.

I have to admit, although I believe in the system, most times I do not bother with wilderness permits because of the bureaucratic bother of obtaining one. But this time I decided to be legal. When I arrived at the back country ranger's office, I found seven other potential hikers waiting to be processed by Emily Edgecomb, the young ranger. At the time she was helping a thoroughly confused Florida couple. The overnight itinerary has to be exactly determined in advance because the backpacker must specify each night's campsite and the ranger has to check a master list to see if there is a vacancy for the specified night. This requires knowing the distances to and characteristics of, say, Mosquito Creek or Hitchens Gulch, before starting out.

After the Florida couple came a man from nearby Fort Collins, who asked, "What happened to the good old days when you just stamped in and the ranger said 'okay'?"

Ranger Edgecomb replied, "This is practically a metropolitan park for the amount of use it gets, if that's any consolation."

It wasn't, but it would have been difficult to argue with her tact. I made my campsite choices, received my permit with eleven rules stamped on the back, and was told 1,300 permits had been issued the previous year for the area I was going into. Once in the mountains I camped at designated sites that I had not chosen, having changed my mind when confronted by the realities of the terrain. In the fall this was not a problem, but in the summer it could have been. As Park Service literature clearly stated, at that time of year "DEMAND EXCEEDS THE SITES AVAILABLE."

Through the lobbying efforts of naturalist Enos Mills, Rocky Mountain National Park was created in 1915. "Without parks and outdoor life," Mills proclaimed, "all that is best in civilization will be smothered." The congressman who sponsored the legislation was Edward F. Taylor, the lawmaker who was later to be the principal force behind changing the name of the Grand to the Colorado River. Earlier, as a Colorado state senator arguing against stockmen having to lease federal lands for grazing, Taylor had said of the rancher, "The streams and the mountains are his. Uncle Sam has been paid a thousand fold already for the land by the blood and bones of these people." There was something in the park bill for both men. Mills got a park, albeit somewhat restricted, and Taylor got water for the ranchers. A clause in the bill allowed the Bureau of Reclamation to "enter upon and utilize for flowage or other purposes any area within said park which may be necessary for the development and maintenance of government reclamation projects." By 1974 the bureau was ready to relinquish this right, but its sister agency, the U.S. Geological Survey, wondered if designation of a large part of the park as wilderness would hinder its water-evaluation studies of the area.

Between the establishment of the park and 1974, the bureau had managed to drive a tunnel 13.1 miles through the mountains under the park in order to siphon water from Grand Lake in the upper Colorado River drainage area into the South Platte River drainage. The clause allowed the bureau to reconnoiter the route from above, and do some exploratory drilling. Both portals of the tunnel, from which all maintenance is performed, were located outside the park. The tunnel passes 3,800 feet below the Continental Divide. The Colorado–Big Thompson Project—the water flows from the Colorado to the Big Thompson River —was a much greater interbasin transfer of water than the older Grand Ditch. But it serves some of the same area on the east slope of the Rockies.

Besides the Grand Ditch, there are other water developments in the national park the Park Service would like to see eliminated. Thirty water rights issued prior to the park's establishment resulted in the construction of four small dams, dating back to around the turn of the century. But even earlier the park could never truly have been considered virgin territory. The Ute and Arapahoe Indians had used it as a transportation route across the mountains, their trail roughly paralleling the present Trail Ridge Road. There was a short-lived gold and silver mining boom around 1880 at Lulu City on the headwaters of the upper Colorado. A post office was opened in 1880 to serve 500 inhabitants, and closed in 1883 when most had departed. A few miles downstream in Kawuneeche Valley, first came the homesteaders, then the dude ranchers, and finally campgrounds serving auto-bound tourists. In 1929 the Never Summer Mountains were added to the park and the Park Service then began purchasing the private inholdings. By the mid-1970s, over 2.5 million visitors were entering the 410-square-mile park annually, most of them attracted by the first mountain barrier encountered on the drive west across the plains.

The hike up to the source begins at Phantom Valley Ranch, the spot where Squeaky Bob, a onetime horse thief and cowpuncher, set up the first dude ranch in the area, a small tent town named the Hotel de Hardscrabble. Bob entertained such notables as President Theodore Roosevelt, Supreme Court Chief Justice Charles Evans Hughes and actress Cornelia Otis Skinner. On a talus slope to the right, not far up the La Poudre Pass Trail, are the black hole and tailings of an early-day mine. The day I hiked up the trail I passed a couple and their blond-haired son.

The father, carrying a metal detector, asked, "How many miles to Lulu City?"

I checked my topographical map. "About a half-hour hike," I replied. "Are you going to find some gold?"

The boy answered, "I hope so."

The Park Service sign at the site of Lulu City has all the breathless drama of Disneyland.

> Gold brought man to this valley. High hopes for a big strike. Hammers and nails built a town bearing a pretty girl's name. Man paid a price in sweat and aching muscle. Picks and powder scarred the land. Dig! Scrape! Blast! But to no avail. The land did not reveal its treasure. Finally the blows of the picks echoed a cadence of despair. Man left Lulu City. His talk and laughter yielded to the sounds of wind and river. Trees and grass returned to the wounded land.

But the real gold, not the metallic but the liquid variety, was in the Grand Ditch, and there is no mention of that on any sign. The ditch, over a period nearing one hundred years, has contributed to the production of many millions of dollars' worth of produce and livestock, but somehow the drama is lacking in the story of irrigation and the West. It has failed to capture the imagination. The West has always been thought of in terms of Indians, cowboys, and gold miners—picturesque folk either riding horses or pulling mules. I cannot recall any movies made about ditch tenders or irrigators.

From Lulu City the trail climbs the west bank of the upper Colorado and passes through Little Yellowstone Canyon, a miniature of its namesake to the north. There was a gust of wind, and a shower of colorful aspen leaves fell to the ground. A mule deer momentarily blocked the trail ahead; then, thinking better of it, bounded off into the woods. The next day I was to see a small herd of the now-rare Rocky Mountain bighorn sheep scamper along the Continental Divide at Thunder Pass, an early wagon route into Lulu City and the companion mining settlement of Dutchtown. Lulu City got its name from the daughter of the developer; property values doubled there in one six-week period. Dutchtown was named for two German prospectors who shot up the place one Saturday night.

The flow of the streams feeding into the upper Colorado had lessened dramatically, and when I came around a bend in the trail, I saw the reason why. It was just before noon when I reached the ditch, a sinuous curve, half filled late in the season of a drought year. Four flatbed trucks passed me, stirring up the dust on the adjacent service road. They would return about an hour later after a crane, located just outside the park, placed one section of galvanized pipe upon each bed. I walked the mile and a half to La Poudre Pass and investigated the meadow below the ranger station, and just west of the divide. The seepage first collected in a small pond a hundred yards below the ditch. It then flowed sluggishly through the meadow to the outlet where, stained a rusty red, it dropped over a few rocks at the point where a timber-and-earth dam had once impounded these first waters. The dam had long ago given way and the water now dropped unimpeded into the Little Yellowstone.

The wind started to blow and thick, fluffy clouds occasionally blocked the sun, cooling the air. I felt disconsolate, not jubilant as I had at the start of the Green River. Perhaps with the bite of autumn and winter not far behind it at the 10,175-foot level, it was time to get out of the mountains. Or perhaps it was the lack of a proper beginning that chilled me.

3

What I remember most about the San Juan Mountains was the rain and a hot spring. I remember the rain not because of the discomfort it caused me, although there was some, but because it was the first solid, sustained downpour of water I had felt in some time, and it eventually led, in the fall and winter of 1977, to the end of two years of drought in the West. During this period, the continued existence of large-scale human habitation in this region sometimes seemed questionable. But water stored in the extensive system of reservoirs along the Colorado River became the sustaining factor for much of the West during this period that ended, at least for me, with the rain in the San Juans. The hot spring high up on the West Fork of the San Juan River at about the 9,000-foot level I remember for its comfort, a warm place in which to soak and watch the cold rain fall. Nude in the warm, saline water, I felt impervious to the elements.

The headwaters of the San Juan River, the third in the arc of the four main tributaries of the Colorado River circling from Wyoming to Colorado to New Mexico, are in the Weminuche Wilderness Area of southwestern Colorado. It is the state's largest officially designated wilderness area and the least known, since it is far from any large cities. Before leaving for the mountains I had asked an official at the San Juan National Forest headquarters in Durango, Colorado, which tributary the Forest Service considered the source of the river. It was a fair question, since the federal agency within the Department of Agriculture administers the area. A senior hydrologist said the West Fork, a junior hydrologist said the East Fork, and a third member of the office thought it was the West Fork. I was told there was an interbasin diversion high up on the headwaters of the West Fork; that was enough to interest me in that source, rather than the more readily accessible East Fork. The pattern of water use promised to be similar to what I had encountered on the upper Colorado.

Two San Luis Valley irrigators, Don La Font and Harley Fuchs, built the original three ditches in 1938 to transfer water from the Colorado basin to the Rio Grande River basin. The ditches run from 12,000 feet to 11,400 feet, well above timberline, with the water being captured south of the Continental Divide in the wilderness area and dumped into Red Mountain Creek, a tributary of the Rio Grande north of the divide. Although they were originally built to benefit agriculture, the Colorado Division of Wildlife purchased the ditches and water rights in 1970 and since then has used the water to enhance hunting and fishing in the upper

Rio Grande area. Like the Colorado, the Rio Grande is a fully committed river, and this diversion is the only water available to prevent fish kills when reservoirs in the Rio Grande National Forest are emptied every irrigation season. "Trans-mountain water," stated a report by the Colorado Division of Wildlife, "is the only water that is free from controls as related to the [Rio Grande] Compact. As a result, it is the *only* means available to fill newly constructed reservoirs that may need to be drained for repairs or fishery improvements, or otherwise increase total basin-wide storage above present levels."

The hike up the West Fork began at the small parking area on the dirt road just below Borns Lake, a private resort. While readying my pack, I encountered a man with a rifle who informed me he was hunting elk and had just bought a new pair of boots for that purpose at Sears. The boots were guaranteed not to leak for one year, he said. The guarantee was about to be thoroughly tested. I informed the elk hunter of my route, hoping not to get shot, and set off up the trail.

The San Juans are the largest subrange of the Rockies in this country, encompassing an area of 10,000 square miles in southwestern Colorado. It is an exceedingly wild area, even in this time of burgeoning population in the West, yet most of it has been combed at one time or another by prospectors and miners. Although I was not to see much of this portion of the range because of the rain and clouds, on previous trips I had seen enough of other portions to be greatly impressed by its dramatic beauty. I happen to feel that the Front Range or central Rockies, the area west of Denver and Boulder which gets the heaviest use, is greatly overrated, and that the more remote ranges, such as the Wind River and San Juans, form a more classical backdrop for the division of the nation's major river systems. The mountain ranges that split the watersheds run in a continuous rank from the end of the Brooks Range in Alaska to where the Andes dip into the sea in Tierra Del Fuego at the tip of South America. Nowhere else in the world is there a more extensive, clearer demarcation of that essential element, running water. At other times I had traveled to these two extreme ends of the Great Divide that also harbor their own arid areas—the Patagonian Desert to the south and the tundra barrens of the North Slope of Alaska.

The extremes of climate at the two ends of the Colorado River system have never failed to amaze me. Here along the Continental Divide in the San Juans there can be fifty inches of rain a year, while at the end of the river in Mexico it is a good year if three inches fall. So the practice has been to trap the water, falling mainly in the form of snow at higher elevations, and store it in reservoirs until it is needed. From the reservoirs the water is diverted onto the land by canals and ditches. The mechanism can be as crude as the unlined dirt ditches, periodically choked by rock and snow slides, at the headwaters of the West Fork. Or it can be as sophisticated as the $2 billion Central Arizona Project, where water will be pumped from the lower Colorado into a concrete-lined aqueduct and transported to storage reservoirs in the Phoenix and Tucson areas. An eventual shortage had been foreseen; indeed, a 1968 federal law had directed that the water supply be augmented, which was one reason why, as I hiked up the West Fork Trail, I could not be sure whether the rain that fell on me was the result of natural causes, as one would normally suppose, or of generators spewing silver iodide into the air.

Everyone—and that constituted mainly the small, tight knot known as western water interests—who listened to the voluminous testimony involving passage of the Colorado River Basin Project Act of 1968 knew there was not going to be enough water to go around to meet all expectations after this last, major diversion of the river's water, so one of the key points on which passage of the act depended was supplementing the river's flows. Diverting Columbia River water into the Colorado system would be costly and was politically infeasible, since Senator Henry Jackson of Washington, chairman of a key committee that had to act on the bill, was opposed to such a plan. Therefore most hopes for more water came to be pinned on old-fashioned rainmaking—known as cloud-seeding, weather modification, or precipitation management in bureaucratic circles.

Wasting no time, in the fall of 1970 the Bureau of Reclamation placed 35 silver iodide generators along the southern flank of the San Juan Mountains near the headwaters, and over the next five winters released 704.66 pounds of chemicals into the air, hoping rainfall would increase over natural precipitation levels. At the time, this was the nation's largest cloud-seeding experiment over mountains. When it ended, it was determined that there was a "potential" to increase normal precipitation by about 10 percent over the San Juans. This meant an additional 159,000 acre-feet of water for the San Juan River basin and 151,000 acre-feet for the Rio Grande. Silver iodide was not judged harmful to the environment —at least not in the short run, although nobody was sure about the longer-range consequences—and most of the small mountain communities worried only about the increased avalanche danger. So the bureau avoided making rain over populated areas and set up criteria to cease operations if the avalanche danger threatened the road over Wolf Creek Pass, near where I was hiking. There was also to be no cloud seeding during deer and elk season, when hunters were in the mountains.

The experimental project ended in 1975, so most likely the rain I experienced came from natural causes, but the bureau was seeking to expand Project Skywater into a much larger, more permanent operation. A few months after I left the San Juans, Reclamation Commissioner R. Keith Higginson testified before a congressional appropriations committee, "Data from the recently completed Colorado River Basin pilot project and other experiments substantiates the estimated potential of increasing mountain snowfall 10 to 15 percent annually by selectively seeding winter storms. Also, the accompanying environmental studies showed no significant short-term adverse ecological impacts. These results remove much of the basic scientific uncertainty. . . ." But others, including the National Science Foundation, were to point out that such "potential" increases in precipitation were "speculative," and the bureau itself admitted in one document that such estimates "are very divergent and in many cases controversial. It is acknowledged that it is not clear under exactly what conditions increases will occur." Such diverse water interests as the Colorado River Board of California and the Southwestern Water Conservation District of Durango, Colorado, wanted a larger demonstration project. There was little doubt it would eventually come about, although proof was lacking as to its effectiveness. More water was needed, and cloud seeding seemed the most painless way of obtaining it.

I camped that night at the hot springs, and leaving my tent and backpack behind, started at seven the next morning for the climb to the divide. The San Juans, compared to the two ranges further north harboring the headwaters of the upper Colorado and Green, seemed lush and, at least on this misty day, resembled the richly vegetated rain forests of the north Pacific Coast. Ferns grew in cool, moist places and fallen trees spanned swollen creeks. I stopped at one vista point to take a picture, and where I would have been walking had I continued, a small rock slide tumbled down the steep slope. Everything seemed to be loosening up with the rain.

The unmaintained trail gave out in places, and it took a bit of guesswork to pick it up again. Just below the crest of the divide ridge I saw the fresh Vibram-soled tracks of another hiker in the mud and soon came upon a lone man huddled in the lee of a fir tree at the divide. He lived in a Denver suburb and usually took his two-week vacation on a different backpack trip every year. This was his second day enroute along the Continental Divide Trail to Silverton. We talked for a while. The rain descended in nearly horizontal sheets and, given the poor visibility, I decided to descend without searching for the ditches. They were described in the guidebook as ugly scars in an otherwise beautiful alpine setting. I would have to leave this description unconfirmed. On the way down I halted for a moment to try to separate out the different sounds of water. I could distinguish the rain hitting the ground and my parka, the squish of water in my boots, the gurgle of a small creek, and the roar of the West Fork. That made five separate sounds, but I was sure there were more. Whatever the number, they all sounded good that day as the drought began to end in the West.

4

By way of contrast, the headwaters of the Gila River, the fourth and last major tributary of the Colorado River system, were dry when I went in search of them one April. Or, perhaps more accurately, they were running underground as does most water in the Southwest, and I had failed to perceive this difference. Granted it was a drought year, but what I found, lack of running water where the source should have been, is endemic to the region. The mistake I made was to lay the humid-area concept of a river on an arid land—in humid areas, rivers have visibly flowing water, while the more likely reality in western New Mexico, southern Arizona, and the northern portion of the Mexican State of Sonora is a dry riverbed.

"Intermittent" is the technical term used to describe the flow of such rivers. But for most of the Gila River system, including its tributaries in Mexico, the lack of any running water is more prevalent than the occasional surface flows from intense summer storms or the spring snowmelt. It was a lesson I was preaching to others—that the West was a different land—but one I had not completely absorbed myself as I drove inland from the West Coast. I assumed the sources of the Gila would be backed up against the Continental Divide, like the headwaters of the three more northerly tributaries. After two days of driving and hiking without an accurate map and only my preconceptions to guide me, I found this simply was not the case. They bubbled up west of the divide, which at this point is more a series of insignificant rounded hills than a sharp demarcation. This is juniper and piñon country, with low mesas and volcanic hills, dark and laden with the history of ancient Indian civilizations and the early Spanish explorers. The grass was dry and golden that spring.

I camped at Wall Lake in the Gila National Forest of western New Mexico. The outlet from Wall Lake, twenty acres of water dammed by the New Mexico Game and Fish Department, was designated on the map as the East Fork of the Gila River. The next day I saw a dead cow athwart the spillway of the small dam, which must have done wonders for the water quality downstream. A rancher stopped his pickup on the road over the dam, eyed the cow, then drove on. I was to spot two more dead cows lying in the stream that day. My plan was to walk up Taylor Creek, the lake's principal tributary, to the source and thus the headwaters of the East Fork.

There was frost on the ground early that morning; it had snowed two days previously. A slight mist hung over the lake as I set off up Taylor Creek. But the morning heated up rapidly, considering it was still spring at the 6,500-foot level. This was cattle and mining country. There were numerous barbed-wire fences to cross. Cow droppings were everywhere; hoofmarks had churned up the landscape. The grass was chomped short and there were those arroyos zigzagging back through the streambank that are found on overgrazed lands. On the south bank a weathered mill stood, and not far beyond, rusted and almost blending into the dry landscape, a mechanical vibrating screen patented in 1925 and built in Racine, Wisconsin, lay on the ground. The instructions read, "Force a small amount of neutral ball bearing grease into bearing housings TWICE A DAY. Be sure grease is clean and free from sand or grit." The armature still moved and I could hear the sloshing of grease.

I looked up to see a man and boy emerge from behind some bushes. The man was carrying a rifle and I thought, Oh, oh, it must be the rancher on whose property or leased land I was trespassing. But it turned out to only be Jesse Judd of El Paso, Texas, and his son, Tom. Jesse sported a plastic baseball-style cap with the label PETERBUILT emblazoned on the front, while his son had a mouthful of braces. The pair were not very threatening on closer inspection. They were hunting bear, black bear, to be exact, and I learned from them that it was also wild turkey season. Jesse rested the barrel of his .270-caliber rifle on the toe of his boot and leaned on the stock as he told me of the powerful CB radio in his pickup truck. Forty miles on a good day without much skip.

I was to meet a number of other Texans in the Gila National Forest. It is their closest escape to greenery and coolness. Not far upstream the flow of Taylor Creek lessened appreciably and I suddenly came upon a dry streambed. While I was thinking about Texans, I overshot the source, so I backtracked a few score yards to a rather sickly, algae-choked pond where the water rose to the surface from an underground spring. The spring was thirteen miles due west of the Continental Divide.

The West Fork of the Gila River, much further west of the divide, is considered the true source, and a few days later I set off up Willow Creek, an extension of the West Fork, to its headwaters at Bead Spring just inside the Gila Wilderness Area. The Indians supposedly made their offerings of beads at this spring. On this day I had difficulty in locating the spring, because the snow on the north slope of the Mogollon Mountains was still deep and covered up almost all running water. In an area corresponding generally to where Bead Spring should be located, according to the Forest Service map, I imagined I was at the source. A swatch of running water, black against the white snow, was briefly exposed. It was fresh and cool at this spot, and I felt this was the proper place for such a river to begin. Ross Calvin wrote of the headwaters of the Gila in *River of the Sun:*

> Through the sunless canyons the myriads of fine tributaries, eating at their roofs of ice, go leaping along. Farther down they bathe the roots of red-twigged willow bushes half submerged in snow, which respond with soft awakening catkins. The waters are fully unchained now, and in glassy flow they go shooting down mountain slopes all over the arid Southwest.

But the waters of the Gila and its tributaries no longer reach their parent stream, the Colorado, at Yuma, Arizona, just as the waters of the Colorado no longer reach their historical outlet to the ocean, the Gulf of California. The surface flow of the Gila is used up before the halfway mark along its six-hundred-mile length, and nothing flows into the Colorado now except an occasional flash flood. Maps for general use, such as the official state road map, mistakenly still show a solid blue line designating a perennial river, while on the ground the river is a dry streambed meandering without visible nourishment through the desert.

It was not always this way. Before the first large influx of settlers and cattle in the late 1800s, although dry during periodic droughts, the Gila normally flowed to the Colorado as a wide, shallow stream, and such tributaries as the Verde, Salt, San Francisco, San Pedro, and Santa Cruz rivers were, at the very least, marshy creeks rich in surrounding vegetation. In 1540 the chronicler who accompanied Coronado on his march across Arizona described the Gila in its middle reaches as "a deep and reedy stream." In 1870, Lieutenant William H. Emory, who made the first scientific survey of the Gila, measured its flow as one-half the Colorado's at their junction. Rowboats and an occasional raft floated down the river, although its flows were never deep or consistent enough for steamboat traffic. Overgrazing and a slight change in the seasonal rainfall pattern at the end of the nineteenth century shrank the flows. With the completion in 1910 of Roosevelt Dam on the Salt River, a tributary of the Gila, and Coolidge Dam on the Gila in 1928, the surface flows withered away until, in the words of a recent Bureau of Reclamation report, "Except for infrequent large floods or an exceptional runoff sequence, outflow from the [Gila] subregion under present conditions of development is negligible."

The headwaters of two tributaries of the Gila deserve special mention because they rise in Mexico and were the corridors through which Europeans first found their way into the West. The San Pedro and Santa Cruz rivers drain about 1,250 square miles in the Mexican state of Sonora. The surface flows of water north across the border into the United States are small and erratic from these most southern tributaries of the Colorado; but here, as elsewhere, there are plans to wring the last drop of water from the river system. Charleston Dam, a part of the Central Arizona Project, would be built on the San Pedro to trap runoff north of the border and transport it by a sixty-four-mile pipeline to the city of Tucson. While the San Pedro River originates in Mexico at about the 5,000-foot level near the town of Cananea, the Santa Cruz actually gets its start in the Canello Hills of the Coronado National Forest some twelve miles north of the border, dips south into Mexico, then, much like the shank of a fishhook, runs north in a basin parallel to the San Pedro, crossing the border just east of Nogales. At one time there were plans to build a dam and reservoir astride the border to serve the needs of the two communities on each side of the boundary. However, since the river flows only between 10 and 15 percent of the time, such a plan was not thought to be sufficiently feasible.

It was by way of the headwaters of the San Pedro River, and later the Santa Cruz, that the first Europeans came to the Southwest sixty-eight years before the settlement of Jamestown, Virginia, and eighty-one years before the Pilgrims landed at Plymouth, Massachusetts. In 1539 a black slave known as Esteban strode into the Colorado River Basin by way of the headwaters of the San Pedro, followed shortly by a Franciscan missionary, Fray Marcos de Niza. In the same year Francisco de Ulloa, also part of the same grand Spanish scheme to find the supposedly rich Seven Cities of Cibola, sailed to the head of the Gulf of California, where he was forced back by the threatening tidal bore, but not without suspecting he had found the mouth to a great river. All three men were scouting the trip Francisco Vásquez de Coronado was to make in 1540. Esteban, who posed as a god, was proved mortal by the arrows of Zuñi Indians, and the priest, who was a bit of a liar, turned back before reaching the goal of the Zuñi pueblo in western New Mexico. But the description he carried back with him was enough to excite Coronado, who, after an exhausting trip across the headwaters of the Gila watershed, found nothing but a collection of mud huts. It was a bitter disappointment for Coronado and his men, but they left their marks upon the West. Their search for riches established the pattern for the exploitation of the region, and they introduced livestock to the West. Colonization by the Spanish came later and was to result in the first use and diversion of the river's waters by white men. Coronado's men had approached the Colorado River in the Grand Canyon, but drew back from the abyss. The interior West would remain unexplored for quite a few years more.

The best perch from which to view the start of the San Pedro and Santa Cruz rivers and the early Spanish route into the Southwest is from the peak of 9,453-foot Mount Wrightson, not too far north of the border in the Santa Rita Mountains. It was early in the morning of a late spring day when I set off on the sixteen-mile round-trip hike. Three or four inches of new snow lay on the ground, the result of an unseasonable storm the night before. The trail first followed a creek past some old mining claims and then climbed steadily to a saddle where a sign, rather dramatically headed "Stranger Take Heed," designated the spot where three Boy Scouts in a previous year had frozen to death in an early-season snowstorm. As the sun rose, the snow quickly melted at the lower elevations. Higher up near the summit the snow was molded in wind-blown forms to the dwarf vegetation; the encrusted forms, dazzling in their whiteness, stood in stark contrast against the sereness of the Sonoran Desert below.

Eventually eight of us, not including one dog, made it to the summit on that Sunday. On the way up I passed three Michigan youths who were trying to dry out after spending a wet, cold night camped on the mountain. At the summit a University of Arizona student related the mistake he and some friends had made when they hiked up the mountain with heavy packs in the heat of the summer. There were only a few signs of water visible from the peak. A thin band of vegetation paralleled the dry riverbeds. There was a glimmer of blue from Lake Patagonia, a small impoundment of water near the Arizona community of the same name. And beyond to the south were the gray ranks of low, serrated mountains that cradled the start of the two river basins in Mexico. From this point the southernmost branch of the Colorado River system bends in a large arc and, after joining the main Colorado, ends back in the silt of the delta plain, bordering this same Sonoran Desert in Mexico.

The Spanish were not the first to march across these lands at the headwaters of the Gila River system. The Indians came first, and the evidence of their civilization gives the region a sense of an ancient presence. The Mogollan culture got its start around 300 B.C., a couple of hundred years before the Anasazi to the north, and flourished along the narrow, shallow canyon bottoms of the upper Gila drainage until it was absorbed by the Anasazi around 1000 A.D. The perennial streams in the mountains sustained the agriculturally oriented Indians. Yet the Mogollon (named after an eighteenth-century governor of New Mexico and pronounced "muggy-own") never became as sophisticated in the use and diversion of water as their neighbors in the desert to the west, the Hohokam Indians. All three early Indian cultures suffered the same fate of abandonment and dispersion due to changing climatic conditions throughout the Southwest in the thirteenth and fourteenth centuries.

The best examples of the Mogollon culture are contained in and around Gila Cliff Dwellings National Monument, astride the river and not far from its source. One day I set off up the Middle Fork of the Gila with Sharon Prell, a ranger at the monument, and two of her friends—Steve, a bicycle-builder from Iowa, and Dave, an elderly dropout from the East who has been doing odd jobs around the monument since the early 1960s. Our goal was some Mogollon ruins Dave had found eight years previously. After numerous crossings of the river and some guesswork we stumbled across the circular pit dwellings sheltered under a massive rock overhang about 250 feet up from the river. Dave posed the question, was there a closer source of water? Steve followed a dry creek bed down a short distance to where the water popped to the surface. Food, water, shelter, and safety were available at this spot for the ancient Indians.

Some years after the Mogollon disappeared from this region, the Apaches arrived in the latter part of the seventeenth century. Following the period of Spanish exploration, a few fur trappers and American explorers like Kit Carson passed through the area, but it was not settled permanently because of the hostility of the Apaches. The first white men through this area had found the Apaches friendly. But a rather nasty trick played on the Indians in 1835 by an English fur trapper named James Johnson changed all that. A bounty had been set on Indian scalps by the Mexican government, so Johnson and his trapper friends decided to give the Apaches a party. A large number of gifts were set out, and when the Indians went to inspect them a small cannon loaded with enough odd pieces of metal to constitute a deadly shrapnel charge was discharged, and some twenty Apaches lay dead. The Apaches retaliated in kind, then retreated to their hiding places in the Mogollon Mountains after their deadly raids. It was not until around 1870 that the headwaters of the Gila were safe again for permanent habitation by white settlers and miners. Colonel H. C. Hooker moved into the area with fifteen thousand head of cattle in 1872, and Mormon settlers populated the wider valleys where the river issued forth from the mountains and built irrigation ditches and diverted the water onto the land. More cattle followed; the consequent massive erosion of the watershed began in the last two decades of the nineteenth century.

The Gila River Forest Reserve was established by President William McKinley in 1899 to bring some order into the use of natural resources, and Theodore F. Rixon, who surveyed the area for the federal government in 1903, made some observations. Corn and alfalfa were being raised in the small farming communities along the Gila and San Francisco rivers. "Agriculture is carried on to a limited extent only along such of the main streams as rarely run dry, no large agricultural areas exist anywhere within the confines of the reserve. With the introduction of reservoirs and irrigation ditches the amount of available agricultural land could be largely increased, but as the market for the products of this district is so distant, being in no instance less than 90 miles away, the cost of putting new land under water would not pay." The primary use of the reserve was grazing, and Rixon noted that livestock, particularly sheep, had created "a barren desert, not a blade of grass being seen and even the roots being entirely destroyed. When the wind blows, the sand and soil rise in vast clouds." The farm products were sent to the mining communities at higher elevations. Logging was described by Rixon as desultory. He criticized past timber harvest practices, recommending that they be confined to smaller areas and made more systematic. But recreation was flourishing, with tourists from all over the country flocking to Gila Hot Springs to take the cure.

This is the region where the naturalist Aldo Leopold spent much of his early career in the Forest Service and first developed the concept of setting federal lands aside as unspoiled wilderness areas. He had seen firsthand what damage could be done to such areas in the Southwest, and later wrote in *A Sand County Almanac:* "In arid regions we attempt to offset the process of wastage by reclamation, but it is only too evident that the prospective longevity of reclamation projects is often short. In our own West, the best of them may not last a century." Primarily through Leopold's efforts, the Gila Wilderness, the first such Forest Service unit, was established in 1924. But this hardly guaranteed its retention as an untrammeled area. A 1975 Forest Service management plan for the wilderness unit stated: "Poor distribution and excessive numbers of domestic livestock in portions of the wilderness have resulted in overgrazing and damage to wildlife resources." Air pollution from the copper smelters in Hurley, New Mexico, and Morenci, Arizona, had drifted into the wilderness. There are 107 miles of barbed wire fencing, plus 19 small rock or masonry dams and 56 earthen dams forming stock ponds in the wilderness.

On a rise just off the highway bordering the west side of the wilderness, with a fine panorama back toward the dark mass, a bronze plaque was attached to a boulder in 1954 by the Wilderness Society to commemorate Leopold's foresight in helping to establish the Gila Wilderness. When I visited it, the plaque had been splattered by three bullets, a rather low average for signs in the West. One had smashed between the *e* and *o* in "Leopold," making it indecipherable to all those who did not otherwise know of his role in establishing the wilderness area.

In 1930, six years after the Gila wilderness was created, the first study was undertaken for a dam on the Gila River. It would eventually become known as Hooker Dam; and the waters it impounded would inundate a small portion of the wilderness area, if it was constructed as originally planned.

Two men who knew the political process intimately, and were not afraid to use their potent leverage, eventually got southwestern New Mexico more water than had been apportioned to it by the Supreme Court in the landmark *Arizona v. California* decision of 1964. New Mexico was disappointed with the court's decision, so it immediately set about to recoup its losses through the legislative process, much as California also did, using Arizona's desperate need for the Central Arizona Project as the point of leverage in the political process.

When Clinton P. Anderson arrived in Albuquerque as a young newspaper reporter, one of the friends he made was Aldo Leopold, who was attached to the regional Forest Service office there. The two men talked, and Anderson became a fervid convert to the concept of wilderness. Many years later, at the dedication of the bronze plaque commemorating Leopold, Anderson—by then a powerful senator—would declare: "Those of us who may visit within the wilderness and who are able to rest and be restored in our peace of mind and body by the quiet that it will always possess have none the less an obligation to see that the work of one generation shall not be sacrificed by those that come after. We have an obligation to make sure that this area may remain untouched for generations and perhaps centuries to come." During the congressional hearings on the Wilderness Bill in the early 1960s, Anderson, then chairman of the Senate Committee on Interior and Insular Affairs, would avow that it was his early contact with Leopold and his ideas which had brought him to sponsor the bill.

But one of the provisions the senator inserted in the bill was a clause permitting construction of water projects in wilderness areas, if the President determined that such facilities better served the public interest than wilderness. Anderson had Hooker Dam in mind and would later claim, as would the state of New Mexico, that Administration support and President Lyndon B. Johnson's signature of the Colorado River Basin Project Act in 1968, of which Hooker was a minor component and the Central Arizona Project a major one, constituted the necessary presidential approval. As an elected representative from an arid state, Senator Anderson had a long history of support for water projects in the West. When the Central Arizona Project came before Congress for serious consideration after the Supreme Court ruling, Anderson was in a key position. Respected by his colleagues and possessing the requisite seniority, Anderson was the chairman of the Senate subcommittee the bill had to pass through on its way to the full committee chaired by Senator Jackson of Washington, with whom he had a close working relationship. Wilderness and a dam at this site were not incompatible in Anderson's mind. He referred to the site as "an ordinary riverbed," and maintained that an equivalent amount of acreage could be added elsewhere to the wilderness area. The senator preached flexibility and balance in the use of wilderness, and was disheartened by the intransigence of the conservationists, who feared the precedent that might be set by this first intrusion into the newly created wilderness system.

Behind every western congressman dealing with water matters is at least one able technician. In this case Anderson had S. E. Reynolds, New Mexico's longtime state engineer, to rely on for advice. The two men set out to get New Mexico more water. Such men as Reynolds, be they state engineers or chief engineers, are common in the western states. They have formed a loose network among themselves, with their separate boards or commissions on one side and their congressional representatives on the other. This triumvirate overlaps with the "Iron Triangle"— the federal bureaucracy, Congress, and water users. There are common elements in both; for instance, a water user can be a member of both a local water district board and a state water commission. The apex of both interlocking triangles—which together form what are referred to as western water interests—is Congress, and the technicians are the principal conduits.

Locked into their positions for most of their careers, they outlast the more transient governors they are supposed to serve. In the process, the technicians have built up their own constituencies among the various local water users, who frequently have the largest stake in the economies of these western states. Few people other than the technicians have the time to master the web of complex and arcane laws and institutions that govern the distribution and use of water in the West. Most defer to them for advice. Sometimes the advice goes beyond the merely technical to the formation of basic positions—western water policy, in other words. This is especially true if a state water commission or a congressman have not taken the time to master the technicalities of water.

For a number of years, Reynolds always had a trusted lawyer from the New Mexico Attorney General's Office close at hand. His name was Paul Bloom. When Bloom left for another job, Reynolds's counterpart in Arizona, Wesley E. Steiner, joked, "How are you managing to keep out

of jail now that Paul has flown the coop?" Novelist John Nichols depicted
the economic and political forces behind the distribution of water in his
novel *The Milagro Beanfield War*. Presenting an account of an incident in
New Mexico that had certain echoes of real life, Nichols wrote of two
characters not unlike Reynolds and Bloom:

> During that time they had weathered the heaviest political storms to
> sweep the state. They had also sweated, plotted, finagled, begged,
> twisted and driven their way to what they felt was their state's fair share
> of Colorado River Basin water; they had made deals with Texas and
> California, with Arizona and Colorado and Utah; and they had created
> lobbies in Washington to have dams built and rivers channeled; they
> had set into motion adjudication suits to determine how much water
> people did or did not have in all areas; they had literally decided how
> the rivers would run and which people must benefit the most from
> those rivers.

When it became obvious that the Supreme Court was not going to rule
in New Mexico's best interests, Reynolds convinced the state to furnish
the Bureau of Reclamation with funds to study the water needs of the Gila
River basin in New Mexico. He wrote Senator Anderson in 1964 that,
with the Central Arizona Project now being considered by Congress, "it
does not seem unreasonable for New Mexico to ask for a small share in
the future development of the waters of the lower Colorado River system." This share would be above and beyond the court's apportionment,
and the price for it would be Anderson's support for the bill. Local
Bureau of Reclamation officials recognized the political reality. They
wrote, "New Mexico feels neither the Central Arizona Project nor the
Pacific Southwest Water Plan can be authorized without the support of
New Mexico." And they were right. Arizona also recognized this reality.
Rich Johnson, then executive director of the lobbying group for Arizona
water interests, later wrote in *The Central Arizona Project*, "As a matter of
fact, in a private conference with Reynolds, representatives of Arizona
had called New Mexico's demand a clear case of blackmail; to which
Reynolds had laughingly replied that he thought extortion would be a
better word for it." California Congressman Craig Hosmer said to Reynolds at a hearing, "In essence you want to hold the lower basin project
for ransom, for some water. . . . You call it equitable apportionment. I
call it ransom."

Whatever it was, it worked. Arizona desperately wanted the bill passed,
so it paid the ultimate western price—the surrender of some water. Although Arizona was bitter about the loss, the two states were essentially
on the same team. After the April, 1965, meeting where Anderson made

his demands clear and a bargain was struck with Arizona's congressional leaders, Representatives Morris K. Udall and John J. Rhodes, the latter noted in a memorandum: "Fundamentally he [Anderson] is with us and he would like to be helpful. Once he stated, 'I hope I live long enough to see this bill passed.' " Hooker Dam was included in the 1968 bill, and New Mexico got an additional 18,000 acre-feet of water with the possibility of 30,000 more when the river was augmented. When conservationists, who were distracted by the more visible dams in the Grand Canyon controversy, objected at the last minute to such a structure in the Gila Wilderness, the phrase Hooker Dam "or a suitable alternative" was inserted into the bill. This action defused the last-minute opposition to this bill which was so crucial to all the water interests in the seven basin states, and it was subsequently passed by Congress.

No suitable alternative had been found for Hooker Dam by the time President Carter challenged the water projects in early 1977. The money to study alternatives was struck from the appropriations bill that year, at the request of the Administration. Although the money was deleted for one year, the dam still remained an authorized unit of the Central Arizona Project. The Carter Administration would have liked to see it deauthorized. But water projects do not disappear that way; at most they just remain dormant for a while. Previous administrations had never been too enthusiastic about Hooker Dam, but that does not mean that someday it will not be built. Already New Mexico was thinking about changing the beneficiaries of its water from ranchers to the copper industry. In the West, time was on the side of the dam builders.

CHAPTER THREE High Country:
Of Cows and Coal

1

The Kendall Dam, if it were ever to be built where the Green River issues forth from its beginnings into the high country, would inundate not only the small patch of habitat of the Kendall Warm Springs dace but also the upper five miles of the Canyon Ditch, a far greater sin in the eyes of the arid West. The Canyon Ditch is the first diversion of water from the Green River. It is the highest man-made interference with the natural flow of the Colorado River system and thus of great, although virtually unnoticed, significance to the seven states in the watershed. From the headgate of the ditch, it is almost 1,700 miles to the last diversion of water from the river—the headgate of a similarly unlined ditch the Mexicans have dug through the sands of the delta to divert the last flow of the river north into Laguna Salada. Between these two ditches, dug with the same knowledge available to the ancients—that water runs safely downhill if the incline is steady but slight—is gathered the most technically complex assemblage of waterworks in the world, run by such complex gadgetry as computers and laser beams and all girdled by a dense network of treaties, laws, and administrative decisions of such talmudic proportions that they are known only to a few.

The Canyon Ditch, of course, is nowhere near as imposing as the vast plumbing system found lower down on the river, but it is a start. Near

Black Butte on the east bank of the Green River, just below the boundary of the Bridger-Teton National Forest, a pile of rocks extends out into midstream. The rocks are a diversion barrier. They channel a portion of the river's flow through the ditch's headgates, raised or lowered by heavy cast-iron wheels. The country around Black Butte is open, a high desert or steppe environment, as is so much of the land between the start of the four major tributaries in mountain wildernesses and the canyonlands that follow after the high country. It is as if the river, faced with the knowledge of another great vertical descent before it, had decided to rest between plunges.

It takes its rest here in the high country where the Green becomes a slow, meandering river, passing through a rolling land dominated by gray sagebrush. The marks of beaver, who have made a comeback, can be seen in the thin line of green vegetation between the river and the high desert. Some of the willow branches have been neatly severed by sharp, exact teeth. It was the beaver in the early 1800s that brought the fur trappers —the first whites to begin the large-scale extraction of resources from the upper Colorado River basin. The resources and the extraction processes were all dependent on the river. Next came the livestock interests, miners, energy producers, and the seekers after the good life—mountain vistas, an elk or a trout, or a second-home site. Almost all their trophies, whether a side of beef or a color photograph, would be transported out of the basin. A fair amount of the water would follow the same route.

The day I was at the headgate of the Canyon Ditch with Harv Stone, rancher and president of the Canyon Ditch Company, a herd of about two dozen pronghorn antelope were warily watching us from a low rise to the northeast. "Pretty sagey meat," commented Harv, as he used a long pole to dislodge the debris that had been sucked into the headgate, restricting the flow of water into the dirt-lined ditch.

The Canyon Ditch Company is a one-man operation, with Stone being the part-time ditch rider and president. For holding a few meetings with the half dozen or so shareholders in the ditch company and looking for washouts along the nine-mile canal in a snowmobile or pickup truck, the rancher gets $1,500 a year. Besides the occasional wages for a temporary machine operator, the company has no other expenses. Near the end of the river, the Imperial Irrigation District has a budget approaching $40 million a year. Regardless of their sizes, such water agencies are the prime governmental institutions in their respective areas because they control and distribute, with little outside interference, the key to all wealth and well-being in the arid West. John Wesley Powell correctly foresaw the need for such institutions when he wrote in 1878, "Small streams can be taken out and distributed by individual enterprise, but cooperative labor or aggregated capital must be employed in taking out the larger streams."

There were a number of attempts to build the Canyon Ditch through the first half of the present century. Such small irrigation schemes were highly speculative ventures. At the end of World War I, a promoter went bankrupt when one of his partners ran off with the money. In 1923 about one mile of the ditch was dug with a steam-driven dredge. Finally, one of the more solid citizens of nearby Pinedale, Curt Feltner, who served four terms as county clerk and surveyed much of the area around the town, completed the ditch; water first flowed in 1952. Feltner owned 200 of the 1,313 acres irrigated that first season. He was a doer. Feltner put together the deal for the city park and the town's first airport. His surveyor's office served as Pinedale's first library. Feltner was responsible for planting the pines on Pinedale's streets and he helped plan and build the town's first water and sewer systems.

It was this drive, this belief in the collective benefits accruing to a specific area and the right to one's own personal enrichment that so typified the Reclamation movement. It was this dedicated persistence in an arid land that got Hoover Dam and the All-American Canal built to benefit the farmers of the Imperial Valley and the developers of southern California. And it was these men, the Harv Stones and the Curt Feltners, and the federal and state water bureaucracies that served them, who were first thwarted, then confused and finally embittered by the change in values—the rules of the game—that were imposed first by the environmental movement in the early 1970s and then by President Carter. By the time Feltner was finished, he had built three canals and owned and operated a bentonite mine. A contemporary of Feltner, a retired rancher by the name of Bill Thomas, commented when I visited him in his home next to the Green River, "The river means a way of life to us, as far as that goes. You know, if we didn't have water we couldn't be here at all."

Although the region surrounding Pinedale is prototypical Marlboro Country, ranching is a very marginal way of life at these high altitudes where a short, sixty-day growing season means just time enough for one crop of hay. The crop is gathered by a mechanized hay baler or the more traditional beaver-slide hayricks made from lodgepole pines that have been cut in the mountains. In these high pasturelands the water runs from the main canal into smaller laterals along the tops of fields from which it is released with very little control over its flow. This is called flood irrigation and is very inefficient. From the bottom of the first field the water collects, to be spread on the next lower field and so on until whatever water remains finds its way back to the river through gullies or arroyos to be used again further on downstream. These are called return flows. The water is frequently allowed to run continually across the fields in unequal amounts. This is called overirrigation and is wasteful. But the water here is cheap, of excellent quality and plentiful. Rarely do ranchers

Rancher Harv Stone at the headgates of the Canyon Ditch, the first diversion of the Green River, near Pinedale, Wyoming.

use any fertilizers on their fields, as their counterparts do, in extensive amounts, lower down on the river. Harv Stone used fertilizer once and swore he got less grass the next two or three years. His idea of the river is of a pure body of water of limitless amount. He and others like him have no concept of the devastating problem of salinity that farmers in the Imperial and Mexicali valleys have to deal with at the end of the river. The Colorado's growing salinity load begins at such sources as Kendall Warm Springs and the first use of water from the Canyon Ditch. The river and its tributaries are seen as a whole by so few. But Stone, whose creased, suntanned neck shows the marks of a lifetime on the ranges of Wyoming and Colorado, felt things were changing. He remarked as we rode back to his ranch from the ditch, "Well, I guess there's just one thing we have to face, and that's too many people. Everybody wants a place somewhere. Everybody has got to be fed and clothed."

The dilemma Stone was facing that second of two drought years (long-time ranchers in the area said it was worse than the Dust Bowl years of the mid-1930s) illustrated the crowded nature of what, at first glance, seems to be an uncrowded region. He and his wife, Lois, homesteaded 320 acres west of Pinedale in 1959, shortly after water began to run through the Canyon Ditch and not long before homesteading, for all practical purposes, came to an end in the West. Ranching on such a small land base, without another source of income, has become a marginal existence. Although the Stones did not plan to subdivide their home ranch, they had divided some additional land they owned not too far away. The sign on this property, along the same dirt road that takes one to their log home, advertises ten-acre "recreation ranches."

The "rural renaissance" has come to Sublette County in a big way. Some of those recreation ranch sites have been purchased and there are more people using water from the Canyon Ditch for purely domestic purposes than for working cattle ranches. In 1970 there were 2,000 subdivided acres in the county. Five years later the number of subdivided acres, mostly being advertised for second homes, had jumped to 12,105. The population of the Green River basin within Wyoming rose from approximately 28,000 in 1970, at which point it had been hovering since 1940, to nearly 45,000 in 1975—an increase of 62 percent. Such a rapid jump in population brought with it reports of organized crime in Rock Springs, slaughtered wildlife on the range, and medallions set in concrete to mark the Oregon Trail being chiseled out. Demands from outside the basin brought an increase in the production of oil and natural gas, extensive searches for new supplies of fossil fuels along with uranium, and the renewed production of coal. Southwestern Wyoming also produced the bulk of the nation's supply of baking soda. Instant mobile homes had taken over from log and frame structures, and coal and its kindred were displacing the traditional ranching economy.

The first large-scale exploitation of the upper basin's resources, the first boom in what were to become chronic boom-bust cycles, was made possible by the rediscovery of South Pass by the fur trapper Jedediah Smith in 1824. The beaver were trapped in the upper reaches of the Colorado River system and shipped elsewhere to satisfy a momentary fashion craze. Smith and his small party made it possible for large numbers of fellow trappers to follow them into the beaver-rich streams of the upper basin; twenty-five years later the gentle, sloping pass that skirts the southeast end of the Wind River Range served as the primary migration route across the interior West to gold-rich California and farm-rich Oregon. The immigrants passed through the inhospitable interior regions as quickly as possible, but some of them were to return as part of the backwash when they failed to find what they were looking for on the West Coast. Smith, like the Catholic missionaries of the previous century, was one of those handful of intrepid explorers who traveled widely through the unknown, fear-inspiring lands of the arid West; like Powell, but without Powell's flair for self-promotion, he faced incredible deprivations in the blazing deserts and frozen mountains. The backpackers that were to follow him into the mountains a century-and-a-half later were equipped with space-age technological innovations and trail guides and maps to lead them. But Jedediah Smith faced the most terrible of all fears—that of the unknown—at both ends of the river system. While still a young man, he was killed by Indians when caught alone in the desert looking for water. Smith stands out among the explorers of the West, not only because of the significance of what he discovered, but also because of his uniqueness and mystique. Primarily a fur trapper and only secondarily an explorer, Smith was the prototype of the good-cowboy image. In a boisterous age, he neither smoked nor swore nor took up with women; he was polite, gentle, religious, yet extremely competent in his craft. Lewis and Clark, who traveled the West before Smith, and Powell, who came later, were professional explorers, scientists with entourages backed by the federal government. Smith, who came from New York, was an amateur and loner who learned his trade quickly while on the job. Both of these types of explorers and exploiters were to contribute to the opening up and eventual crowding of the West.

Smith and his men, having heard of an easy crossing of the Continental Divide from some Crow Indians, traversed South Pass in the second week of March, 1824, on a day when they could find no running water to drink and had to melt snow. They had the good luck to kill a buffalo, which they ate raw. Most of the men had not had any food for four days. Smith was not the first to use the low pass—fur trappers returning from the Northwest in 1812 had passed that way; but that discovery had been forgotten by Smith's time. Dale R. Morgan, Smith's biographer, wrote of the rediscovery: "In retrospect this crossing of South Pass is a high moment in American history. Others had traversed South Pass before him, but Jedediah Smith's was the effective discovery, the linking of the pass in the lines of force along which the American people were sweeping to the Pacific." Smith not only forged an easier east-west link, but also a north-south connection within the Colorado River system. Escalante had gone looking through the interior West for a route from Sante Fe to central California in 1776 and in his travels came upon the Green River, which he named the Rio de San Buenaventura. He guessed that the river ran into a large interior lake. Later cartographers supposed the outlet from the lake flowed into the Pacific Ocean near San Francisco, causing such people as Smith considerable grief, because there was no such easy transmountain route to the West Coast. In his 1826 travels the fur trapper disproved this cartographic myth, and when he reached the lower Colorado River by way of the Great Basin and Virgin River, Smith had linked Escalante's route up with that of another Franciscan missionary, Francisco Tomás Garcés, who had crisscrossed the lower basin. When Smith rejoined the Colorado River system along its lower length he thought he had found his old friend, the Green River; in a sense, he had, although in its lower reaches the Colorado was more reddish in color and swollen with the waters of many subsequent tributaries.

In the peak year of 1850 about fifty thousand immigrants traveled over South Pass on the Oregon Trail, bound for the gold fields of California. One of those who came a few years later was Mark Twain, who wrote of his stagecoach trip over the pass, "We bowled along cheerily, and presently, at the very summit (though it had been all summit to us, and all equally level for half an hour or more), we came to a spring which spent its water through two outlets and set it in opposite directions." This was a rather fanciful, yet graphic, description of the Continental Divide. At 5:00 P.M. the next day, after crossing the Green, described as "a fine, large, limpid stream," Twain arrived at Fort Bridger, the first permanent settlement in the Green River basin of Wyoming. Founded by trapper Jim Bridger in 1843, the fort was similar in purpose to the present-day Little America motel complex on Interstate 80 not far to the east—a place to get some food, fuel (fresh horses), and sleep before quickly resuming the cross-continent journey. It was near Fort Bridger and the Mormons' Fort Supply, all in the same area, that the first irrigation works, the first man-made diversion of water from the river system in the upper basin, were built in the early 1850s. Bridger described the fort's ideal placement: "The fort is a beautiful location on Black's Fork of Green River, receiving fine fresh water from the snow on the Uinta range. The streams are alive with mountain trout. It passes the fort in several channels, each lined with trees, kept alive by the moisture of the soil."

By 1900 the water from Blacks Fork and Smiths Fork was over-appropriated, and three years later the newly formed Reclamation Service surveyed the area for dam sites. Meeks Cabin Dam was completed in 1971 and a proposed China Meadows Dam was not built because of environmental objections. The dam would have inundated 340 acres of mountain meadow on the north slope of the Uinta Mountains, an area heavily used for recreational purposes. Instead, an alternative site was chosen for the Stateline Reservoir on the East Fork of Smiths Fork and construction was started. The Lyman Project, of which these two dams were a part, was reviewed by the Carter Administration in 1977 but not challenged, since it was already two-thirds completed. The project will supply irrigation water for about 36,000 acres in the Bridger Valley. Up to 1970 the valley, whose economy was primarily based on cattle, had been losing population; but then this trend was sharply reversed, and over the next four years there was a 69 percent gain. Power plants and coal mines were responsible for the sudden influx of immigrants. One of the power plants was given the name of Bridger.

Besides having been immortalized by a coal-burning power plant, the memory of the mountain men has been kept alive in other ways. It was in February of that drought year that I rounded a curve on the dirt road to the campground at Brown's Park and came upon thirteen tepees pitched in the dry grass beside the Green River. I had no idea what was happening, but soon discovered there was to be a wedding, mountain-man style. Most of the participants, who called themselves the Rocky Mountain Men and were from the suburbs of Salt Lake City, were clad in buckskin and toting huge muzzle-loading rifles and lots of clinking paraphernalia, such as powderhorns and knives. There were black bowler hats, stovepipes, and fur pieces atop shaved, scrubbed, suburban faces. Children and squaws, as they were referred to, were attired in similarly authentic clothing, much of it homemade after careful research. A bottle of Black Velvet whiskey, a substitute for home-brewed liquor, was being passed around, and some chili was simmering in a huge, blackened cast-iron pot set over a fire.

Then somebody yelled, "Here he be comin'."

Through the riverbank willows rode J. D. Waddle ("The Circuit-Ridin' Preacher") leading a pack horse. Out came the Instamatic cameras as

Waddle, with black bowler and waxed mustache, rode up to the group and asked, "Who's in charge of this?"

"He be over there," answered a tall buckskin-clad figure.

"I hear they be trouble around here and I come to settle this matter," said Waddle, as if reading from a script.

An older man stepped forward. "I think it's about time you married my daughter." And then he shot off both barrels of his gun.

In this manner Lee Robertson and Alice Oliver were married by Waddle, who produced an accordion from his pack. The crowd sang "Shall We Gather at the River." At the end of the ceremony there were three shouts of "Hip, hip, hooray." A volley was fired into the air over the river as Lee and Alice departed downriver in a fiberglass canoe, only to reappear that night at a dance in the nearby one-room schoolhouse. For Lee, it was his second marriage. His first wife had not enjoyed buckskinning, as it is called, but that was how Alice and Lee had met. When not preaching at a Baptist church or giving demonstrations of western lore to Sunday-school children, Jesse Waddle was a salesman for an oil company.

Part of Brown's Park, just over the line from Wyoming in the northwest corner of Colorado, is a national wildlife refuge. The 12,838-acre refuge lies within a thirty-five-mile long valley just below Flaming Gorge Dam, surrounded by mountains. Before the dam was built in 1962, the Green River used to flood annually, creating marshes and backwater sloughs in the valley that were ideal for migrating geese, ducks, and other waterfowl, such as sandhill cranes and whistling swans. The idea was to restore some of the marsh habitat, much as one would irrigate a hayfield, by building dikes and ditches to divert and spread the water pumped from the river. Water from dams on Beaver and Vermillion creeks was to create additional habitat. This is called intensive water management, in this case for wildlife and in other cases for coal-burning power plants or cattle.

The fur trappers held rendezvous in this lush valley from 1826 to 1840, and after they left, a few settlers arrived, to be followed by cattle rustlers and outlaws, among them Butch Cassidy, who had traversed much of the interior West on the Outlaw Trail. The park was a perfect hideaway, isolated and virtually sealed off from the outside world. As matters turned out, however, the park's isolation served the purposes not only of the outlaws but of the settlers, when the latter decided to take the law into their own hands and rid the valley of rustlers. A gunman was hired; he shot one rustler while he ate lunch and another when he went to relieve himself. The second rustler was Isom Dart, one of the West's few black outlaws. Needless to say, such precipitate action had its effect, and the outlaws disappeared from Brown's Park in 1900. Then the serious grazing of cattle and sheep began. As many as 100,000 cattle wintered in the park at one time, and it was not too long before the effects of overgrazing were evident.

But the legacy of lawlessness and vengeance remained. During the Administration of Theodore Roosevelt the federal government attempted to impose some controls over the use of public lands, and the cattlemen in Brown's Park reacted by initiating a campaign of small-scale guerrilla warfare in 1906. When Forest Service rangers tried to collect a grazing tax from one Brown's Park rancher, Ora Haley, he refused to pay; he also scattered his cattle so they could not be counted. When rangers rounded the cattle up, Haley and his men sabotaged their efforts by stampeding the herd through the rangers' camp. Haley finally hired a gunman, Sheriff Bob Meldrum of Baggs, Wyoming, and in a showdown on a street in that small town, Meldrum faced Forest Supervisor Harry Ratliff, who outdrew the gunman. Meldrum's gun never left his holster, and Ratliff simply took it from the surprised gunman and walked away. It was an uncharacteristic display of federal toughness toward western economic interests.

The West is, first of all, predominantly an arid or semiarid land; the flip side of that primary characteristic is that a fair amount of its population is watered by the Colorado River. Its second unifying factor is the dominance of the federal government in land and water matters. It is a dominance the West has welcomed, in terms of benefits, but struggled against in terms of concomitant regulation. In the first 100 years of this country's history the federal government either acquired or was ceded vast tracts of unsettled lands. In the last 175 years the federal government either granted or sold over 1 billion such acres. What remains as public domain is 760 million acres, or about a third of the nation's 2.3-billion-acre land area. These lands are national parks and monuments, national forests, wildlife refuges, Indian reservations, military reservations, and a category called national resource lands, which generally appear to be nondescript desert but constitute 60 percent of the total federal holdings. The Bureau of Land Management, an agency within the Department of Interior, administers the national resource lands. It is on these lands that energy and agriculture have come into greatest conflict in the West. The Forest Service, within the Department of Agriculture, has the next largest chunk of federal lands to administer, 24 percent of it, all at higher, more densely vegetated levels. Within the eleven western states, around 60 percent of the surface-water supplies originate on federal lands. About half of the public domain is in Alaska, and the vast majority of the remaining 350 million acres are in the West—that area stretching from the Rocky Mountain States to the Pacific Ocean and encompassing the Colorado River basin. In the upper Colorado basin, 60 percent of the land is federally owned; in the lower basin, 52 percent. However, when Indian trust lands, military reservations, and state and municipal property are added to the federal total, they leave only about 20 percent of the lands within the basin for private owners. Besides being an oasis civilization, the West is an archipelago of small, privately owned islands within a vast federally owned sea.

It is difficult to imagine the vastness of federal land holdings in the West, especially when traveling by automobile or airplane across it. The terrain seems to be an indistinct, desolate mass punctuated by a few snow-capped peaks and wrinkled by some canyons. There is little indication of who owns what or who would even want to own some of the more arid stretches. The answer, in most cases, is the federal government. Eighty-six percent of Nevada is public domain, an area twice the size of the state of New York. The public lands of California are equivalent to an area eight times the size of Massachusetts. Florida would fit onto the public-domain lands of Utah, if they were of one piece. There is another perspective to this immense scale, and that is the relatively small number of people living on these lands. Nine of the eleven western states have population densities substantially lower than Maine, the most lightly populated state east of the Mississippi River. Again, the statistics of an arid land.

Of the 350 million federal acres in the West, nearly 250 million are administered for grazing, most of them by the Bureau of Land Management and the remainder by the Forest Service. The public lands used for grazing purposes constitute an area larger than the fourteen Atlantic seaboard states. But only 5 million head of cattle, representing 8 percent of the total number of beef cattle in the nation, make use of the public domain. The federal lands account for but 3 percent of all forage consumed by livestock in the country and 12 percent of the feed consumed in the eleven western states.

Simply put and not often emphasized, no other activity uses so much land area in the West, or for that matter in the entire nation, as cows eating grass. This is to satisfy the nation's hunger for red meat. Per capita beef consumption was expected to hit 150 pounds in the year 2000, nearly half a pound per day for every citizen. While other less fortunate countries have sought the bulk of their protein requirements in grains and fish, America has been historically a nation of beef eaters. No other single activity or combination of activities has contributed more toward altering the shape and texture of western lands, or the wildlife that is dependent upon them. Nor does any other activity consume more water. For instance, in the near-normal-water year of 1975, before the drought,

Irrigated fields on a high-country ranch in western Colorado are surrounded by sagebrush.

irrigated agriculture, whose principal crop is alfalfa, accounted for 7.5 million acre-feet of water used within the basin. On the other hand, municipal and industrial use, including power plants and mining, siphoned off only 630,000 acre-feet. Slightly more than 50,000 acre-feet went toward recreation, fish, and wildlife. Evaporation from reservoirs consumed 2.3 million acre-feet, an amount exceeding the total use of the four upper-basin states plus Nevada. Seven million acre-feet were diverted out of the basin, an amount which includes the water used in the Imperial and Mexicali valleys for agriculture. More water is diverted from the Colorado River watershed than from any other river basin in the country. Clearly, the Colorado River and the West were different.

Although the grazing of cattle is the one activity that extends from the headwaters of the Colorado River to its end—I have seen cattle eating grass in the Bridger Wilderness Area and just north of the Mexican village of El Golfo on the gulf—it is most closely associated with the high country, particularly Wyoming. Jim Bridger did not spend much time at the fort he founded on Blacks Fork, and he sold it to the Mormons in 1855. The Mormons burned it to the ground in 1857 at the approach of U.S. Army troops, sent to Utah to reassert the supremacy of the federal government. With the Army came William A. Carter, a Virginia gentleman whose job it was to supply the troops. Carter settled at the fort and imported, along with a Steinway piano, the first shipment of Texas longhorn cattle into the Green River Basin of Wyoming in 1868. Three years later a New Mexico cattleman by the name of George Baggs trailed 900 steers in from southern Colorado to Brown's Park, and the cattle boom in the upper basin had begun in earnest. For the next thirty years, until a crippling cold winter and a drought, the first major land-use revolution in the West would result in fertile grasslands being turned into eroded gullies and a permanent change in the type of vegetation, along with a corresponding jump in population. It was an ecological and social change, governed by economics, that was not to be matched in the interior West until the "rural renaissance" of the 1970s. As always, water followed the predominant land use and allowed it to flourish.

The first sheep, cattle, hogs, mules, and horses—variously estimated at between 1,500 and 6,500 animals—were brought into the West in 1540 from Mexico by Coronado. Neither Coronado nor anyone else in his large entourage had any idea of the consequences of the introduction of these exotic animals into the West. How the livestock were first dispersed is not known, but Walter Prescott Webb theorized, "Perhaps the horses were stampeded by Indians or by herds of buffalo; but it is more than likely that some were set free because they became too poor or footsore or crippled to be of further use to their masters." The same was probably true of the cattle. And so began the eating of the West. Later Spanish and Mexican missionaries and settlers brought more livestock into the Southwest, and after the revolt of 1836 the Texans moved into Mexican territory and appropriated more cattle. The great cattle drives then began from Texas; but it was not until 1850 that the first drive traversed the Colorado River Basin on its way to California. Between the years of 1850 and 1860, the increase in Texas cattle was over 1,000 percent and after the Civil War they began showing up in greater numbers in the Colorado River Basin.

In the early 1870s two Wyoming governors invested in the cattle business and took the lead in promoting its development. In 1883 a Wyoming governor could report: "Stock raising is the chief industry, comparing with all others about as 90 percent to 10. . . . Cattle by the thousand roam in every valley and drink from every stream in the territory." The boom crested in Wyoming in the mid-1880s, when there were 1.5 million cattle in the state. On an average in Wyoming, forty acres are needed for one cow. The range was becoming very crowded. A later federal report noted: "Not an acre of grazing land was left unoccupied, and ranges that for permanent and regular use would have been fully stocked with a cow to every 40 acres were loaded until they were carrying one to every 10 acres. . . . No one provided any feed for the winter, the owners preferring to risk the losses. Gradually the native grasses disappeared. As fast as a blade of grass showed above the ground some hungry animal gnawed it off." Inevitably, disaster struck. The winter of 1886–1887 was unusually harsh—like a dry year, just part of the normal weather cycle—and a lot of cattle died. At the Chicago stockyards, prices for the starved cattle which flooded the market hit record lows, and in 1888 the governor reported to the Secretary of the Interior, "This was the turning point in the history of Wyoming." Cattle, coal, oil, natural gas, copper, and gold were to boom and bust in Wyoming through World War II, but the state stagnated in the 1950s and early 1960s.

For the Southwest, the situation was much the same, except the bust in the livestock industry did not come until the drought of 1893. The Jesuit and Franciscan missionaries who had settled in the Santa Cruz River Basin south of present-day Tucson in the 1700s brought the first large herds of livestock into the basin following Coronado's brief visit more than a century and a half earlier. In the Spanish and Mexican periods, ranching alternately flourished or became dormant, depending on the whim of the Apache Indians who periodically terrorized the countryside. At one time eight million sheep and one million cattle roamed southern Arizona under Spanish rule. Again in the 1830s and early 1840s under the Mexicans, herds blossomed, then dropped dramatically when the Apaches raided and the inhabitants fled, leaving the cattle to wander untended about the country. The 1849 gold rush in California brought a great demand for beef; and large herds were trailed from Texas across Arizona. Some of the cattle and cowboys dropped out along the way, and the first ranches to be established under American rule sprang up along the Santa Cruz River in the 1850s. There had been no great change in the vegetative cover yet (although the early Indians had periodically set fire to the grasslands) except for some spreading of mesquite shrub into grassland areas. But the invasion of mesquite was not yet particularly dense, as it was to become. The grass was still plentiful and many of the streams ran perennially through unchanneled, open, marshy river bottoms. There were beaver and trout in the San Pedro River, a fact now hard to believe. Rarely now are there any significant amounts of water in these streams, except when there is a sudden, fierce rainstorm. Their channels are normally dry, as is the channel of their parent river, the Gila.

Following the Civil War, ranching boomed and spread throughout the Arizona Territory. As was happening elsewhere in the West, easterners and Englishmen invested heavily in livestock operations. With absentee owners, profit, not care for the land, was the primary consideration—not unlike the attitude of energy companies headquartered outside the basin in the 1970s. In the mid-1880s there began to be talk of overcrowding on the range and, although the severe winter late in that decade had little effect on the Arizona cattle industry, the depressed prices being paid by the Chicago stockyards signaled that all was not well. There were at this time between 720,000 and 1.5 million cattle in the territory. As in the 1970s, two dry years occurred back-to-back, and by the spring of 1893 the dead cattle lay so thick that it was said a rock could be thrown from one carcass to another. From one-half to three-quarters of the herds were lost in Arizona.

Something else was lost, too, something more permanent and meaningful. While the cattle would return again in more moderate numbers, the land would never again be as productive. At no time in this country's history has a landscape been so quickly, so unalterably changed. The Colorado River, nicknamed "Big Red," had always been a silty river—after all it was the first Spanish explorers who named it for its predominant color—but the millions of hooves that destroyed the vegetation hastened the process. Nowhere has this been better documented than along the Santa Cruz River.

In their remarkable book, *The Changing Mile,* James Rodney Hastings of the University of Arizona and Raymond M. Turner of the U.S. Geological Survey traced the alterations in the landscape. They stated, "Taken as a whole, the changes constitute a shift in the regional vegetation of an order so striking that it might better be associated with the oscillations of Pleistocene time than with the 'stable' present." Up until about 1890, the landscape had remained pretty much intact. Then, very quickly, a dramatic change took place, not dissimilar to what had occurred eight hundred years earlier in Chaco Canyon. But this time the inhabitants would not have to leave, at least not yet, because they could dig deeper for water. With the plant cover removed, the severity of flooding increased. The result was arroyo cutting—the lowering of the rivers below the surrounding terrain. The marshy, flat riverbottoms and perennial streams disappeared, to flow underground and only intermittently above ground through a drier, more crackly landscape. The country became incised. Using a set of matched pictures taken in 1890 and 1965, the authors dissected the changes. The grass cover disappeared, to be replaced by shrubs spread by cattle ingesting, then eliminating, the seeds. One cow chip, it was determined, contained 1,617 undigested mesquite seeds. Ocotillo, turpentine bush, desert broom, and rabbitbrush invaded the grasslands. The zone of oak woodland moved uphill. In some places, such as the region now included in the Saguaro National Monument, there were not as many giant cactus. Overgrazing was to blame, as was a drying trend in the climate. But given enough time, the climate would once again become wetter. The authors felt, nevertheless, that the damage had been irrevocably accomplished: "The suspicion remains that the desert grassland, by and large, is a thing of the past and that, short of spraying them with diesel oil or uprooting them with a chain and bulldozers, the shrubs are here to stay. There is no evidence that the elimination of grazing can bring about their disappearance, once they have been established."

As in the high country, the first change in land use had been established. As time went by, that change was to be cemented. Throughout the West, in the name of "range improvement," chains were dragged by bulldozers, and airplanes and helicopters sprayed herbicides, including 2,4,5-T and 2,4-D of Vietnam fame, on many millions of acres (3.5 million of Forest Service lands alone) in an effort to rid the range of such invaders as sagebrush; more nutritious types of grass were planted in their place. The result was vast monocultures, such as thousands of acres of the exotic crested wheat grass. But the cows loved it, although the wildlife suffered. All of this was done to produce more red meat.

Around the end of the nineteenth century a number of factors leading toward large-scale federal involvement in the water business coincided. Ranchers saw that they could not allow their cattle to run untended, that the livestock would have to be fed during the winter in order to be certain to survive into the spring. This meant growing hay and getting the cattle off the pastures of the home ranch during the short summer growing season and onto federal lands at higher elevations. The 1890s were also a time when the most easily dammed and diverted small streams had already been utilized and larger amounts of capital and expertise than could be found locally were needed for those bigger water projects that were deemed necessary to provide water to grow feed. Thus was born the Reclamation movement, whose main beneficiary was to be the cow. In a sense it was the bailing out of the West from what threatened to be economic stagnation.

The philosophers of the movement did not have much interest in cows. In his report on the arid lands of the West, John Wesley Powell made some scattered references to livestock destroying the native grasses and trampling the ground; but Powell was more interested in the social-engineering aspects of furnishing water. That great boomer of the Reclamation movement, William E. Smythe, had little sympathy for cattle, which he called "dumb brutes," and foresaw irrigation as mainly benefiting fruit and vegetable farms. Smythe said of the predominance of the cattle industry in Wyoming, "It is a pursuit which does not develop the higher possibilities of the country, either in a material or a social way. . . ."

If the smart money was going where public officials were putting it, then in the 1890s irrigation schemes (many of which went broke before completion) were the most attractive investment. Wyoming Governor William A. Richards had a hand in an irrigation project, as did Senator Joseph M. Carey, the author of a bill designed to give federal aid to private irrigation projects. Even William F. Cody, better known as Buffalo Bill, put his money into an ill-fated private irrigation scheme. As the century faded, irrigation projects, not cattle, were the hottest investment prospects.

Not surprisingly, Wyoming was the first state to take advantage of the Carey Act of 1894, which provided for the federal government to give up to one million acres of land to a state, if those lands were reclaimed and settled. Even with this incentive, Wyoming did not develop particularly fast, although the Carey Act and the Reclamation Act of 1902 set off a boom in water rights and canal building. Every stream in the state that could possibly be tapped was sought after and the necessary filing made for a water appropriation. "Never before in the state's history were so many irrigation enterprises launched, nor were there ever so many miles of new ditches constructed in the same period of time in the state," reported Wyoming State Engineer Elwood Mead, who was later to head the Bureau of Reclamation. Buffalo Bill's project was not successful as a private venture, and the federal government completed it after passage of the 1902 act—the state's first federal irrigation project. Wyoming historian T. A. Larson wrote of it, "Although the Shoshone project has

not lived up to expectations, it may well be regarded as the most successful federal reclamation project in Wyoming."

With Theodore Roosevelt in the White House and a conservation ethic
abroad in the land, the Forest Service took over rangeland management
on forest reserves in 1905, an action that caused rebellions similar in
intent if not actual violence to the Brown's Park incident. Gifford Pinchot,
chief of the Forest Service under Roosevelt, issued this muted warning:
"In new forest reserves where the livestock industry is of special importance, full grazing privileges will be given at first, and if reduction in
number is afterward found necessary, stockmen will be given ample opportunity to adjust their business to the new conditions." The Forest
Service is still trying to cut back on the amount of grazing, with varying
degrees of success. But the Forest Service had jurisdiction over only a
small portion of the public domain. The remainder went unregulated.
These unregulated lands were the vast tracts of sagebrush desert and
rolling, shrubby terrain that seem omnipresent throughout the West.
Most of the choicest parcels, like national parks and forests, were carved
out of the areas now called the national resource lands at an early date;
or the better lands were virtually given to settlers under homestead laws
and other legislation, such as the Desert Land Act and the Carey Act.
With the grazing conditions worsening during the drought years of the
early 1930s, the Taylor Grazing Act was passed in 1934, creating the
Grazing Service within the Department of Interior to regulate rangeland
use on national resource lands. The problem was that the Taylor Act was
a weak law and the Grazing Service quickly became a captive of the
livestock industry that it was supposed to control. By this time, all the
early federal water projects on the Colorado River system had been built,
mostly in the high country where they benefited livestock interests.

In 1936 the Forest Service, in a bid to take over all of the federal
government's range programs, published a report entitled *The Western
Range*. It was a document of doom and gloom, to which the livestock
industry replied in a publication called, *If and When It Rains:* "The stock-
man is not a despoiler of the range. He is necessarily a conservator. His
livelihood depends upon feed for his livestock. . . . Dust storms come and
dust storms go, and bureaucrats build up the need for big appropriations
based thereon, while sensational writers harvest a big crop telling about
it, but when the rains come, the grass grows again and all is well on the
range." That argument would be heard again after the rains began to
come in late 1977, ending the most recent drought.

In 1946 the Grazing Service and the General Land Office, created to
sell off public lands, were merged into the Bureau of Land Management,
the most important federal agency dealing with the West after the Bureau

of Reclamation. The first BLM director, Marion Clawson, wrote, "The political influence of the range livestock industry differs in no essential respects from the political influence of other economic groups. It is simply more powerful in relation to the number of people involved." Clawson noted that the "oldest and best" families in the West, who had the best political connections, were associated with ranching; and western congressmen "often dominated legislation regarding federal lands because they were the best informed and most interested in it." A similar view involving water interests was voiced by William E. Warne, a contemporary of Clawson's within the Department of Interior. "The old school tie of Reclamation," said Warne, "binds together a loyal but unorganized clientele of considerable importance to the bureau." Frank Gregg, the head of the BLM during the Carter Administration, would state: "The political attitudes in the western states congressional delegation, with the exception of California, are more attuned to grazing than any other issue." Gregg saw a stronger BLM, bolstered with a new law giving it a larger mandate and the increasing importance of energy issues, as factors that would erode the power of the grazing interests. But during the "War on the West" it was the livestock industry, not the Carter Administration, that had its way. Congressmen from the western states form the majority on those obscure subcommittees where water and grazing bills are shaped, then passed with little debate by the full committee and frequently enacted by a voice vote of Congress, before being sent as a *fait accompli* to the president for signature. This was the pattern of politics developed by a region that has found it has special needs about which outsiders are likely to be uninterested, ill informed, or unsympathetic.

No other writer so stirred the ire of the West, and particularly his home state of Utah, as Bernard DeVoto. As I concluded an interview with Calvin L. Rampton, who served twelve years as governor and is considered Utah's elder statesman, he remarked, "Well, you know those writers who are hardest on us were born here and then left, like DeVoto. But he had some personal disappointments here." And that comment was made nearly twenty-five years after the death of the novelist, historian, and journalist. After a western trip in 1946, DeVoto, who had previously left Utah for Harvard, wrote a series of impassioned articles for *Harper's* magazine and other publications about the threat of livestock interests taking over the public domain. Wallace Stegner, also a Utah inhabitant at one time, wrote of DeVoto in *The Uneasy Chair*, ". . . he heard in 1946 a tune he had heard all his life: the ambivalent clamor for more federal subsidies and federal aid discordantly fused with complaints about absentee federal landlordism."

In the process of flailing away at the attempt to take over the public lands and other conservation issues in the decade following the end of

World War II, DeVoto pinpointed the essential ambivalence of western attitudes toward natural resources—the region's great reliance on and great dislike of outside money and controls, specifically those imposed by the federal government. "It shakes down," he wrote, "to a platform: get out and give us more money. Much of the dream of economic liberation is dependent upon continuous, continually increasing federal subsidies —subsidies which it also insists shall be made without safeguard or regulation. . . . It is the forever recurrent lust to liquidate the West that is so large a part of Western history." If DeVoto could have lived to witness the energy boom and "rural renaissance" of the 1970s, he would have felt vindicated.

DeVoto mounted his campaign to save the public domain forty years after the Forest Service attempted to cut back on grazing permits. In the late 1940s the livestock interests had suggested to the sympathetic House Subcommittee on Public Lands that permit holders get permanent grazing rights and be able to purchase such lands at a minimal cost. At a 1977 hearing on grazing fees before the same subcommittee, a New Mexico congressman asked, "Why didn't they [the federal government] keep their promise to turn the lands over to the states in the West as they did in Texas and Oklahoma and Virginia and Maryland and all the rest of them? No, they have to be greedy, and own one-third of all the land of America, and it is all in the West." A very important statement, said the subcommittee chairman from Wyoming. The other congressmen sitting in on the hearing that day represented Colorado, Montana, and Nevada. The Carter Administration wanted grazing fees on public lands raised to what they were worth on the open market, a level substantially higher than what they have traditionally been set at. The bill approved by the subcommittee and later signed by President Carter contained provisions for a fee somewhat less than the fair market value and allocating an extra $360 million to make the rangelands more productive. The money would buy more water developments, more fencing, more chaining, and more spraying of herbicides in the West.

Nowhere has the sense of uneasiness pervading the western range the last few years been more noticeable than in Nevada and Arizona, the two most overgrazed states in the West. These two states are also the most arid, and possess large amounts of federal lands—only 18 percent of Arizona is privately owned, while 87 percent of Nevada's land area is federally owned. Outside of the glitter capitals of Las Vegas and Reno, Nevada is hard-rock mining and ranching country. The Great Basin in Nevada is a closed-off society that tends to do as it pleases, regardless of what the federal government says. There is more illegal grazing on public lands, politely termed trespassing, in Nevada than any other state. It is the home and main operating base of John J. Casey, the king of trespassers, who has successfully defied the legal talent of the federal government for more than twenty years, in part by just ignoring it. Casey used a motel he owned in Reno as a mail drop. Frequently the certified letters the BLM sent him with notification of trespass by his cattle and other legal matters were returned unopened. Asked in court by a federal prosecutor if he had a problem receiving his mail, Casey answered, "I don't have much luck having it forwarded to where I am." The federal government has spent over $1 million to prosecute Casey over the years, and he has said he plans to spend a similar amount fighting the government. And still Casey's cattle roam the public range.

Nevada, partially as a result of the "War on the West," was the source of the sagebrush rebellion of the late 1970s, a movement within the western states similar to the attempt in the late 1940s to take over the public domain. The state's congressional delegation, which played a leading role in forming a western coalition of lawmakers when President Carter challenged the water projects and other western economic interests, wore large orange-and-black buttons proclaiming "Welcome to the West. Property U.S. Govt." The state legislature passed a bill allocating $250,000 to do legal battle with the federal government over the public domain. From Nevada the movement spread elsewhere. Utah Senator Orrin G. Hatch termed the revolt "the second American revolution, destined to lead the Western states to the most dramatic development in our history since entering the Union: emancipation from economic and political control by the federal government." What was billed as a rebellion was mostly rhetoric, since a condition for joining the Union had been for each state to disclaim "all right and title" to federal lands. To espouse the rebellion was politically expedient at that time in the West. And besides, who knew? Perhaps it would put pressure on the Administration to soften what was perceived to be its antiwestern stance. It was recalled that President Carter had not won a single western state in his 1976 election. (He would lose even more decisively in the West in 1980. Ronald Reagan is a westerner—one of us, it was thought, in fact as well as image.) The same feelings that were bound up in the rebellion were what caused Gary McVicker trouble in Kingman, Arizona, that drought year, and the McVicker incident had a clear precedent in what John Wesley Powell had suffered at the hands of western congressmen nearly a hundred years earlier.

Regardless of which way Kingman is approached on U.S. Highway 66, the scenery in this part of northwestern Arizona is typical of BLM lands. Hot, dry, scrubby desert is mixed with jagged heaps of desolate mountains running north and south between broad alluvial valleys. This is country where one cow needs a hundred acres of grazing land. Driving west from Kingman, the traveler drops off the Colorado Plateau and begins the gradual descent into the summer furnace of the lower Colorado River and Needles, California—frequently the hottest place in the nation. Kingman is between the south rim of the Grand Canyon and Los Angeles, and the only reason for the traveler to stop is to grab some gas and a quick meal and rest in air-conditioned comfort. In January of 1977 McVicker arrived in Kingman to begin work as the new BLM area manager, and John Neal, whose family had been ranching in the area since the 1880s, was on hand to meet him that first day. The two were to become bitter antagonists in a classic struggle with many precedents in the West.

McVicker felt no hint of trouble when he first arrived, although he later said the ranchers, led by Neal, had been out to get him from the start. The first indication that things were not going smoothly came in the spring at a meeting with the ranchers when they complained he was being unsympathetic to their needs. McVicker, who represented a new style of BLM employee more interested in firm regulation than polite diplomacy, had replaced an area manager with a reputation for leniency. The BLM was in the process of going through the necessary motions that would probably result in decreased grazing, and this, along with all the other pending changes, such as the debates going on in Washington, heightened the ranchers' fears of a reduced livelihood. All of this was descending from the outside, and McVicker was an outsider, a federal employee, *the* federal government in Kingman. In a sense, he was doomed from the start.

The complaints about McVicker centered on his attitude. He was accused of being curt on the telephone, possessing "total negativism" toward the rancher's plight, and "lacking sympathy" or being "uncooperative." Those who had worked under both McVicker and previous area managers said the newcomer lacked a sense of camaraderie and that he questioned their proposals closely, but acted fairly. They said he was more of a resource manager than a diplomat. The ranchers were more familiar with the latter type of area manager. By June the situation had deteriorated to the point where the Mohave County Livestock Association formally requested that McVicker "be replaced by someone who will work with the ranchers," namely the former area manager. Copies of the request went to the Arizona congressional delegation. The state's freshman senator, Dennis DeConcini, had already visited Kingman and listened to the ranchers' complaints. On July 7 he wrote Interior Secretary Andrus asking that McVicker be replaced with someone "who will better reflect the intent with which our governmental employees are to serve." Assistant Interior Secretary Guy Martin replied that McVicker would stay where he was unless an investigation showed he was not doing a good job. Additionally, Andrus wrote an inquirer, "Employees acting within the law for the good of the general public will receive my full support." The state BLM director investigated, found McVicker was doing a good job, and recommended he be retained in Kingman. A short time later, however, the decision was made to transfer McVicker, and it was announced in June of 1978.

McVicker's superiors, with the area manager himself agreeing, decided that the controversy had so tainted the atmosphere that it would be extremely difficult, if not impossible, to carry out the BLM's program in the area. Besides, Senator DeConcini thought he had a commitment from one of Andrus's aides for McVicker's removal, an assumption that had some merit. Interior Department officials did not want to challenge the senator, since he had just gained stature with the White House by casting a key vote for the Panama Canal treaty. When the decision was made public, there were few BLM or Forest Service employees who did not see its significance. BLM Director Frank Gregg commented, "It was a sticky case. It is going to raise a signal that we are not going to support our field people. There is going to be some damage." For their part, Neal and the other ranchers thought it was a complete victory that could be duplicated elsewhere in the West, if necessary.

In the high country of Wyoming there had been periodic energy booms in the early years of the century, but nothing to match the change in land use, and thus the use of water, in the 1970s. It was to be the second major alteration in the West's landscape during its short history. Water projects that benefit agriculture have always been, at best, of marginal feasibility in southwestern Wyoming. Hay is about the only crop, and the growing season is extremely short that far north and high up. Yet, between the political influence of the livestock interests, the willingness of the federal government to fulfill those demands, and the preoccupation of Wyoming with the mystique of ranching, water has flowed toward cows. As the time of large-scale and perhaps continuous energy extraction began to dawn in the late 1960s, this predominance was threatened.

Another factor centered attention on the diversion of Green River water for use in energy developments throughout the state. Other rivers in Wyoming either did not have the additional supplies of water or were restricted by compacts from diverting it outside of their basins. Not so the Green River, on which there was no such restriction. The Green also had a surplus of water since Wyoming, the slowest of all the Colorado River basin states to develop its share of the river's waters, was only using about 30 percent of its allocation under the Colorado River Compact of

1922 and a subsequent division of the waters by the upper basin states in 1948. The state's water plan noted, "Thus, Green River water can be the most flexibly used of Wyoming's presently available surface water resources." Like other upper basin states, Wyoming was anxious to put its share to use before California or Arizona gobbled it up or environmental and water-quality constraints restricted its use.

In the late 1960s, coal—once mined here for the railroads but in decline since diesel engines had made their appearance—began to make a comeback on the Wyoming scene. This time coal was to be used in power plants serving the West Coast and Middle West. The state had stagnated since World War II. Immediately following the war, agriculture, meaning livestock, was the state's leading industry. In the 1950s oil and gas developments and tourism replaced agriculture, and gradually Wyoming became more dependent on mineral extraction. Population was being lost from the state, as elsewhere in the rural areas of the West, and the federal government was pumping more money in than was going out in the form of federal taxes—a sure sign of colonial status. In the late 1960s, Governor Stanley K. Hathaway, like his counterpart in Utah, launched a promotional campaign to attract industry; it happened to coincide with the message Reclamation Commissioner Floyd Dominey brought to Wyoming in 1968. Let's start planning to supply more water to the energy companies that are showing a great deal of interest in developing the state's abuandant coal lands. Dominey knew it was becoming more difficult to get straight irrigation projects past the President's cost-conscious budget watchers. The predominant thinking was expressed by the Wyoming Natural Resources Board: "Wyoming is working vigorously to attract new industry and help already existing industry to expand. Water is the key to much of this development. . . ."

The swing toward large-scale diversions of water for energy use climaxed in 1972 when the Bureau of Reclamation released its report, *Alternative Plans for Water Resource Developments, Green River Basin, Wyoming.* The report had a clear bias against the further use of any significant amounts of water for irrigation projects. It noted, "Accordingly the only new depletions from irrigation anticipated for the future are minor depletions from supplemental irrigation that could be undertaken in conjunction with certain storage developments for industrial [energy] use." This was a slap in the face to the bureau's traditional clientele, the livestock interests, and there was the predictable outcry. The bureau, caught in the midst of changing times, would be accused of tailoring its projects for the expediency of the moment. Also predictably, the bureau retreated and faded back into the woodwork after this momentary indiscretion. It had once again abdicated its decision-making role.

In the early 1970s Wyoming was flooded with a dizzying number of energy proposals, many of which did not get built because there was not enough water to go around, and a reaction set in after the first flurry. The Casper *Star-Tribune* editorialized, "We don't want to sacrifice irrigated agriculture for industrial development. Those irrigated farms are a part of our environment too, much more pleasing than strip mines and coal slurry pipelines." But nobody could prevent the energy interests with their large bankrolls from buying up agricultural water rights. The *High Country News,* a bimonthly environmental journal, warned at about the same time in 1973, "But water is the key element in future industrial development, and what happens to Wyoming water will determine what happens to Wyoming." Wyoming boomed as it never had before in the mid- to late 1970s. Income and population took incredible jumps and Hathaway, the former governor who spent a short time as Interior Secretary before returning to Wyoming to work as a lawyer for energy companies, prophesied, "By the turn of the century, we'll be the number one energy state." The only dark spot in the general glow was agriculture. For three years running, its labor force had declined by 8 percent, while mining increased by 17 percent.

Harv Stone dealt with subdivisions. Floyd Bousman dealt with a rather massive and potentially devastating energy project. Both ranchers, who lived near Pinedale, Stone to the west of town and Bousman to the east, came out of the same traditional mold to find themselves separately confronted by the two contemporary trends in the West that have displaced the importance of ranching. Stone, born in South Dakota, began cowboying in Colorado before coming to Wyoming in 1938, where he watched over livestock in their national forest summer range for a cattle association. He worked as a hired hand for a number of years before homesteading his own ranch in the early 1950s. Bousman served as an aerial navigator during World War II and emerged from that conflict with some savings that he used to buy a ranch in 1948. For the first fifteen years, he supplemented his ranching income by packing dudes into the mountains on horseback, an occupation he found very demanding and nowhere near as satisfying as tending cows. The June I visited with Bousman, his ranch was a sweep of green fields set against the dramatic backdrop of the Wind River range in one direction, and a flat expanse extending into the high desert in the other. It seemed set at the breaking point between greenness and aridity. Bousman, who was secretary-treasurer of the Boulder Irrigation District, was irrigating his fields on this day. A stubble of gray beard softened the angular lines of his face, set between a checked shirt and well-creased, broad-brimmed hat. He was the classic picture of the high-country irrigator. With a shovel in hand,

Bousman walked along the raised dirt bank of the ditch. He leaned down to check the headgate. Behind him the water ran fully onto the field that stretched into the seemingly limitless high desert. The water was a silver sheen of fingers seeking the easiest route downhill and back to the river from which it had first been diverted, to be used again and again. Bousman, too, was cognizant of the high price of land, driven up by urbanites seeking a place of refuge. He did not think he or any other working cowboy could ever again take a small amount of savings and buy a ranch.

I had talked with the Swifts, Bousman's neighbors from the Chicago meat-packing family, that previous winter in their home overlooking Boulder Creek with a view toward the mountains. Along with other residents of the Pinedale area they were involved with Bousman—who led the fight—in trying to stop Project Wagon Wheel from becoming a reality. Project Wagon Wheel was the Atomic Energy Commission's scheme to set off a string of five consecutive underground nuclear explosions to unlock natural gas supplies near Pinedale in the mid-1970s. Most of Pinedale and surrounding Sublette County—with a total population of 3,755—wanted no part of Wagon Wheel. Pushing for the last remnant of the Ploughshare program, an attempt to promote the peaceful use of nuclear energy for construction and other purposes, were the AEC and El Paso Natural Gas Company. It seemed a very unequal match, but people like Bousman were extremely believable and well informed. There was a rationality and immediateness about their opposition lacking in most other environmental confrontations at the time. Along with local resistance, there was a growing regional concern about littering the West with more radioactivity, and, as a result, Project Wagon Wheel never came about. There had been two previous nuclear explosions within the Colorado River basin for the same purpose and a third, besides Wagon Wheel, was scheduled. At the time, Wagon Wheel was the granddaddy of them all, with not one but five blasts contemplated. By 1981, one AEC study suggested, there could be as many as 370 such explosions in the upper basin. It seemed that a bureaucracy was about to run amok with a rather dangerous new toy. The staff of the Federal Power Commission, however, was much less enthusiastic and noted that the commercial application of such technology had not yet been proven, although there had been some limited technical success with the previous two explosions.

It is hard to believe now that the plans for the detonation of those nuclear devices existed, what with the subsequent concern about nuclear safety and the health dangers of low-level radiation; but the interior West had a history of being a radioactive province. It is here that most of the world's uranium has been mined and milled, to be shipped elsewhere to be used as fuel for commercial nuclear reactors and in military weapons. New Mexico and Wyoming are, respectively, first and second in the production of uranium. Workers in underground uranium mines have experienced a higher than normal rate of lung cancer. The wastes from the milling process, called tailings, have been left at twenty-two abandoned mill sites to be washed into the Colorado River system or blown as dust over rural settlements. In addition, tailings from active mills have accumulated in large piles. The tailings are relatively high in radioactive content and can exceed health standards when confined in poorly ventilated spaces. Over the years they have been carted away from the abandoned tailings piles to be used as foundation material in homes, businesses, schools, churches, and fire stations in Salt Lake City, Denver, and Grand Junction. The first atomic bomb was exploded in the desert of New Mexico and there were proposals in the late 1970s to bury nuclear wastes in that same desert. The post–World War II weapons-testing program in the Nevada desert exposed thousands of soldiers and residents in nearby southern Utah to potentially harmful doses of radiation. Large amounts of fuel from the end of the nuclear fuel cycle are currently stored in the West, and more storage is planned. As the decade was ending, President Carter announced plans to locate the MX missile system in Nevada and Utah, a project that could affect up to 22,000 square miles and bring tens of thousands of new people into the area. So it was no surprise that the AEC decided to set off such a nuclear blast near Pinedale. The only surprise was that it never came off.

As head of the Wagon Wheel Information Committee, Bousman was concerned about the genetic effects on his grandchildren and the safety of Boulder Lake Dam, whose reservoir was the source of irrigation water for about 10,000 acres of pasturelands. At $1.25 an acre-foot, the lake's water is incredibly cheap, and there is as much high-quality water as any rancher wants during the two-month growing season. The present dam is the third such, previous ones having washed out. The Carey Act made possible the first diversions from Boulder Creek, a tributary of the New Fork River, around the turn of the century. Early settlers cleared the slopes by building fires atop large rocks and then pouring cold water upon them. They split open, literally exploding in some cases, and the small pieces were picked up and carted away. The new community that grew up around the newly irrigated lands was named Boulder. Ranching is the traditional use of land around here, and grazing on the nearby national forest has come to be regarded as a historical right. The present dam was built in 1966 and a proposed enlargement was part of the Bureau of Reclamation's 1972 scheme to transport Green River water over the Continental Divide and into energy-rich but water-poor eastern Wyoming. Bousman and others opposed such plans. It is people like these high-country ranchers who have been the bureau's traditional clientele, but Bousman felt there was little future in ranching, what with the incursion of second-home subdivisions and energy projects that would eventually drain off new water supplies.

The river, always the river. Below lower Green River Lake it spreads out and begins to meander through the high country. There are sandhill cranes and an occasional beaver and mallards at this time of year. Above all else, there is a sense of freshness about the Green River at this point, still preserved from its pristine origins in the glacial ice not too far distant in the mountains. The small creeks that come tumbling in—the Roaring Fork, Moose, Wagon, Beats Me, and Klondike—speak of tangible matters, of things observed by literal-minded people. Where the rocks of the Canyon Ditch's diversion structure jut out into the current, the Green becomes more a river and less a large mountain stream. The creeks have added some water, and the stream has gained some majesty just in time to lose a little of it to the ditch. Not too much further, Horse Creek and Cottonwood Creek enter from the west, and from the east comes New Fork River, into which Boulder Creek flows. But it is not long before the sense of freshness is lost, that the greenness lessens and the dun color of the high desert begins to intrude and the Green becomes a true river of the West, traversing dry lands for most of its length before ending at the junction with the Colorado River. In this high desert there is a fine sense of horizontal openness, the vastness of western spaces. The sky seems to sit upon the land, forcing one's perspective toward the far horizon. The clouds—close, tangible essences—trail rain across the landscape, precipitation that frequently evaporates before it reaches the ground. But the accompanying wind rustles the sagebrush.

Aspen, a tree of the mountains, has given way to cottonwood, a tree of the intermediate altitudes, which in turn will give way to tamarisk, the predominant plant growth along the lower, hotter stretches of the river. But here it is cottonwoods, huge yet wraithlike and fluttering, always indicating the presence of water and shade in what is frequently an otherwise reliefless terrain. Cottonwood trees are a very welcome sight. They have a nice combination of qualities. There are a great many creeks and canyons named Cottonwood in the West. But for some, cottonwoods have an undesirable quality: they use water in a water-short land. There have been periodic state and federal programs to cut them down, along with the tamarisk and other vegetative consumers of water along the lower river. The competition for water is very fierce. The story goes that a crew of Apache Indians was hired by a federal agency to cut down some cottonwoods as part of what was termed a "phreatophyte control program." The white boss came back after a while to check the crew's progress, only to find them sitting under the shade of a cottonwood. No trees had been cut. The boss asked what had happened and an Indian replied, "Apaches can't kill cottonwoods. Something bad might happen."

Oil and natural gas are produced near the river where, a century earlier, the crude oil from natural seeps had been applied to the axles of immigrants' wagons. One crossing of the Green River on the Oregon Trail was at the northern end of what is now Fontenelle Reservoir, a meandering body of blue-gray water this day, set against the light sandstone colors of the desert. "Improbable" is the first thought when such large, artificial bodies of water come unexpectedly into view in a desert setting. Fontenelle Dam is the first of any consequence on the Green River. It was part of the 1956 package of congressionally approved water projects for the upper basin that included Glen Canyon Dam. Fontenelle was first conceived of as a project to benefit irrigation, but that never worked out because the land that was to be irrigated proved to be too alkaline. So the water was put to use for industrial purposes, such as energy production. There are plans for more such uses of the water stored in Fontenelle, and a Wyoming Game and Fish Commission publication warned, "Although present conditions of reservoir operation provide sufficient releases to meet minimum flows, future conditions may not."

Below Fontenelle Dam, the Big Sandy River comes in from the east, having drained the southern end of the Wind River Range and crossed the desert to the Green River. From the west, draining the northern slope of the Uinta Mountains, come Blacks Fork and Henrys Fork into Flaming Gorge Reservoir, also part of the Colorado River Storage Project Act of 1956, which authorized more than forty potential water projects. Flaming Gorge is the second impoundment of Green River waters. By this time the salinity of the water has increased markedly since the pristine beginnings. From the Big Sandy come 510 to 830 tons of dissolved solids daily, Blacks Fork contributes 618 tons, and Henrys Fork adds another 359 tons. At the point the Green River leaves Wyoming, its salinity is about one-third of what it will be at Imperial Dam near the Mexican border. Most of the salinity on the Green is from natural sources, the rest from return irrigation flows. The mineral content comes from such natural sources as Alkali Creek and the irrigated fields around the tiny communities of Eden and Farson along the Big Sandy. As the salinity increases, so does the sediment load from the heavily grazed lands. Irrigation water is used to raise feed for cattle whose grazing contributes to erosion which is deposited in reservoirs, thus diminishing their capacity to store irrigation water. It is a vicious cycle. More than 2,000 acre-feet of sediment is dumped into Flaming Gorge each year.

It is surprising how much of southwestern Wyoming is a desert—a high, cold desert to be sure, since the elevation north of the Utah border does not drop below 6,000 feet—but a dry desert just the same and, at least in one remote area, complete with those shifting sand dunes that fit most peoples' concept of what desert should look like. Northeast of Rock Springs lies the Great Divide Basin, 3,916 square miles of separateness that can be penetrated only on a few primitive dirt roads. No streams flow out of the basin, and few flow for any distance within it. Although it is self-contained, the Great Divide Basin—within which lies the Red Desert —is usually lumped with the Green River Basin in state and federal water studies. Topographically it shares a greater resemblance to the lands west of the Continental Divide than to those to the east. There are magnificently sculptured badlands, dramatic mesas, flat sagebrush plains, the Oregon Buttes near South Pass, a chain of small lakes, and dry, eroded gullies. Livestock, particularly sheep, have heavily grazed the basin. Some oil and natural gas and uranium have been extracted, and as the decade ended with an energy crisis, there was a lot of feverish activity by prospectors looking for more such mineral resources in the basin. Toward its western end are vast ranks of sand dunes, some 100 feet high, and stretching over 170 square miles—a visible symbol of aridity not to be matched along the length of the Colorado River system until the Imperial sand dunes near the Mexican border in Southern California. The difference between the two dune areas is the perennial ice buried at the base of the Wyoming dunes.

The Green River Basin in Wyoming is roughly triangular, with a base of 213 miles and an apex extending north 168 miles. From Knapsack Col at the river's start in perpetual ice to the sands of the Great Divide Basin, the scenery is a microcosm of the drama within the larger Colorado River basin. Wilderness, ranches, recreational-home subdivisions, mines, power plants: the river has gotten off to a well-used start on its journey toward the sea as it crosses into Utah.

2

Not far from the Wyoming border, in fact only forty road miles south of Baggs across the rolling sagebrush terrain, lies Craig in northwestern Colorado. Wyoming's predominant image is one of ranching, with liberal doses of mining in the past and returning in the present; in contrast, Colorado's history was first associated with mining, although ranching has been a constant. In Colorado, people go looking for ghost towns with a history of gold mining, only to find them garishly rejuvenated into modern tourist traps specializing in separating the traveler from his paper dollars. Not many tourists visit Craig; if they do, it is only because they find it a convenient place to stop and get gas while driving from Steamboat Springs or Denver on to Dinosaur National Monument and eventually Salt Lake City on Highway 40, a two-lane alternative to the interstates to the north and south.

Craig had just crossed the line, the transition from rural agricultural to urban industrial, when Pinedale was approaching urban status at the end of the last decade. It was not a visitor-tourist enclave, as were such Colorado communities as Steamboat Springs or Aspen, oriented more toward the metropolitan areas of Chicago and Los Angeles than their immediate milieu. Craig was more a part of the traditional oasis civilization of the West than the newer enclaves. But Craig was like a Steamboat Springs and Aspen, and other once-small communities along the western slope of Colorado, in that it was booming in the 1970s. Coal, uranium, oil and natural gas, and electricity were being exported from northwestern Colorado. Immigrants who had come to work in the new industries had taken over, were in fact dictating many of the decisions; and there was hope, doubtless the same hope that existed during past booms, that Craig would not revert to being just another small ranching center, that the boom-and-bust history of the West would not repeat itself. There was a certain feeling of smugness—This time we've got it licked—mixed with desperation, since there was no past history of stability in such communities as Craig. In the state capital of Denver there were grave concerns about where the remaining water would go, especially in that drought year when President Carter's "hit list" seemed to single Colorado out for special persecution.

Craig sits on the north bank of the Yampa River, whose flow, as Colorado River tributaries go, is relatively unimpeded by large dams. That is not to say there were not future plans for such structures, because there were, but no major impoundments were yet in place as Craig began its metamorphosis shortly after the start of the 1970s. Although there were no major dams, there were, as elsewhere in the West, plenty of small impoundments. Their number in this one region gives a good indication of the intensity of water use elsewhere in the basin. There are about 122 small reservoirs within national forest boundaries in northwestern Colorado, 550 stock ponds and erosion-control structures on other public lands in the area, and 5,700 small reservoirs on private lands. Additionally, there are 8 reservoirs of 500 acre-feet or more used for irrigation or municipal purposes.

The Yampa River begins along the Continental Divide, marked here by the 12,000-foot peaks of the Park Range. The first trickles grow into Elk River, Spring Creek, Fish Creek, Green Creek, Silver Creek, Morrison Creek, and Bear River. There is also a Coal Creek. They join to form the

Yampa, whose drainage basin takes in nearly 10,000 square miles, mostly in Colorado but some in Wyoming. In the mountains there can be as much as fifty inches of precipitation a year, in the high desert about nine inches. The growing season for the irrigated hayfields can be as long as four months. Flowing across the high desert, interspersed with low mountains and hogback ridges, lying between the 6,500 and 7,500 foot level, the Yampa dips into Bear Canyon before merging with the Green River at Echo Park in Dinosaur National Monument. In the early 1950s there were plans for a dam in Echo Park, but conservationists succeeded in blocking it and the site was switched to Glen Canyon.

About one-third of the way along its journey to the Green, the Yampa passes Craig; here its meandering course through the flood plain is marked by green fields, cottonwoods, and willows. It is prime pastureland, good for raising hay. In the Yampa River basin about 85 percent of the land area is used for grazing; less than 2 percent is irrigated, and of this watered 100,000 acres, almost all is used for raising livestock feed. But energy developments are beginning to impinge on the river near Craig.

There is not much to say about Craig's esthetics, because they are mostly lacking, a fact that has probably saved it from the western kitsch that predominates elsewhere. It is a typical high-country town, actually a small city, with streets laid out in the four compass directions. The main street is Highway 40, which bisects the town; the other east-west streets are numbered, while their north-south counterparts bear such stalwart names as Lincoln, Pershing, Colorado, Green, and Yampa, the last-named being the main road north to Baggs. There are the usual amenities such as a fairgrounds, armory, golf course, and city park, closed when I visited Craig because the energy workers who could not find housing had overrun it with housing on wheels. And there is an unusually large number of bars to serve the needs of the inhabitants. Craig's western flavor, if it ever existed, has been erased by the preoccupation with making a living off the surrounding natural resources and the pursuit of progress. What exists is actually more typically western than the false-fronted buildings selling western wear to tourists in resort towns. The preferred architecture is Safeway and Western Auto Stores moderne. Craig is remote, being, as the locals put it, one mountain pass from everything, including the nearest urban areas of Denver, Grand Junction, and Salt Lake City.

The land surrounding Craig does not vary greatly. Sagebrush dominates the low, rolling hills, and grows where other plant communities find insufficient moisture in the soil. There are four different species—basin big sagebrush, Wyoming big sagebrush, low sagebrush, and black sagebrush. One large patch covers 1,500 square miles in north central Moffat County; there are scattered patches elsewhere. Sagebrush prefers fine, deep soils, while piñon and juniper, which together form another type of vegetative community found over large areas of the interior West, grow best in coarse, porous soils. In northwestern Colorado sagebrush starts giving way to piñon-juniper, which predominates along the tops of mesas and plateaus in the Southwest; the colors begin to shade from the light gray of the sagebrush into the dark greens found further south. Around Craig some oak brush and serviceberry grow on the slopes where the snow remains longest. It is a semiarid steppe, with an average temperature of 42 degrees and an average rainfall of 13.4 inches.

Most of the land has been overgrazed, this being the reason for the large amount of sagebrush—an invader of former grassland areas. The dominant type of mammal is the rodent. Open range predominates. There is some dry-land farming, a little irrigation and oil production, and strip mining for uranium and coal. There are two coal-burning power plants in the area and more are planned, along with more coal mines, railroads, highways, and dams. Craig was an energy boom town, a species that had come to dominate the interior West in the 1970s as the country experienced two energy crises and three presidents, all of whom emphasized there must be more domestic production of fuels. Energy policies like Project Independence meant boom times for the interior West, where most of these resources were located. What happened to Craig was similar to what happened to Rock Springs, Green River, Rifle, Moab, Farmington, Page, and a dozen or so other small communities in the Colorado River Basin that had the good fortune or bad luck, depending on how one viewed this second major land-use change, to be near a vein of coal or oil shale, a deposit of uranium, a pool of oil or natural gas, or a newly constructed power plant dependent on one of the above. Only the tourists-visitors, be they skiers or second-home owners, could match this sudden influx of people concerned with the extraction of natural resources. Certainly the livestock industry could not, and ranching began to be displaced, pushed aside; but its image, or what was thought to be its style, was emulated by the newcomers who bought sheepskin-lined coats and wide-brimmed cowboy hats or played mountain man with down jackets and freeze-dried food. It was all part of the "rural renaissance," and the advertisers of cigarettes and ski areas quickly capitalized on the cowboy image.

It was surely boom times for the city of Craig, which had only gained 5.5 percent in population between 1960 and 1970. Between 1970 and 1976, there was a 75 percent increase. For the three counties that make up northwestern Colorado, there was a slight loss in population between 1950 and 1970, but a 70 percent increase in the next five years. With the decline in coal production after World War II and the migration of people to the cities, Craig and the surrounding region fell on unspectacular times. Craig settled into its role as a ranching and trade center, and little else. An oil refinery, located there during the war, shut down in 1948 —and that seemed like the end of any new industry. The longer-term inhabitants were not unmindful of their past history. A study commissioned by a coal company in 1974 determined that "The erratic nature of mineral-related development has left residents skeptical regarding growth. Prior to 1966, for instance, there were 19 operating coal mines in the area. Due to changing economic conditions, only one was in operation as of 1966; there are now three operating mines. Although economic growth is seen as beneficial to the area, there is the fear that mineral-related growth is temporary, and that after a limited period of expansion, the economy once more will experience a decline." Craig's history of booms and busts was similar to the experiences of other high country communities.

After the fur trappers cleaned out the area by the early 1840s, the

A coal strip mine near Craig, Colorado.

explorer-soldier John Charles Frémont passed through the Yampa River Valley in 1844 on his way from Steamboat Springs to Salt Lake. Frémont saw little use for the land around Craig, although he did note the outcroppings of coal. Other travelers at the time also saw little of worth in the area, a not uncommon view of utilitarian-minded travelers and explorers from the humid East. A few years later the Grand Canyon was to be judged equally valueless. Following the removal of the bothersome Ute Indians, a number of small communities sprang up in the 1880s in northwestern Colorado. In 1889 the Craig Land and Mercantile Company was formed, and among the first purchasers of land was the Reverend W. B. Craig. Cattle was the first bonanza, then the cattlemen tried to ward off sheepmen, homesteaders, and the Forest Service. Eventually all came to live together in harmony. The livestock industry remained economically dominant to the 1920s, when mineral production began to take over; it remained politically dominant into the 1970s. When the railroad reached Craig in 1913, there was hope of a coal boom but it never materialized. The railroad never did very well; one of the reasons was a high mountain pass it had to cross, so it was thought a tunnel through the Rockies would help. The Moffat Tunnel was built, and again hopes soared. A series of U.S. Geological Survey publications promoted the mineral wealth of the area, but still nothing much happened. Oil was produced in the Rangely area in the 1920s. There was a land-development scheme, called the Great Divide Homestead Colony Number One, but dry farming in the high desert never proved very productive and the settlers eventually drifted away in the drought years of the 1930s. In the intervening years the region just puttered along on its own with lessened expectations until the 1970s. A historical study of northwestern Colorado, written for the BLM, states that "This region's history was one of constant 'booms' that never worked out. From fur to mining, cattle to oil, coal to oil shale and back to coal, the northwest corner of Colorado showed great promise, but for various reasons, it has never been able to fulfill the dream that seemed so real." It had all changed by that watershed year when the dream, already fulfilled, threatened to turn into a nightmare.

What Craig and northwestern Colorado were trying to deal with in 1977, and constantly falling behind on, was the impact of construction workers from the Yampa Project, a nearby coal-burning power plant, and the advance guard of numerous proposals for coal mining. The proposals, not all of which would come to fruition, included five mining plans for already leased federal lands, thirty-three competitive lease applications, fifty-four nominations for additional leased lands, eleven for preference-right leases, and one planned right-of-way for a short rail line to haul coal. What this could all amount to in 1985 was 62 percent of the coal production in Colorado and 2.5 percent of the national total—25 million tons of coal from seven mines. There would be three new power plants, twenty-six miles of new railroad track, fifty miles of new roads, two hundred miles of new powerlines, and a population increase of nearly ten thousand people. This, of course, was all conjecture by the planners. The reality that Craig was dealing with was a serious breakdown in the traditional social fabric that had held the community together for the previous ninety years. Streets were congested, schools were on double sessions, the small hospital was overcrowded, crime was way up, ministers and mental-health workers were dealing with heavy caseloads, and the suicide rate was about 200 percent above normal. Drug use was out of control, and about 80 percent of the arrests were alcohol-related, not a surprising number if one made a tour of the crowded bars on a late Friday afternoon after paychecks had been distributed. As was happening elsewhere in the interior West, mobile homes had proliferated. In the first four years of

Mobile home subdivision near Craig.

the decade, 590 new housing units were erected in Craig—85 percent of them mobile homes. It was no longer a static society, and in the midst of the uncontrolled growth Craig's mayor, Doyle Jackson, quit after issuing a rather prophetic warning:

> In a community where rapid growth requires all available resources to be directed at basic services, subsequent problems result which are lumped into what is now known as the "boom town syndrome." Educational services, recreational facilities, and other systems are all attacked and all suffer from the inability to perform adequately in meeting the demand of the populace. This, in turn, forces a decline in quality. Lack of and/or poor quality services and amenities, in turn, result in upswings in crime, alcoholism, suicides and suicide attempts. Social problems and negative signs of urbanization (congestion, higher prices and fear) contribute to feelings of alienation and loss of the old, better way of life. The highest toll extracted from boom towns is the literal destruction of a community which has in the past sustained and nurtured its people. The decline in the quality of life has stolen a community from its people.

If the social impacts were to be the most devastating, which they were, then Craig, as so many other small communities in the West during that decade, was to shed its rusticity and emerge an urban, industrialized center, which it did. Actually, the amount of acreage to be converted from grazing to mining was infinitesimal, even if all the mining plans became a reality. Areas actually mined would account for less than 1 percent of the total land area in the three counties. The real competition would be for water. As a BLM study noted, "Perhaps competition for water could be more significant because of the large consumptive rates generally expected for power plants." Before the coming of power plants, about 90 percent of the water consumed in the Yampa Valley went toward irrigation of the bottomlands for hay production. Coal mining requires very little water and a power plant a moderate amount compared to agriculture. Colorado's three northwestern counties accounted for about 2 percent of the total value of the state's agricultural products. While agriculture was of little importance in the total economic scheme, its advocates would voluntarily surrender very little water.

Not only would the use of water shift toward other purposes, but there were other, more subtle ways increasing energy production would affect the traditional use of land. Cumulatively, they could be more decisive than the more direct changes. Favorable political decisions, like water in the West, flow toward the interests with the most clout and money. When I visited Craig, two of the three county commissioners had close ties to energy companies. One commissioner had helped a coal company obtain options on the land it sought because he was on the board of the utility company that would use the coal. Another received payments from a coal company that had an option on his land. Condominium development around the nearby ski resort of Steamboat Springs, as well as coal, had its effect. Second-home subdivisions drove up the price of surrounding acreage, and some ranchers sold out. Rarely was such land again used for ranching, except as a rich man's hobby. Packs of dogs killed sheep and had to be killed themselves. The dogs were brought into the area by urban newcomers and either abandoned or let loose to run wild. The increased population trespassed on ranchers' lands, when hunting or snowmobiling, and fences were torn down and other improvements damaged. Ranchers went to work for energy companies to supplement their incomes, and became more financially stable and less dependent on the vagaries of beef prices. Besides the land directly disturbed by mines, transmission lines and pipeline corridors cut up grazing lands. There was an increasing shortage of farm laborers, the legendary hired hand; ranchers could not meet the wage levels of the energy companies. This drove ranchers toward increased mechanization and more debt. As one rancher put it, $100,000 worth of machinery had to be purchased to put up $15,000 worth of hay. Ranchers, a proud lot, having come from some of the best families with the longest histories in the area, had once dominated community affairs. They now found themselves being treated as second-class citizens, with all that new energy money attracting most of the action. There was a temptation to get in on it, to mine their own ranch lands. "The coal is there, and I don't see any reason I shouldn't make some money off it, do you?" asked Glen Miller, whose ranch is eleven miles north of Craig on the road to Baggs.

I drove north on that road one winter evening on my way to a meeting in Baggs, just over the state line in Wyoming. The population of Baggs had jumped from 150 to about 300 in the six years following 1970, with most of the new population being transients involved in exploration for oil. But Baggs was still small enough that drought year so that if one blinked while driving through, as the saying goes, he would miss it. There was one restaurant in Baggs, the Drifter's Inn, and I sat near a long table that had been set up in the middle of the room. Sitting at the table were about twenty-five men, most with sportcoats and a few with suits and neckties or bola ties. They were the Bureau of Reclamation personnel up from the Grand Junction office and their clientele—local ranchers, water district officials, and an attorney. It was all very chummy, and I remember eating chicken-fried steak and overhearing snatches of conversation from the long table. They were talking about the arrangements for the meeting and wondering how it would turn out.

The public hearing on the Savery–Pot Hook Project was held in the Baggs High School gymnasium. Pickup trucks were parked outside; inside, a screen had been set up under a basketball net. Basketball is big in these small high-country communities, since it only takes five to make a team and outdoor sports are hampered by the long, cold winters. Baggs had a class 3 Wyoming state championship team in 1973. Starting with "Good evening, ladies and gentlemen, we are glad to be with you this evening," bureau personnel, using an overly simplified phraseology and reading agonizingly slowly, launched into their factual presentation.

Then a Craig attorney, James M. Pughe, an advocate for the project and one of the diners, took over the meeting. Pughe represented an interweaving of various public and private water interests, such as coal companies and water boards, and a genre peculiar to the West—the lawyer who

A crowd at a basketball game in Craig.

is expert in the byzantine ways of western water law, which has its roots in England but has undergone an almost complete transformation to fit the realities of the arid West. His occupational roots go back to such persons as Delph Carpenter, the brilliant Colorado attorney who was both a water statesman and strong Colorado states-rights person at the 1922 compact negotiations, and Northcutt Ely, who went from a top Department of Interior post into practice as a high-powered, sometimes imperious lawyer-lobbyist for southern California water interests during their epic battles with Arizona during the 1950s and 1960s.

This evening, Jim Pughe was telling the 150 or so assembled ranchers and their families sitting on folding chairs that if they did not approve the Savery–Pot Hook project in an upcoming election, the water would most likely go to energy use or, worst of all offenses, continue to escape downstream to light up Las Vegas. And the Bureau of Reclamation people smiled and nodded assent, and when the one identified environmentalist got up to ask them a question, he was heatedly berated for being an obstructionist, although he was only seeking factual information.

But Savery–Pot Hook could not bear much scrutiny and still survive as a project of much merit. The meeting was in January of 1977. One month later Savery–Pot Hook would be under intense study by the new Carter Administration, much to the consternation of those same career bureaucrats who attended the meeting. Eventually the Administration would recommend it not be funded and be deauthorized; the latter step is unprecedented and has yet to be acted upon by Congress. On paper Savery–Pot Hook did not look good. Costs outweighed benefits. The growing season was short, and grazing would be eliminated or reduced on 21,750 acres while only 14,650 acres would be irrigated by the project. Sage grouse and antelope populations in the area would suffer. And there was always the threat, hovering in the background, that energy interests would make off with at least part of the water to be impounded behind the project's two dams. Pughe played upon that threat, at the same time assuring the rural audience that the governor would have to approve the transfer of water use. He also played upon the historic fears of a transbasin diversion, that if the water was not used and used quickly on the west slope of the Continental Divide, it would be transferred out of the basin to Cheyenne, where some of the water from the headwaters of the Little Snake River was already going. Cheyenne was almost as unthinkable an alternative as Las Vegas. The audience clapped when Pughe declared, "My message is, Little Snake River water is to be used here."

East slope versus west slope, agriculture versus energy versus second-home subdivisions versus recreation versus wildlife—these were some of the issues that Colorado and the rest of the basin states were trying to deal with in the 1970s when it finally became obvious that the Colorado

River was not going to serve everyone's needs forever. Colorado found itself in a position similar to Wyoming; the only significant amounts of surface water remaining for future use in the state were in the Colorado River Basin, and that water was escaping downstream to be used by others, like the insatiable Californians and their counterparts in Arizona who were about to tap into the system in a big way with the Central Arizona Project. The upper basin had been slow to develop, slower to use its share of the river's waters than the lower basin, and the only schemes it had in mind at the time were such marginal projects as Savery–Pot Hook or Fruitland Mesa further south in Colorado. The state's share of the river's waters was a promise sealed in interstate compacts signed in 1922 and 1948. Projects like Savery–Pot Hook had been dreamed up thirty or more years ago and set in motion by Congress in 1956 as part of a political trade whereby the upper basin got Glen Canyon and a number of smaller dams after the faster-growing lower basin—read California here—already had Hoover Dam in place. The conquest of the river was always the dominant theme, it was what was being voiced in Baggs that night and it was what the National Academy of Sciences commented so eloquently upon in its 1968 report on the Colorado River:

> The management of water sometimes is advanced as an end in itself. Flowing from a basic tenet of occidental philosophy that nature, distinct from man, exists to serve him, is a strong and pervasive view of man as a manipulator of nature, and this view places a premium on technical proficiency in regulating water volume and quality. Arguments for controlling the flow of the Colorado and for transporting water long distances to make the desert bloom reflect the view that an uncontrolled resource is a wasted resource and that if man has the capacity to control and completely utilize the waters of a river he should do so. . . . Especially in the arid West, this aim or motive is present and pervasive; elsewhere it may not be as obvious, but it should not be ignored as a motivating factor in both public and private efforts to manage water.

Savery–Pot Hook was certainly not the world's most needed water project, but one would have thought the challenge to it and other projects in Colorado had assumed the proportions of a Middle East crisis, judging by the angry, fearful reaction of water interests and elected officials in Colorado following President Carter's announcement. The battle to retain the Colorado water projects was led by Governor Richard D. Lamm, who had been elected in 1974 on a strong environmental platform. Both Lamm and Senator Gary Hart came into office the same year during a liberal landslide that seemed to bring fresh voices and outlooks into

Colorado politics. The state was at a critical point, questioning the old growth policies and looking for new directions. No more nuclear explosions to unlock natural gas supplies were wanted, and Coloradans successfully voted down a proposal to stage the Winter Olympics in their state—an extravaganza that would have been eagerly sought a few years earlier. There was a strong antigrowth, or slow-growth, movement afoot in the state, the reaction to the helter-skelter development during the boom of the late 1960s and early 1970s when the influx of new immigrants from the West Coast and the East climbed to 28 percent. Rural subdivisions and condominium developments sprang up in the mountain counties of the west slope. The immigrants were younger; their urban backgrounds led them to be more protective of the natural environment. They did not particularly want too many others like them following the same migratory path, since this would destroy the very amenities, such as clean air and clear water, that brought them to such communities as Steamboat Springs, Aspen, Vail, Crested Butte, Telluride, and Silverton on the west slope and the Boulder-Denver metropolitan area on the east slope of the Rockies. And they were instant voters; it did not take them eighteen years of living on a ranch to entitle them to vote for a county commissioner.

Lamm, like Hart, was very much a candidate of this new wave. He was elected and then reelected after he personally reminded President Carter, in a panel discussion, of what John Gunther had written in *Inside U.S.A.*: "Touch water in the West and you touch everything." As governor, Lamm could safely defend Colorado's right to its water allotment, but the ticklish decisions were where and how it would be used within the state. So, in order to keep Colorado's water in Colorado, Lamm and the state legislature and almost everybody else except the most dedicated environmentalists favored more dams on the Colorado River system. "Every prognostication indicates that Colorado is likely to put to use its remaining undeveloped water—whether by private, state, or federal sponsorship—sooner or later," stated Lamm. "The question is not so much whether this water will be developed, but only *how, when,* and *for what purpose.*" The same viewpoint had been offered from a different perspective in 1909 when D. C. Beaman—a vice-president of Colorado Fuel and Mining Company, the state's single largest industrial firm—remarked, "I have listened with much interest to the gentlemen who have been so earnestly sounding the foghorn of danger from the consumption of our natural resources. . . . A very large proportion of the conservation movement proceeds from misinformation. It is worse than that, it is ninety percent wind."

Lamm was perhaps aware of another John Gunther quote, to the effect that "Water is blood in Colorado," because the governor, like all politicians of the interior West, was a blood brother. He had not always been so. In 1966, Lamm, then vice-president of the Young Democrats of Colorado, had written Congressman Wayne N. Aspinall, who represented the west slope of Colorado and was the powerful, autocratic chairman of the House Interior and Insular Affairs Committee. Lamm was objecting to dams in the Grand Canyon proposed in a bill authorizing the Central Arizona Project and five small irrigation projects in Colorado. "The Colorado Young Democrats are proud of you as our Congressman of greatest influence; yet we seriously question whether you fully appreciate the ever increasing threat to our scenic resources, the ever increasing need to preserve some of America in its natural state, to save some from those who would plow, pollute and pave under (as well as dam up) our major scenic resources." (Aspinall replied tartly in a letter to the group's president, "I have always believed the Creator placed natural resources to be used as well as viewed, and to be used by the people generally.") While Lamm would be elected in 1974, propelled mostly by those new forces who were flexing their muscles in Colorado at the time, Aspinall would be ignominiously defeated in the Democratic primary in 1972 by a law professor who was subsequently beaten by the Republican candidate in the general election. Aspinall was a victim of the same forces that elected Lamm. The veteran congressman, then seventy-six years old and a member of Congress since 1949, had been redistricted by an unsympathetic state legislature dominated by Republicans. He was perceived as being too old, too entrenched, and one of the "dirty dozen" congressmen singled out for defeat by environmentalists. The voters in the rejiggered district were younger and more urban-oriented than in previous elections. He did not fare well in such west-slope counties as Pitkin, of which Aspen is the county seat. Aspinall, who gained his first elected office in 1920, had not known when to quit.

It was the end of a remarkable career, almost an era of the West, and an end barely noticed outside of Colorado and Washington, D.C., and quickly forgotten if ever really understood in the state's urban areas. But in the mining and ranching communities of western Colorado there were few who did not know Wayne Aspinall by either sight or name. His colleagues in Congress called Aspinall "Mr. Reclamation." He was certainly "Mr. Water" as far as Colorado was concerned. In the East he might have been thought of as that venal western congressman who cheated President Kennedy out of his wilderness bill (President Johnson subsequently got it when Aspinall was ready to release it from his committee), but to the West he was scrupulously honest and fair, and above all else, knowledgeable. Aspinall could challenge the technicians with equal expertise, a rare ability, since most politicians were dependent upon them.

If Aspinall had a basic weakness, it was his unrelenting attitude toward conquest of the river—not being able to discern where all this might lead other than to more fruit orchards, hayfields, and mines; knowing the water was going to run out before everybody's wishes were taken care of but still writing those wishes into law. The denial of water, for whatever reason, was not Aspinall's forte, nor was it the strong point of any western politician. His departure from Congress coincided with the end of the era

Wayne N. Aspinall of western Colorado, who, as chairman of the House Interior and Insular Affairs Committee at a crucial time, had much to do with dividing western waters.

of big dams and aqueducts; now would be the time to build the smaller, more marginal projects like Savery–Pot Hook and clean up the water-quality problems, at least partially caused by all this construction. But it was Aspinall who had seen to the authorization of those smaller projects and some of the bigger ones, and in 1974 when the House Subcommittee on Water and Power held hearings on how to provide Mexico with better-quality water, Aspinall, by then a private citizen, was invited to sit with the committee. It was the ultimate compliment, but one aptly paid, since Aspinall had always believed Congress to be more representative of the people than the more transient presidential administrations. He had tenaciously held onto this belief in the field of natural resources during the time the presidency was becoming more imperial. Aspinall was aided by such crises as Korea and Vietnam, and presidents who were more eager to make their mark in foreign and urban affairs, rather than the hinterlands of the West. That heritage—of Congress being the essential arbiter of what happens with western lands and water—was to remain unchallenged until the Carter Administration.

Although he was a water statesman, Aspinall was first of all a man of his congressional district. He consistently brought home the bacon, be it low grazing fees, continuance of an outdated mining law, or that ultimate prize, nearly $1 billion worth of water projects. During his long tenure as chairman of the House Interior and Insular Affairs Committee, a post he held during four presidential administrations, Aspinall had more to say about the governing of the West than any other individual, including the many governors, the Secretaries of Interior, and his counterparts in the Senate, where the discussions of western land and water use in the Interior committees were never as thorough as in the House. Interior Secretary Stewart L. Udall, who served during a portion of Aspinall's reign, frequently deferred to the committee chairman. His brother, Morris K. Udall, who was to succeed Aspinall as committee chairman, followed his chairman's lead. When Morris, as an Arizona congressman, begged Aspinall to appoint him to the conference committee considering the crucial Central Arizona Project bill, he promised, "You would be my leader, always." Aspinall's fellow congressmen, who lived outside the interior West and knew little or nothing about complex water matters, came to trust him. One such congressman, when the 1968 bill came to the House floor, simply stated, "I have never heard or learned of Chairman Aspinall bringing a bill before this body unworthy of consideration. . . . If it's good enough for Wayne Aspinall, it is good enough for Ed Willis."

A staff aide for Senator Carl Hayden of Arizona, in a memo assessing the chances of passing a Central Arizona Project bill, said of Aspinall, "He is probably objectively judged as a sincere, capable, knowledgeable

One of the water projects provided by Wayne Aspinall on the Gunnison River in western Colorado.

technician—dedicated to placing before the House sound projects that have been scrutinized in every aspect. He is deliberate and sometimes seemingly arbitrary and immoveable." Aspinall was nowhere near as charitable toward the venerable Hayden, judging him unknowledgeable about water matters. But then, Hayden had committed the unpardonable sin of attempting a legislative end run around Aspinall. The last comment in the memo was probably in reference to Aspinall bottling up the Wilderness Act in 1963, despite personal phone calls from President Kennedy, until he was assured that mining would be permitted in wilderness areas and that Congress, not the President, would be the final arbiter of what new wilderness areas could be added to the system. When he got these assurances after President Kennedy's assassination, the bill was passed by Congress. Publicly environmentalists castigated Aspinall, but privately they respected his legislative talents. After the two dams in the Grand Canyon were removed from the 1968 bill and a compromise affecting all factions in the West had been molded by Aspinall, the Sierra Club's Southwest representative, Jeffrey Ingram, wrote Aspinall, "Appreciation may be too weak a word, for as I reflect even over the short time I have been involved, I find my feeling to be one of awe at your accomplishment."

When Aspinall left Congress, he returned to his home town of Palisade, which has the same neat orderliness about it as Aspinall's work habits. Palisade, just east of Grand Junction, is where the upper Colorado River issues forth from the high mesa and plateau country of the western slope onto the high desert. Aspinall's family moved to Palisade from Ohio in 1904, when he was eight years old. As a boy he was raised with a shovel and a hoe in his hand and picked peaches from trees watered by one of the first Bureau of Reclamation projects. The view from Aspinall's basement office in his home, crammed with the mementoes of a long congressional career, was toward the Colorado River. Late in life he recalled, "The river means the West to me. It is the heart of my West. My whole ambition was to see western Colorado, no, excuse me, I mean western water made available for the development of the West." And more than any other single person, Aspinall had succeeded in spreading the waters.

But such diverse politicians as Aspinall and Lamm were constrained by certain realities in Colorado, the primary one being that during the first part of the Tertiary Period some 70 million years ago a great series of north-south mountain ranges began to rise through almost the whole length of the continent. The glaciers during the last 3 million years gave the finishing touches to the creation of the Rocky Mountains, and by the time the first humans arrived in Colorado some 20,000 years ago, the rivers were already flowing west and east from the Continental Divide. With the first permanent settlements springing up after 1858 along the eastern slope, or what is now known as the Front Range, the state's primary land-use pattern had been established—80 percent of the people strung along the east slope, and 70 percent of the water and most of the mineral resources on the west slope. For a politician, this situation called for a delicate balancing act, one that would become more delicate as the remaining water became scarcer.

The Rocky Mountains divided the state into four watersheds; the snow that fell at the higher elevations melted and flowed down into the Colorado, the Rio Grande, the Missouri, and the Arkansas river basins and then left the state. The Rio Grande, Missouri, and Arkansas basins were east of the divide and contained the least amounts of water, although the latter two flowed through the most populous regions. A state report noted, "East of the Continental Divide, it is clear that there is little additional water left to consume, given physical and legal limitations and present consumptive uses." There was, at the most, an additional 1 million acre-feet of Colorado River water that could be used in the state, the exact amount subject to the river's flows and various interpretations of the 1922 compact; unfavorable rulings and droughts could lower this amount. The report continued, "In short, Colorado water supplies are limited, and their availability is subject to fluctuation and uncertainty." Uncertainty would be the key word for the remaining amount of water in the whole river system as the 1980s began. Of Colorado's remaining 1 million acre-feet, if that was the true figure, energy developments might want from 250,000 to 620,000 acre-feet, agriculture a maximum of 713,000, and diversions an additional 400,000 over the next twenty or so years. Clearly, the competition for water was going to get exceedingly intense in Colorado.

Agriculture was already losing out, or at least not gaining, in comparison to energy uses whose economic return was far greater. Energy producers, such as the large oil corporations that were spreading out into coal and uranium production, could easily afford to buy agricultural water rights, and the water in the 1970s began to flow toward them. A 1975 Bureau of Reclamation report, referring to the five water projects Aspinall had secured for the west slope in 1968, stated, "A major authorized purpose of these projects was irrigation. Federal policies with respect to support of irrigation projects were currently under review. With the increasing pressure for energy development in the area, these projects may be reformulated to provide additional municipal and industrial water with corresponding reduction in irrigation water." It was true the governor had a qualified veto over any transfer of water use, but he would be subject to some of the same pressures. In some Front Range areas, urban communities had purchased 5 to 10 percent of the total amount of water being consumed for agricultural purposes, and this figure was bound to rise as people kept flooding into the state. Other communities were aggressively pursuing plans to divert their remaining amount of Colorado River water or pushing other water programs, such as increased pumping of groundwater. Agriculture was getting squeezed. Large irrigation projects were less popular with the federal government and too costly to be tackled alone; the west slope had lost Aspinall, its great provider; and the emphasis had shifted to industrial and municipal water uses. All of this did not mean that western Colorado was going to dry up, because most of the large energy projects were located in that region, but it did mean that the bucolic style of ranch life was about to be eclipsed by boom towns like Craig and enclaves like Aspen.

For the east slope it would mean more cities, more smog, more subdivisions, a process that had been started with the first diversions of water from the Colorado River basin. Around the time that the Grand Ditch was constructed in the 1890s, Louis G. Carpenter, who would come to be known as the father of scientific irrigation in Colorado, was studying the Big Thompson Project, which would eventually result in another diversion from the basin. Carpenter, a professor of engineering at Colorado Agricultural College in Fort Collins, went on to serve as State Engineer in the early 1900s. Most of the construction activity under his direction was on the more rapidly expanding east slope, but he was not unmindful of the importance of Colorado River water. In 1908, when Colorado had more land under irrigation than any other state, Carpenter declared, "The farmers go into the mountains and in numerous cases, bring water across the summit of mountain passes at elevations of many thousands of feet, construct ditches on the precipitous slopes, bore tunnels, all for the sake of bringing water to the lands which need it." Irrigation ditches at this time were more than a hundred miles long and as much as sixty feet wide. An unlimited, prosperous future seemed to be in store for anyone who shared in this water, and Carpenter typified the enthusiasm and optimism of the time. "It is an example of the enterprise of our western people, the boldness and the confidence with which they proceed to develop the country when assured the protection of the laws and of the recognition of the right to water," he stated. The protection of the laws was what those interests on the populous east slope needed to divert the water, while at the same time dealing with the rising acrimony on the west slope.

Everyone wanted a piece of the river. The east slope got a large chunk of it with such diversion projects as the Colorado–Big Thompson Project on the upper Colorado River and the Fryingpan-Arkansas Project, not. too far from Aspen on two forks of the Roaring Fork River. Altogether there were some thirty points of diversion where water was carried from west to east either across in open ditches or under the Continental Divide in tunnels.

It was a clear example of the extractive principle. The river was the one unifying geographical feature in the West—in fact, its lifeblood, to paraphrase John Gunther. For the five years beginning in 1971, an average of 480,000 acre-feet a year was diverted in Colorado. During the 1978 water year, 630,000 acre-feet were diverted out of the whole upper basin. But the system lacked a certain efficiency. More water was lost to evaporation from upper-basin reservoirs than was diverted out of the basin. From the headwaters of the Little Snake River in Wyoming, where Cheyenne was seeking to increase its diversion from a roadless area in Medicine Bow National Forest, to a point near Mexicali at the end of the river, where the Mexicans were building a seventy-six-mile aqueduct to transport water over the mountains to the border city of Tijuana, the search for and transportation of water in the West knew no bounds.

By the 1970s the enthusiasm for moving water from one basin to another was being seriously questioned in Colorado. Water equalled growth and more water equalled more growth, and no longer was that a concept that was unchallengeable. The law professor from Denver University who had defeated Aspinall was subsequently appointed regional administrator of the Environmental Protection Agency. Alan Merson's opposition to the Foothill project, part of one more large diversion scheme, was based on restricting Denver's growth by not increasing its water supply. Said Merson, "Denver is getting to be unlivable. The air is dangerous. We can't allow subdivision after subdivision to spread like cancer."

It was the energy boom in the upper-basin states that was fueling Denver's growth. There is a great deal of coal, considerable oil shale, and some uranium and crude oil and natural gas in northwestern Colorado. In the mid-1970s about 4.6 million tons of coal were being mined each year. The expectations were that if all the proposals for new and expanded coal mining in the region went forward to production status, seven times that amount of coal could be dug by 1990. The hopes for oil shale were even greater. In January of 1974 the Bureau of Land Management held a prototype lease sale in Denver of a 5,089-acre oil-shale tract in the Piceance Basin of northwest Colorado, and two oil companies successfully bid what was then considered the tremendous sum of $210 million. It was estimated there were more than 4 billion barrels of oil contained in Tract C-a. Suddenly it seemed like old bonanza times were here again. Water, it was agreed, would be the only limiting factor to virtually unlimited growth, and the planners set about figuring out how it could be diverted to this new source of wealth.

Although in comparison to irrigated agriculture, mining consumes very little water and disturbs much less land than the sustained grazing of cattle, its symbiotic accouterments—power plants, oil-shale processing facilities, coal-gasification plants, and coal slurry lines—are large consumers of water. There were many proposals for such facilities floating around the West in the early 1970s. Through the decade, presidents Nixon, Ford, and Carter tried to promote energy self-sufficiency, which in large part meant western coal and oil shale, but except for a momentary flurry in the first half of the decade, mostly the result of plans made by the utility companies in the 1960s, nothing much happened. What did occur was a very modest boost in coal production, not on the gigantic scale desired by energy planners in Washington, but enough to create a dozen or so boom towns. This was enough to make one pause to contemplate what the growth pains would have been like under the planned agenda. Oil shale never did get off the ground during that decade. Like the big gold and silver strikes of the preceding century, a sustained energy bonanza seemed to be always just around the corner, the 1980s for sure now, and the water and land kept being reserved and committed for that purpose.

It was western history and its own particular brand of resource politics repeating itself once again. Whatever the issue—grazing, the creation of national parks and wilderness areas, water projects, or mining on public lands—western interests or those groups of people interested in the West, whether intent on preservation or extraction, had created a special niche for themselves in the laws that were supposed to govern the public domain according to the needs and wishes of the nation as a whole. Of course, not everyone was interested in or knowledgeable about the interior West, so what were termed special interests—more likely advocating single use of the land rather than multiple use—dominated the decision-making process. Livestock grazing was by far the most widespread activity in the West. But mining had the greatest potential for future growth. An 1872 law gave it precedence over all other uses. There were mining claims staked—but mostly not producing—in national parks and wilderness areas, national wildlife refuges, grazing lands, timber lands, and on top of other mining claims. Perhaps 6 million mining claims had been filed between 1872 and 1962. Investigators from the General Accounting Office visited 240 randomly selected claims and found that only one was being mined. The real importance of mining to the West was debatable. The saying went that more money had been put into the ground than taken out, and Otis E. Young, Jr., a professor of history at Arizona State University, wrote in the book *Western Mining* of the period up to 1893,

"Economically speaking, it would seem that in the long run, despite its image as a bonanza frontier, the mining West scarcely broke even." But the first mining boom did bring people and payrolls and agriculture trailing along behind it into the West.

The origins of the mining laws in this country can be traced back to Germany in the Middle Ages, a 1586 opinion of the Royal Solicitor in England, ancient Spanish mining codes used in Mexico, Cornish mining traditions, and certain practices in the thirteen original colonies and in the Middle West in the first half of the last century. The general thrust of these laws, as they gradually evolved in the western portions of this country, was that the federal government owned the public-domain lands, but the search for minerals upon them was to be encouraged by free access, nominal royalties, and if a mining claim proved to be productive, transfer of ownership for a nominal payment. By the time of the California Gold Rush in 1848, when every miner was a trespasser on the public domain and the federal government was most conspicuous by its absence, the miners had formulated their own laws, which later became institutionalized by the state legislature and Congress. The guiding principle was free mining on public lands.

It is interesting to note that the California Gold Rush also gave birth to the West's novel water laws, based on prior appropriation rather than the English tradition of riparian rights used in humid areas to the east. The practicalities of climate were the prime motivating force in making this basic change. Under the riparian water laws of humid areas, location next to the many rivers and streams meant everything. In the water-short West, the most important factor was putting water to beneficial use—often at a great distance from the source. Not many miners or ranchers were fortunate enough to be located adjacent to one of the infrequent watercourses, so water had to be transported far beyond riverbank properties by flumes and ditches. First in time, first in right was the basis for the arid-region doctrine of prior appropriation that grew out of the Gold Rush experience. As a corollary, water was also regarded as free.

From California, miners fanned out all over the West, taking with them their attitudes toward the use of land and water. To this day they are the most intractable of all western interest groups when they feel their historic rights are being threatened. It is impossible to travel the remote regions of the interior West without sighting a long-abandoned pile of mine tailings, the ruins of a ditch, or the remnants of some log or plank shack erected by a miner at the site of his diggings. The first Europeans to explore the interior West were interested in mineral riches. Miners guided the first tourists into the Grand Canyon, were the main impetus for the brief period of steamboat traffic on the lower Colorado River, and were the first to live briefly but intensively in such upper-basin settle-

ments as Lulu City. Their nature fit their place in western history—
secretive, elusive, and fleeting.

Congress began to deal seriously with the issue of the mining free-for-
all on public lands after the end of the Civil War. Two blocs developed
in Congress, eastern and western, not an unusual split on matters pertaining to the development of natural resources in the West. Nor was the
outcome, or the means used to achieve it, that much different from what
came later with the politics of water. The eastern bloc, small but idealistic,
was headed by Representative George W. Julian of Indiana, the chairman
of the House Committee on Public Lands. Julian, like his fellow midwest-
erner John Wesley Powell, was a strong advocate of the orderly, rational
development of the West and a champion of the little man. He wanted
the land surveyed first, then sold at public auction at minimum prices so
small miners could afford to purchase it.

Opposing Julian's concept and heading the western bloc was Senator
William M. Stewart of Nevada, a hell-for-leather but knowledgeable mining lawyer who was later to wreck Powell's career by blocking the appropriation for his irrigation survey. Stewart once used a derringer to cross-
examine a witness whose veracity he doubted. Whether it was miners or
irrigators, the senator was unflagging in protecting Nevada's interests
during an eighteen-year senatorial career. "His political opportunism,
deference to the railroads, and myopic view of national issues did much
to give Nevada a reputation as a 'rotten borough' and to cast doubt on
western politics in general," said one writer. To another, "Stewart's
political success was due in large part to the fact that he was bright,
articulate, and painstakingly well prepared."

Unquestionably, Senator Stewart was a quick learner. When he arrived
in the Senate, he perceived that the real power for a western congressman
lay in the obscure committees that dealt with public lands. So, there being
no Committee on Mines and Mining, he got one created and had himself
appointed to it. The senator drafted a bill opening up mineral lands to
free exploration with the right to later obtain title to a mining claim, a
procedure called patenting, for a nominal fee. Stewart's bill passed the
Senate in 1866, but was effectively bottled up in the House by Congressman Julian, who favored his own rival measure. The western bloc then
went to work and gutted a bill that had been passed by the House and
awaited action in Stewart's committee. Stewart's mining bill was attached
to this measure, under the improbable title of "An Act Granting the Right
of Way to Ditch and Canal Owners over the Public Lands, and for Other
Purposes." It was passed by the Senate and referred back to the House
for final action. Julian's motion to have it referred to his committee was
resoundingly defeated by voice vote and the bill passed, with a large
number of congressmen absent at the time of the vote. In Julian's words,

"The clumsy and next to incomprehensible bill then became a law, and by legislative methods as indefensible as the measure itself."

With only a few additions, none of them substantive, the 1866 bill became the Mining Law of 1872. Its author, a California congressman, assured his colleagues, "This bill simply oils the machinery a little, it does not change the principles of the [1866] law. . . ." Needless to say, Senator Stewart also supported the 1872 act. It has remained the basic law governing hard rock mining for more than a hundred years, despite continuing criticisms dating almost from the day it was enacted to the present time. It was hard to find a government study commission or agency that had to administer the law that was not critical of it, be it the Aspinall-dominated Public Land Law Review Commission or the General Accounting Office, an investigatory arm of Congress. There have been attempts to change it since 1969, but none got beyond the key House subcommittee. On leaving government in January, 1969, Interior Secretary Stewart L. Udall stated, "After eight years in this office, I have come to the conclusion that the most important piece of unfinished business on the nation's natural resource agenda is the complete replacement of the Mining Law of 1872." The Carter Administration tried unsuccessfully to change the law, and Interior Secretary Andrus testified before a hostile House Subcommittee on Mines and Mining in 1977. In reply to a statement from a congressman from Andrus's home state that "we have been able to develop a great nation by this system," the former Idaho governor replied,

> Yes, in 1872, if I had been a member of Congress, I would have voted for the mining act. We gave away land to the railroads to bring two ribbons of steel to connect the coasts. We gave away land to develop the West. We were trying to develop a nation. We have developed the strongest nation on earth, but we cannot continue to give away the public resources. We have accomplished what we set out to do. Now 105 years later I say it is time we look at a different approach to the management of the public resources. That is probably why we are seated here today.

But despite the secretary's plea (Andrus said later of the importance of the mining industry to western states that "when that industry sneezes, too often an entire state runs a fever") the subcommittee failed to act on any bill, whether written by the Administration or the mining industry. So the last significant bastion of *laissez faire* management of public lands remained in effect, in large measure because the chairman of the full committee, Morris K. Udall, was facing a recall movement by small miners in his district and because the copper industry in Arizona was depressed at the time. Udall had once been a strong advocate of mining-law reform. But at the hearings he came out in favor of mild changes proposed by the mining industry, thus effectively killing any chance for meaningful reform.

The exploration for and extraction of uranium are governed by the 1872 law. Both activities boomed in the upper Colorado River basin during the 1970s when the price soared from around $6 per pound in 1972 to over $40 four years later. The need for uranium for nuclear weapons had touched off an earlier boom after World War II, then through the 1960s uranium production subsided. But in the 1970s there was a shortage of fuel for nuclear power plants, and the uranium industry blossomed once again. Most of the nation's and a great deal of the world's supply of uranium came from the Colorado Plateau area, particularly the Four Corners region and the interior basins of Wyoming; so, in one more instance the raw resources of the Colorado River basin would feed the appetites of the rest of the country. The region had a long involvement with radioactive materials, dating back to the extraction of radium in the early 1900s.

The transgressions on the public domain allowed by the 1872 mining law were best illustrated by the activities of uranium prospectors in the Great Divide Basin of Wyoming and the seekers after the myth of riches in the Superstition Mountains of Arizona. The Great Divide Basin, also known as the Red Desert, is a sort of dry no-man's land, an unfilled cup, straddling the crest of the nation. In the lee of Oregon Buttes, a landmark on the old Oregon Trail, I once saw stack after stack of new claim stakes. The unweathered markers were part of the uranium claim-staking operation of the Denver-based Fremont Energy Corporation, whose activities had been detailed in testimony before the Subcommittee on Oversight and Investigations of the House Committee on Interstate and Foreign Commerce. It took a subcommittee not dominated by western mining interests to reveal widespread violations of state and federal mining laws. Within the Great Divide Basin alone, Fremont had staked some 18,000 claims totaling 360,000 acres—at the time one of the largest contiguously staked areas in the nation. Elmer L. Gibson, the foreman of Fremont's staking crew, which consisted of Mexican aliens who could not speak English, testified that none of the 30,000 to 40,000 claims he had staked over a period of seven years for Fremont in the states of Wyoming, Colorado, and Utah had been legal. In addition, said Gibson, the chairman of the board of Fremont, a Casper motel owner named Ralph L. Schauss, had tried to induce him to salt some uranium claims—that is, to plant some uranium to make them look valuable. This practice was not uncommon during Gold Rush days, when miners used to load their shotguns and scatter gold dust over their claims to fool tenderfoot investors from the East. Now it was tenderfoot representatives from a Northwest utility company who took an option on Fremont's uranium claims in the Red Desert. The testimony revealed the claims had been illegally staked. The claims had not been properly surveyed, nor had an adequate number of stakes been pounded into the ground. Also, the required annual maintenance work had not been performed. There was no evidence actual salting had taken place. The subcommittee concluded there had been "widespread illegal practices involving the staking and maintaining of uranium lode mining claims on public lands." It was clear this was not only the case in the Great Divide Basin, but also elsewhere in the West.

The myth of quick mineral wealth has been one of the more enduring themes of the western states, Coronado having been the first and certainly not the last seeker after it. Nowhere was the myth being more actively pursued than when I rode a horse into the Superstition Mountains of Arizona to find prospectors divided into armed camps and shooting each other up because of the legend of the Lost Dutchman Mine— one of the West's more persistent illusions. Already it had cost some fifty persons their lives along the way. None have ever found any riches, but that had not deterred them from staking the convoluted, lava-strewn hillsides and cactus-studded valleys of the Tonto National Forest with thousands of claim stakes. One Los Angeles man and his crew had staked 10,000 acres and planned to add another 20,000 to their mining claims when I visited the wilderness area.

The heat was sometimes fearsome during the midday hours, and I wondered if this did not affect the minds of those who worked there. Mary Austin, who knew about arid lands and their attractions, wrote of prospectors in *The Land of Little Rain,* "There are many strange sorts of humans bred in a mining country, each sort despising the queerness of the other." The ride with a Forest Service ranger, who had about as much chance of enforcing the law as the federal government did during the California Gold Rush, was into dry heaps of mountains; back into time —not geologic time or Wild West time, but back into that time and place where one had the freedom to be a little bit queer if he chose to be so.

First of all, there was Robert S. Jacobs, who preferred to be known as Crazy Jake. That was his trademark, along with a hair-trigger temper, a sawed-off shotgun, and a .44-caliber magnum pistol he sometimes loaded with dum-dum bullets. Jake's camp in the mountains, from which he made occasional forays out to search for the mine, was an armed fortress surrounded by a five-foot-high stone wall, just high enough to shoot from behind. Crazy Jake sat at a table, drank coffee endlessly, and chain-smoked in his home at the foot of the mountains. A pistol lay on the table next to the coffee cup. There was also a tape recorder on the table. Jake was not taking any chances with the interviewer.

I never had the impression Jake's eyes focused on me, although he did stare quite a bit in my direction. His eyes had a near-far, intense look about them. He leaned over and confided, "I never have shot anybody. All I have to do is look them in the eyes and I scare them to death." It was true.

Jake's nemesis was Harry van Suitt, the man whose crew had staked 10,000 acres of the wilderness area with mining claims. Van Suitt had shot one of his workers in both wrists with his .32-caliber pistol. After being shot the worker was reported to have stated, "If he had shot me once more, I would have quit." Van Suitt, whose trademark in the desert mountains was a fur-lined cap with ear flaps, later fired the man. When a free-lance photographer visited van Suitt's camp after the shooting incident and would not stop taking pictures when asked to, van Suitt smashed his camera and wrapped the lanyard around the photographer's neck. He then started to draw his gun, but thought better of it: there was a ranger present.

Both Crazy Jake and van Suitt, who had an intense hatred for each other, had nothing but the utmost respect for Al Morrow, a gentle older man who had lived in the mountains for twenty-one years, all the time searching for the Lost Dutchman. To keep occupied when not digging in his network of tunnels, Morrow had twice copied the Old Testament in longhand, changing the wording of the Gideon Bible into the vernacular. When working on a particular north-south earthquake fault, Morrow swore he heard music and telephone conversations in the wilderness. (When Jake was out of earshot, Morrow whispered to me, "Just between us, Jake's got his eye on the treasure on this claim. If there is any treasure on my claim, I will find it eventually. Treasure is a will-o'-the-wisp." Jake later told me Morrow had some of the best information.)

The old man could remember when Celeste Marie Jones, a black woman who said she was an opera singer, was in the Desolation Mountains with an entourage looking for the legendary mine. That was the time there was a shootout between Miss Jones's worker, Robert St. Marie, and a rival miner named Ed Piper. St. Marie was killed in 1959 after being outdrawn by Piper. His last request was for a drink of water. Piper gave it to him, testifying later, "Well, I stood there and watched him for a while. Then I walked off and left him when he quit breathing." Piper died penniless of cancer a few years later. Miss Jones wound up in Arizona State Hospital. And when van Suitt visited me a few months later in Los Angeles, he said Al Morrow had been killed when one of his tunnels caved in.

While there was to be no change in the mining law governing the exploration for and extraction of such hard rock minerals as uranium, gold, silver, copper, and molybdenum, it took only forty-eight years after the passage of the 1872 act to put the energy minerals of coal, oil shale, crude oil, and natural gas into a different category. When Congress passed the Mineral Leasing Act of 1920, lands thought to be bearing these minerals were put under a leasing system, competitive in some cases and noncompetitive in others. The key word was leasing. Control and ownership over these lands were retained by the federal government, and there was some monetary return for their use. Those who advocated a reform of the 1872 mining law during the last decade favored a similar leasing system for hard rock minerals. When Presidents Theodore Roosevelt and William Howard Taft withdrew some oil and coal lands from entry, and the Supreme Court upheld their action, balky western congressmen were then pressured into voting for a bill that incorporated provisions for leasing those energy minerals. Needless to say, the provisions of the bill were not too unfavorable to the West. What revenues there were, and they were usually minimal, were divided thus: 10 percent to the federal treasury, 37.5 percent to the states where the leasing occurred, and 52.5 percent to the Reclamation Fund, so dams and aqueducts could be built in those same western states. The 1920 act was to remain the basic law governing the extraction of coal, geothermal energy, oil shale, crude oil, and natural gas on the public-domain lands of the West. The Carter Administration would attempt to make bidding more competitive and revenues less minimal, while at the same time boosting the production of domestic energy sources. In 1971 the Department of Interior halted its coal-leasing program because no coal was being produced on more than 90 percent of the leased lands. Critics of the program contended that not only was there little production from leased lands, but most of the leases were being held for speculative purposes. The revenues were too little, said the critics, and the government exercised few controls. Large oil companies were acquiring these new energy sources on public lands. A report done for the Federal Trade Commission in 1978 stated that four of the top ten coal producers were subsidiaries of oil companies, and oil companies accounted for half the uranium production and held half of the uranium reserves, also mostly located on the public domain. The only thing the oil companies did not own was the water, and they were moving fast to rectify that situation.

Water was all set to serve oil-shale development in the 1970s. There were great hopes for energy self-sufficiency, what with Project Independence and everything. Perhaps 250,000 barrels of crude oil a day could be produced from the federal oil-shale program in the 1970s and a really significant contribution, like 1 million barrels a day, could be made to the nation's energy supplies during the 1980s. But this was not to be, at least not yet. By the end of the decade there had been no commercial production of crude oil from oil shale. There were technical problems and the economies of production were not yet right. Oil-shale fever struck again under the name of synfuels after the second energy crisis in the spring of 1979.

In retrospect, the determined push to promote oil shale within the Department of Interior during the early 1970s seemed to be self-fueled. It fed on itself and the wishes of the oil industry, with little other input. It was a good example of the revolving-door syndrome. Common Cause, a consumer-action group, did a study titled *Serving Two Masters* and found 35 percent (twenty-three out of sixty-six) of top Department of Interior officials had previously held jobs with energy-related firms and about half of the twenty-three had worked for firms with contracts or leases issued by the department. An agency within the department responsible for assessing the mineral potential of public lands, the U.S. Geological Survey, was canvassed, and Common Cause found that almost 60 percent of its top officials used to work for energy-related firms. About half of those thirty employees, or 27 percent of all top officials, had worked for firms with leases or contracts with the department. When these employees returned to private industry, they frequently held positions necessitating contacts with the Department of Interior on business matters. No laws were broken, but it just did not seem a proper way to regulate public lands and water. During the Carter Administration, a number of conservationists went to work for the department, raising the same issue of propriety in the minds of others.

One of the persons who was intimate with the Department of Interior officials at the Denver lease sale was Hollis M. Dole, head of Atlantic Richfield Company's oil-shale program and previously the assistant secretary of the interior for mineral resources. Dole took credit for inserting the reference to oil-shale in President Nixon's first energy message. He saw his role as "trying to develop a stronger domestic mining industry." Dole's opinions and those of the oil-industry-dominated National Petroleum Council were much the same, according to a wire-service report, and when he left the department in March, 1973, he was praised by Interior Secretary Rogers C. B. Morton for being "the one man who had done more than anyone else in the nation to alert us all to the energy crisis we are now facing." Before leaving and going almost directly to work for Atlantic Richfield, Dole had been instrumental in bringing Jack

Rigg of Grand Junction to Washington to work as his assistant in the oil-shale program. Rigg, active in Colorado mining matters for some twenty years, had been executive director of the Colorado Mining Association. Working with Rigg on Interior's oil-shale program was Reid Stone, formerly with the U.S. Geological Survey and before that with Atlantic Richfield's oil-shale program. Dole and another assistant secretary, Harrison Loesch, who oversaw the BLM, were responsible for selling the oil-shale program to Interior Secretary Morton. Loesch, a lawyer, came from the west-slope community of Montrose. After leaving the Interior Department, he went to work for Peabody Coal Company, which had numerous coal leases on public lands, as its vice-president for governmental relations. From his Washington office, Loesch pushed for an expanded coal-leasing program.

There was one other Colorado connection in Washington at this time. Dr. Beatrice E. Willard, formerly president of Thorne Ecological Institute, an environmental consulting firm with many mining-industry clients, had been appointed by Nixon to one of the three posts on the President's Council on Environmental Quality. It was the council's job to review environmental-impact statements, which supposedly were key decision-making documents. Dr. Willard's firm had participated in oil-shale studies funded by the state, the Interior Department, Atlantic Richfield, and other oil companies—which showed up in the impact statement on oil shale. She was personally acquainted with Rigg and Dole. Dr. Willard was prominently featured in an Atlantic Richfield film that sought to allay the fears of environmentalists about oil-shale development. It was produced when she headed the institute but received wide distribution while she was on the President's council.

These informal contacts, so important in government, helped Colorado state officials keep track of the oil-shale program as it was shaped and enabled them to influence its direction. Their counterparts in Utah and Wyoming felt left out of the process. Others, such as Colorado environmentalists, were incensed by the close ties between industry and government. There were some mutterings in Congress. Representative Charles A. Vanik of Ohio declared, "This is, quite possibly, another case of the 'revolving door' in which oil men come into government, have access to special, detailed and technical information and pass their information on to their colleagues who are rejoining the oil industry." A Wisconsin congressman discovered that members of the National Petroleum Council, which advises the Secretary of Interior on oil matters, had contributed $1.2 million to President Nixon's reelection campaign. To Senators Henry M. Jackson of Washington and Lee Metcalf of Montana, the terms of the oil-shale leases appeared to provide the oil companies with large subsidies, but few meaningful controls. None of these congressmen were from the basin states where the oil shale was located.

Although oil shale did not prove to be the bonanza that many thought it might be in the 1970s, there was plenty of other energy-related activity in Colorado. In Denver twenty-seven large office buildings were being constructed in the downtown area, and there were two thousand energy-related firms and more than four thousand geologists trying to figure out how to unlock the region's natural resources by the end of the decade. West of the Continental Divide, the predominant feeling was one of a great deal of energy being expended searching for energy. There was a sense of vibrancy, of throbbing, incessant traffic on narrow, dusty town streets. New pickups sported "Colorful Colorado" mudflaps. Their snow tires hummed. Strapped in back were toolboxes and extra barrels of fuel for long drives into remote areas for exploration and production. The place could be Craig or any other center of frenetic activity.

Craig and its lookalikes were one type of boom town in the high country. Another was Aspen and its progeny—Vail, Steamboat Springs, Breckenridge, Crested Butte, and Telluride, all on Colorado's western slope. These enclave resort communities had little to do with what took place around them in the interior West. They related more comfortably to Chicago, Dallas, and Los Angeles. Aspen, on the Roaring Fork River, had passed through the pains of growth and transition before the other enclaves. Having survived it, and emerged in somewhat different form, Aspen sent its more restless immigrants off to fill in the other shells. In the process Aspen emerged as a bit of Scottsdale or Beverly Hills in the Rocky Mountains—perfectly ordered but prohibitively expensive. The cosmetic orderliness made one yearn for the throbbing realities of Craig.

There were some interesting statistics concerning Aspen and the surrounding county of Pitkin in the late 1970s. For instance, it was more expensive to live in Aspen than in New York City, Chicago, Palm Springs, Houston, or Denver, New York being the second most expensive place and Denver the least. Aspen also handily beat out other west-slope communities in the cost-of-living index, scoring sixteen more points than Vail and twenty-eight more than Craig. In 1934 there had been 206 ranches on 80,655 acres in Pitkin County, but by 1974 this number had shrunk to 48 ranches on 48,785 acres as agricultural lands were repeatedly subdivided and built upon with urban developments. At the height of the silver-mining boom in 1890, Pitkin County had nearly 9,000 permanent residents, a level that was not to be reached again until the mid-1970s. By 1978 there were nearly 11,000 permanent residents, defined as those who had lived in the area for six months or more. This number did not include transient skiers; by the middle of the decade 1.3 million skier visits were being recorded yearly at the four ski areas in and around Aspen. The biggest influx of skiers came in the late 1960s, when yearly increases of about 50 percent were recorded. These were the boom years, from 1967 to 1973, when yearly retail sales jumped from $19 million to over $60 million. By state the people came from California, Illinois, and New York, in that order. By region, again in sequence, it was the Midwest, South, and the East and West Coasts. Aspen, and other resort communities like it, were true enclaves of alien cultures within the Colorado River basin. It had not always been that way.

Immigrants founded Aspen, and the few that remained, after the hordes of miners had departed, settled in for the long haul to spawn natives. Aspen's history has been one of extraction, whether it was taking minerals from the earth or money from tourists. The first white man camped on the banks of the Roaring Fork River in 1859 to trap beaver.

Gold-mining ghost town on the Roaring Fork River near Aspen.

Then came the miners, the spillover from the gold and silver strikes elsewhere in the West. By 1892 Aspen had a booming population and an annual silver production of $10 million. At the height of the boom, which lasted from 1885 to 1893, there were forty-three saloons, twenty-five houses of prostitution, twenty-three hotels, and twenty boarding houses serving a population that was as transient as the skiers who would descend on the area nearly a century later. In 1893 it all went bust with the repeal of the Silver Purchase Act. A few miners stayed on in the decaying Victorian structures to eke out a living. A Department of Interior official visited the community in 1922 and suggested that a ski lift be built. No, said the inhabitants, most of whom were on welfare, we prefer to wait for the next silver boom. They were wrong about the metal, but correct in their cyclical expectations. By the late 1940s the central portion of the town could have been purchased for $150,000. When I visited Aspen in 1979 I was told that one of the few new condominium units to be built under strict new growth laws would be priced at $1 million.

Eastern capital had financed the mining and it would be eastern money that would bring the town alive once again. Following World War II, Aspen could boast of a brief silver boom and being the birthplace of "Rough House" Ross—Harold Ross, the editor of *The New Yorker*—and little else. There was some ranching, some potato raising, a little mineral production, and about 1,000 inhabitants. Then Walter Paepcke, chairman of the board of the Container Corporation of America, teamed up with an Austrian ski instructor, Friedl Pfeifer, to put Aspen on the dual track toward a resort community catering to the mind and the body—the former being represented by the Aspen Institute for Humanistic Studies founded by Paepcke and the latter being the mountains and all the varied luxuries of the after-ski life. D. R. C. Brown, a state senator and local rancher whose family owned some key mining claims on Aspen Mountain, became president of the Aspen Ski Corporation in 1957, and Brown and the corporation sought to retain a conservative grip on the community through successive waves of ski bums, beatniks, surfers, and hippies. It was the surfer image in the early 1960s that soured Walt Disney on Aspen as a possible place to invest in a ski area. But Hollywood would gain a foothold in due time.

The clubbish Ivy League atmosphere that had predominated in Aspen up to the early 1960s changed when jet planes, safety bindings, more lifts, and intermediate runs brought the large crowds. The eastern orientation was diluted. Now the emphasis would swing more toward the West Coast, meaning California and particularly the southern portion of the state, with its burgeoning population and peripatetic skiers. These were people who lived at the end of the aqueduct system that delivered Colorado River water, originating as runoff in such streams as the Roaring Fork.

A Southern California real-estate developer, Bill Janss, was responsible for changing the face of western skiing with the opening of Snowmass-at-Aspen in December of 1967. Janss came from a family that was responsible for developing Westwood Village, surrounding the campus of the University of California in West Los Angeles, and the sprawling suburban community of Thousand Oaks to the north in Ventura County. He had also run a cattle feedlot in the Imperial Valley before turning his attention to ski areas, to which he was not a complete stranger; he had been a member of the 1940 U.S. Winter Olympic Team and a director of the U.S. Ski Association. Janss also had Aspen ties, having skied there since 1950 and being a director of the Aspen Institute.

What Janss brought to Aspen, and later Sun Valley when he acquired it, was the concept of a ski area as being primarily a real-estate development. Skiing would become secondary, simply a lure to promote real-estate sales, where the real money was to be made. It was modern, computer-based, systems management. And it had to be cost-effective. The American Cement Corporation joined Janss as a partner, and one of the first to have a home built at Snowmass was Janss's friend, Defense Secretary Robert S. McNamara, then the very model of the hard-eyed, balance-sheet-oriented businessman turned government administrator. A self-contained, instant village, complete with heated road to melt the snow, was plopped down on former ranchland along with accompanying homes and condominiums. The ski business would never be the same again. It was a successful example that was followed elsewhere; condominium fever spread to wherever there was a ski area, or a hint of such a thing, in the West.

The growth rate began to cool off in Aspen after 1972 because of tight money, high interest rates, a slumping economy, and the wishes of the newly arrived immigrants who wanted to slow things down so that many of the amenities, such as clean air, for which they had sought out Aspen, could be preserved, at least to some degree. These were the type of people who voted against Wayne Aspinall, the very personification of the western growth ethic. These were the years of great concern about the deteriorating environment, and antigrowth candidates began being elected to the Aspen City Council and the board of county commissioners. (*Rolling Stone* writer Hunter S. Thompson, running on a different platform, failed in his bid to capture the county sheriff's job.) The community polarized between the newer arrivals and the old-timers, who were more conservative in their outlook and tended to have roots in ranching and mining. But as the decade waned Aspen became more homogenized.

What Aspen had wrestled with in the early 1970s was the problem that was to later plague Craig—how to control rampant growth. Borrowing from the Petaluma experience in California, the city and county adopted a Growth Management Policy Plan, which limited future growth to 3.4 percent a year instead of the 10 percent increases in new housing starts recorded annually during the boom years. It was not a practical solution for many other Colorado River basin communities, where zoning was frequently regarded as being something vaguely Communist, or at best a Socialist plot. But it could work in Aspen, the enclave, "a trendy Rocky Mountain leisure world," as one magazine described it. The end of an era came at the end of the decade when the Aspen Ski Corporation was purchased for $48 million by 20th Century–Fox with the proceeds from *Star Wars.*

The aspen tree, for which the resort enclave is named, achieves its greatest glory in the San Juan Mountains of southwestern Colorado. For me,

Condominiums at Snowmass.

there is no other high country vista in the West so striking as the view from Dallas Divide which stretches northwest from the San Juans. It is just far enough outside the mountains to grasp their totality, yet close enough to absorb their classic drama. The San Juans are snow-capped and serrated, as true mountains should be. This is John Wayne country, so much so, as a matter of fact, that it was around the small town of Ridgeway, at the bottom of the grade from the divide, that his movie *True Grit* was filmed.

Above Ridgeway and other small communities in the San Juans, such as Ouray, Silverton, Rico, and Telluride, the aspen trees achieve an almost iridescent brilliance in the early fall and a tinkling softness in the late spring. Not many trees in the West change color; but the aspen, growing here taller than its eastern counterpart, comes close in one species to making up the total deficit. It is known as the quaking or trembling aspen because of the constant fluttering of its pale-green leaves. It is the tree of the western mountains; it dominates here, just as the sagebrush, piñon and juniper, cottonwood, and tamarisk dominate their particular locales. The aspen is distinctive, no matter what its color phase, and even when leafless. When bare it is an elegant white-stemmed stalk among dark, seasonless conifers. When clothed, it is a friendly presence. The aspen tree, like water, is much sought after in the West. Everyone wants a mountain homesite with a grove of aspen and a stream running through it.

I was attracted to the Dallas Divide area because it resembled those dramatic mountain backdrops in western movies I saw as a child growing up in New Jersey. The image of true mountains I was raised on in the East was one of the European Alps, and the San Juans bear a striking similarity to that classic range. The area was also near Telluride, a decaying mining town ringed by mountains, which was about to undergo an astounding metamorphosis into a ski resort complete with film festival and high prices for subdivided lots and old Victorian-style homes.

This was in the early 1970s, when Telluride had not yet changed that much. The main street was still dirt. But the direction it was to take was inevitable, as was my reason for being there. I was part of the "rural renaissance." At that time I had a dream of escaping from Los Angeles into the interior West. To help that dream along, I looked up a friend in Telluride who sold real estate. We spent days in her four-wheel-drive vehicle combing the area. Much of the former ranchlands had already been subdivided. Lots were staked, roads bulldozed, and even a few vacation homes had been erected. We finally found a remote forty-acre lot on Dallas Divide. It possessed all the tangibles I was looking for—that remarkable vista, a spring, and aspen trees.

There was also a thin, sinuous line drawn through the property on the map my friend had given me. She said she did not know what the line meant, but did not think it important. I thought about that line as I camped for a few days on the property, to experience it properly before making a final decision. I decided I had better make some inquiries. At the Tri-County Water Conservancy District office in Montrose, they said the answer to my question could best be found at the regional Bureau of Reclamation office in Grand Junction. There I was told that the line was a hundred-foot-wide right-of-way for a feeder canal and maintenance road to serve the Dallas Creek Project, one of the water projects Congressman Aspinall had obtained for his district. The canal would tap the runoff from the San Juans and transport it to a reservoir near the parcel I was considering. From there it would flow down Pleasant Valley Creek where another canal would draw off enough water to serve a recreational home subdivision being planned for twelve thousand people. I learned at the county courthouse where I stopped on my way back from Grand Junction that an oil company owned geothermal drilling rights on the property. I felt crowded, and gave up any further search for land in that region.

The river: as it gathers force in its descent through the high country of Colorado, the competition for its waters intensifies. Aspen, with Telluride not far behind it: the playground of the rich and beautiful in the village sublime. Water: it was along the banks of the Roaring Fork that the first trapper plied his trade in 1859, silver mills were built, and the institute and condominiums were to cluster. Suddenly it seemed there was not enough water for everyone's purposes in the 1970s. Pitkin County commissioners filed suit. They wanted the water to stay in the river for fish, wildlife, and esthetic reasons. Farmers and cities in southeastern Colorado, who had used up all the other surface water in the Arkansas River basin, wanted to divert the same water through a tunnel under the Continental Divide and a system of six dams and reservoirs, called the Fryingpan-Arkansas Project. The upper Colorado: from Hunter Creek to the Roaring Fork River to the upper Colorado near Glenwood Springs, the Blue and the Eagle Rivers having joined the Colorado above this point, and from the Never Summer Range in Rocky Mountain National Park to the Utah state line, the river has dropped nearly ten thousand feet through many separate enclaves, each with its own claim upon the water. Before the river descends into the depths of the canyonlands, the Mormons and Indians have their claims upon it.

CHAPTER FOUR　Separate Nations:
Of Mormons and Indians

1

No group of people in the West since the coming of the whites has been more aware of the importance of water, more cohesive and diligent in searching, capturing, and distributing it, or more suitably adapted to preserving and perpetuating a water-dependent culture in an arid land than the Mormons, those members of the Church of Jesus Christ of Latter-day Saints who make up the vast bulk of Utah's population. Like the Indians, the Mormons have their own distinctive relationship to the waters of the Colorado River system. The Mormons have shared in it greatly, the Indians minimally. Unlike the disparate, transient whites that make up most of the West's population, these two groups (not really minorities since they dominate and rule their own lands to a great extent) possess separate identities and unique institutions that have survived despite constant encroachments from the outside world. To wander through their separate nations is to feel like a stranger in one's own land.

The Mormons wasted no time in establishing their own special relationship to water that was later to be emulated throughout the West. Early on the morning of July 23, 1847, the advance party of Mormons, who had taken part in the best-organized mass movement of people in American history, arrived at the site of what was to become Salt Lake City in the Great Basin, just to the west of the Colorado River basin. A committee was appointed and by 11:30 A.M. a suitable plot for farming had been staked. By noon the first furrow was plowed. At 2:00 P.M. work started on the first dam and ditches to divert the water from City Creek. The next day, warm with fleecy clouds, a five-acre potato patch was plowed and planted, and then the water was turned onto the land. After some trials and failures, the desert bloomed, although it was not complete desert; evidence exists of there being abundant grasslands along the watered slopes of Salt Lake Valley before the Mormons arrived. The pioneer Orson Hyde remarked in 1865, "I find the longer we live in these valleys that the range is becoming more and more destitute of grass; the grass is not only eaten up by the great amount of stock that feed upon it, but they tramp it out by the very roots; and where grass once grew luxuriantly, there is now nothing but the desert weed, and hardly a spear of grass is to be seen."

The Mormons, of course, did not invent irrigated agriculture, that most consumptive of all uses of the river's waters. They were just the first whites to practice it on a large scale in the West. Before arriving in the valley that morning, the Mormons had never diverted water to any measurable extent; but, being very thorough people, they had studied irrigation and were organized to practice it on arrival. Various individual Mormons had visited the Southwest, where Indians and Mexicans were irrigating their crops, and the Middle East. In fact, that very same Orson Hyde had visited Jerusalem in 1841 and observed that the overflow of the pool of Siloam, was "sent off in a limpid stream as a grateful tribute to the thirsty plants of the gardens in the valley."

The Mormons would always feel close to the Middle East, in part because of the great dependency upon scarce water supplies. I recall driving toward Salt Lake City late one night in that drought year and hearing an interview with an Israeli scientist over a Mormon-owned radio station. The scientist was discussing the parallels in the two desert cultures. A few days later on that same radio station the president of the American Farm Bureau, who was visiting Salt Lake City, attacked the Carter Administration's water policies and put forward two solutions to the drought—cloud seeding and not allowing militant ("radical" is the word used most often by Utah natives) environmentalists to stop any further dam projects. That was the year twenty-six Utah counties, representing 90 percent of the state's land surface, participated in cloud-seeding programs, and a state water official, Daniel F. Lawrence, said of the Administration's criticism of the Central Utah Project, "A one-month cursory review of such a complex and vital water project by inexpert and uninformed people could only be superficial and not a well-founded basis for concluding it was unneeded, undesirable and should be discontinued."

The Mormons were ideally suited for establishing a culture based on irrigation in the desert. The Salt Lake Valley was remote, undesirable to other whites, and watered by streams from the surrounding mountains. Here the Mormons could find prosperity, and escape the religious persecution they had suffered elsewhere. Second to being extremely religious, the Saints were farmers. They had gotten to where they were by discipline, cooperation, and organization—the qualities most needed to divert more than a minimum amount of water. Elsewhere in the West, irrigation typically passed through four phases: individual efforts, private speculative ventures, publicly owned local districts, and federal government projects. The Mormons skipped the first phase, experienced little of the second, and passed quickly through the third to the fourth phase. The emphasis was always on voluntary cooperation and group cohesiveness, both distinguishing Mormon characteristics. Thomas F. O'Dea wrote in *The Mormons:*

> From the time of the ancient empires of Egypt and Mesopotamia, communal water control had always been accompanied by and presumably always demanded authoritarian leadership and coordination in a tightly knit social organization. The Mormons had both of these, and they had the religious conviction and the experience of common effort and shared suffering and hardships to give them intense group consciousness.

From the start it was the Mormons conquering the desert, ceaselessly seeking prosperity and progress in an inhospitable locale—what was to become the second driest state. There was no private ownership of water. The church controlled its distribution and made the decisions on where and how to capture it. The family was the basic social unit, and farming, not mining, was the preferred occupation. With the coming of the energy shortages in the early 1970s, all these basic Mormon tenets would be challenged during the mad scramble for Utah's mineral wealth. But until then, in the small Mormon towns scattered like frontier outposts throughout the Colorado River basin in the remote eastern and southern portions of Utah, it would still be an almost completely closed, agrarian society, one which viewed the goings-on in the metropolitan area of Salt Lake City as being somewhat akin to those of ancient Babylon. The vision, however, remained intact. Again and again, as I heard on the radio that night in 1977, the Mormons likened their experience to the opening words of the thirty-fifth chapter of Isaiah: "The wilderness and the solitary place shall be glad for them: and the desert shall rejoice, and blossom as the rose." This vision, lacking the same social structure and religious fervor, was adopted by other struggling communities in the West in the latter part of the nineteenth century. One hundred years later the rose was beginning to wilt. It had been overwatered.

The well-ordered Mormon village, still existing to this day and bearing little resemblance to Craig, was another contribution of the Saints to the West. In a region where the virtual absence of land planning reigned supreme, it was an oddity. The typical village was laid out and governed in a thoughtful, rational attempt to best utilize the surrounding land and water resources. Such villages were generally successful, in a rather low-keyed way, and came to personify the Jeffersonian ideal of the yeoman

A Mormon couple after their wedding at St. George Temple, Utah.

farmer transplanted from the lush, humid countryside of Virgin harsh realities of an arid climate. The wide streets were laid ou north-south, east-west rectangular patterns, and farm structures as well as homes were built in town. The fields were for farming, and the land was kept clear of as many structures as possible and divided equally among the inhabitants. The farm lands, and sometimes the village lots, were irrigated. The towns were compact, and exuded orderliness and a sense of efficiency.

The Mormon villages were closer to New England and European concepts of communal living and efficient use of resources than the sprawl of towns and isolated ranches found in neighboring western states. In a classic study entitled *The Mormon Village,* Lowry Nelson characterized the relationship between Mormons and their habitat thus: "Their hundreds of irrigated oases throughout the intermountain states today, with their effective social institutions and satisfying way of life, are living monuments to the intelligent and efficient application of human labor to the natural resources." To be sure, there were precedents for this pattern of living and working in the West: The Anasazi Indians lived in clusters and went out into the adjoining fields to work. As the roads improved and four-wheel-drive vehicles became prevalent, even the ranchers around Craig started moving back into town. But, again, because of the strength of their religion that extended into secular life, the Mormons had enlarged upon this pattern.

Escalante is a good example of such a Mormon village. The streams that drop precipitously into the Escalante River originate as springs fed by snowmelt just below the rim of the towering Aquarius Plateau. The springs issue from volcanic and sedimentary rock at a mere trickle or a gushing 450 gallons per minute. They form Birch, North, Pine, Boulder, and Deer creeks and then tumble into the river, which flows—not at a great rate but usually perennially, which is enough to gain river status in this country—ninety miles through the ever-narrowing, serpentine canyons of slickrock country to end in the placid waters of Lake Powell. Before reaching its end, the walls of the canyon have risen 1,500 feet above the river. Edward Abbey described their color in *Slickrock:* "In the flat sunlight of midday the mighty cliffs appear bluff-colored, a pale auburn, but at morning and evening when the sun's rays come slanting in at a low angle the rock takes on an amber tone, with a glow like the bead of good bourbon. At sundown the coloring deepens still more and smolders—you can feel the heat—through the twilight in all the hues of hot iron cooling off."

This was, and still is to a great extent, the remotest region in the

contiguous states. The Escalante was the last river to be named. When
the geologist Clarence E. Dutton first looked down upon that region from
the plateau he described it perfectly: "It is the extreme of desolation, the
blankest solitude, a superlative descent." The river was named after the
first Spanish explorer to traverse the plateau and canyonland province of
southern Utah, now termed "Color Country" in promotional literature.
But the town-site of Escalante was so inaccessible that Silvestre Vélez de
Escalante and his party came no closer than one hundred miles. Deep
chasms, plateaus soaring to twelve thousand feet, cliffs, mazes, crazy

Escalante, Utah, and town ditch.

convolutions, box canyons, heat, and aridity combined to delay the finding of the river and the settling of the townsite. A. H. Thompson, who named the river in 1872, described the difficulties in getting there: "Travel was exceedingly slow and difficult. Our progress was often barred by a canyon, along whose brink we were compelled to follow till some broken-down slope afforded a way to descend, then up or down the canyon until another broken slope permitted us to ascend, then across a mesa to another canyon, repeating the maneuver a dozen times in half that number of miles."

Thompson and his party finally made camp in a "beautiful meadow" with a cold stream. The topographer, who was working for his brother-in-law John Wesley Powell, studied the surrounding countryside. From the base of the cliffs Thompson saw a long, narrow plateau that he estimated extended some sixty miles to the southeast and the Colorado River. Thompson also named the plateau: "The Indian name for a small elevation near the north end is Kai-par-o-wits (Canaan Peak), so we called the whole plateau by that name."

Kaiparowits, Escalante, the state of Utah, along with the Department of Interior and some out-of-state utility companies, would become intertwined in the 1970s in a wrenching experience that would jolt the Mormons out of their state of insularity and the assumption they could control their own destiny. Outside forces would be dominant within Utah as they had never been before. The search for energy, in this case a large coal-burning power plant to be built on the Kaiparowits Plateau, easily penetrated the geographical remoteness and threatened to tear apart the social fabric that had been woven so carefully over so many years. The Mormon missionary Jacob Hamblin was the first to venture partway down the Escalante River in 1871, thinking it was the Dirty Devil River. It was another Hamblin 103 years later who would write to the editor of *The Southern Utah News* in Kanab expressing concern over the vast amounts of money and the influx of strangers the power plant would generate. "Even if we have plenty left over and our school system becomes rich, will our students become better people? Will they become more grateful for what they have? What good is a new auditorium with broken windows and slashed seats? New school buses with slashed tires and sugar in the gas tanks? Do we need more money? Is there anyone in our area who doesn't have the necessities? Is there anyone who doesn't have a few luxuries? We do have ball games without knifings. Dances without police present. There are very few areas left like ours." It was true. Southern Utah was a world apart, but outsiders now wanted in.

The utility company eventually dropped its plans for the power plant. Those prime symbols of meddling outsiders—environmentalists and the federal government—were burned in effigy in Kanab, and the crowd roasted hot dogs in the fire. Southern Utah went back to being what it had always been; only a residue of bitterness remained, along with some new plans to tap the huge coal resources of the area.

Escalante typifies the ties of a small Mormon Village to water down through the years. When six Mormon men traveled over the mountains to the Escalante River in the winter of 1875, they determined that there were about 1,000 acres in an otherwise rockbound, unwatered region that could be made to bloom. The first six reported back, and it was not long before six more brethren, seeking a milder climate than that offered in nearby Panguitch, dropped over the rim with two wagons and descended to the six-thousand-foot level of what was then known as Potato Valley. The six were without families. A meeting was called, not unlike the conference held that first morning in the Salt Lake Valley, and it was decided to locate the townsite on the north side of the Escalante River. The townsite was switched next year to the south side, thus freeing up

Watching the horse races sponsored by the Lions Club at Kanab.

a larger block of land for farming. Work started in April, 1876, on the first irrigation canals. The first two earth dams washed out but the third, made of concrete, held. The farmland was laid out in 160-acre blocks, to be subdivided into 22½-acre holdings when more settlers arrived. The townsite was divided into eighteen 5-acre blocks and each block was split into four lots of 1¼ acres. The first family arrived in the spring of 1876. The first structure was a dugout, sort of a cellar covered with brush and willow poles, where choir practice was held. Soon there were a dozen families. On the nation's Centennial, 336 years after Coronado marched into what would become Arizona and 289 years after the first English settlement at Jamestown, a Navajo blanket was hoisted to the top of a pole, an American flag not being available.

The first settlers faced a series of dry years, as did the inhabitants a hundred years later, and the church historian wrote, "At this time [1876] there was only a very little water in the Escalante Creek—scarcely sufficient for the people to irrigate their garden spots, to say nothing of their farms—but in a most marvelous way the water commenced to increase, though there was considerable scarcity of water until 1882." These were the years of the boom in livestock on the western range, and isolated as Escalante was, it was not immune to large numbers of cattle and sheep, which quickly decimated the native grasses. With the livestock, as elsewhere, came the first large increase in population. Remote Escalante experienced the same oscillations in population as the rest of the interior West. By 1882 there were 441 inhabitants. A high of 948 was reached in 1894, whereupon it dropped by a couple of hundred, not to hit the 1,000 mark until 1920, where it hovered until 1940. Escalante then lost its inhabitants to the cities, as was happening to other rural areas, and by 1970 the population numbered 638. It began increasing in the 1970s to around 900 by mid-decade.

To a great extent, people followed cows and coal in Escalante. Livestock flourished until drought struck in 1893. There followed a series of dry years, and by 1896 about half of the stock had perished. A 1931 study by the U.S. Geological Survey of the Kaiparowits region stated, "There is no doubt that the Escalante and Paria Valleys and the Kaiparowits Plateau have deteriorated as pasture lands during the last decade, and it seems unlikely that they can be restored to the state existing during the period 1875–1890." The study went on to note, "The one known mineral resource of potential value is the coal of the Kaiparowits Plateau." Since 1901 local coal had been used to heat homes. It was cheap and relatively ash free.

Escalante slumbered through the years of rural attrition in the mid–twentieth century under the Mormon agrarian doctrine, enunciated by a bishop as follows: "A little farm well cultivated near homes, I know, is your doctrine and it is mine and ever was." The irrigated land base remained the same, alfalfa being the principal crop. The church and family were the primary units of social cohesiveness in Escalante. The town government, livestock association, business groups, and two irrigation districts were dominated by Mormons, who constituted 80 percent of the population in the mid-1970s and more before then. Although church and state in Escalante were nominally separate, the town was actually an informal theocracy, especially when it came to water. The church partially financed construction of the area's largest reservoir, the 2,300-acre-foot Wide Hollow Reservoir. It was constructed in the late 1950s but its usefulness was hindered by high evaporation and siltation rates that lessened its storage capacity. In a normal year 8,000 acre-feet of water were spread on about 5,500 acres of farmland. By the late 1960s all of the surface water had been fully appropriated.

One hundred years after the first irrigation ditches were constructed and the first water diverted, water was still the basic factor in deciding what went on around Escalante—whether it was the traditional use for irrigation, to support any finds made by oil, uranium, and coal exploration crews, or to supply the recreational needs of tourists and backpackers who were making their way in increasing numbers into the remotest corners of the region. Environmentalists wanted the water untouched for wilderness values, farmers desired more for their fields, a Utah utility company had plans to dam the Escalante River, and the Kaiparowits consortium of out-of-state utility companies had obtained rights to 102,-000 acre-feet of water to be pumped from Lake Powell. This represented 20 percent of Utah's remaining share of Colorado River water. Suddenly things were getting very tight.

Then, with the April, 1976, announcement that there would be no Kaiparowits project, at least in the form of a three thousand-megawatt coal-burning power plant, Escalante's great expectations were suddenly shattered by outsiders. Prosperity had seemed just around the corner for an economically depressed region heavily dependent on welfare. Parents had been hoping that children would return home to find a job and families could be made whole again. All of this, it was also hoped, could be done with little disruption to the existing social fabric. It was a vain wish, as evidenced elsewhere in the West, but it was not to be immediately tested in Escalante. The bitterness was directed at the environmentalists, not the utility company, Southern California Edison, which had decided to abandon the project mostly on economic grounds. Cars with California license plates parked at trailheads leading into the Escalante wilderness were vandalized, apparently in the mistaken belief that all Californians belonged to the Sierra Club, which had opposed the project.

In Escalante eight months later, with California license plates on my car, I was told in no uncertain terms by four men in the local cafe, while I was eating my breakfast, that I was not welcome in town. Said one, "I don't speak to environmentalists. I don't want anything to do with them. If I saw one and he was belly up in the desert with nothing to drink and a mouthful of sand, I wouldn't offer him any water." That seemed to be the ultimate form of rejection in this arid region and, being a stranger in a separate nation, I thought it best to move on.

The Mormon village experience heavily influenced the Reclamation movement of the late 1800s and was eventually written into federal law as a vain attempt to limit the size of landholdings receiving the benefits of federal water projects. That intrepid promoter of the Reclamation ethic, William E. Smythe, constantly cited the Utah example, and Horace Greeley applied it to his colony in Colorado, which eventually drew on Colorado River water from the Grand Ditch. John Wesley Powell traveled through Utah and, like others, was impressed by the way the Mormons avoided monopolistic control of water and banded together in cooperative efforts to disperse the water. More clearly than anyone else at the time, Powell saw the relationship between land and water in the arid West. Water determined what happened on the land, and who controlled the land would control the water.

To people like Powell and Smythe the ideal solution was to divide land into small parcels and put a benevolent authority that would be publicly accountable, rather than a grasping private speculator, in charge of the procurement and division of water. In time this would come about, but it would not do away with the monopolistic control of water in the West. An oligarchy or community of interests would come to dominate water policy in the West. It was the control of water by default, since others, including the conservationists, did not organize or retain interest long enough to affect basic policy. It is interesting that Powell's ideas, given focus by his travels through Utah, were aired in the reform Administration of President Rutherford B. Hayes. The Carter Administration, particularly in its early policies toward the western states, exuded the same aura of reform. But the policies of both Powell and President Carter would be undone by the tenacity of western interests.

More than any other state, with the possible exception of Nevada, Utah would come to abhor federal regulation of land and water resources, especially after the Kaiparowits experience. But, in the best tradition of western schizophrenia, it was the recipient of a great deal of federal largess, the latest being the massive Central Utah Project and the first being the Strawberry Valley Project. Throughout the West initial irrigation projects were small, possible for individuals and small groups, such as the first Escalante settlers, to undertake. These diversions were relatively easy to accomplish and from obvious locations, such as high-country creeks surrounded by rich bottomlands. By the 1880s most of these locations had been utilized and larger projects demanding more capital were needed. Privately financed, speculative irrigation ventures then began to predominate. As with the mining ventures of a few years earlier, most of the capital came from the East. Many of these ventures were failures, being underfinanced and poorly planned and constructed. By 1900 more than 90 percent of the private irrigation companies were either bankrupt or close to it. Another boom had gone bust. With the federal government entering the scene with the passage of the Reclamation Act of 1902, irrigation was given another boost, one that would go unchallenged for the next seventy-five years, and Utah was one of the first to line up at the federal door with a project in hand.

The Strawberry Valley Project was first conceived of in 1900, and no sooner had the federal Reclamation Service, the precursor to the Bureau of Reclamation, been formed by the 1902 act than it was petitioned by farmers from Utah Valley to study the feasibility of diverting water from Strawberry Valley in the Colorado River basin through a 19,500-foot tunnel to the Spanish Fork River, from which Mormons had been diverting water to farmlands in the Provo area since 1851. The streams on the Great Basin side of the divide were not always reliable; frequently there was not enough water for late-season use. The total project was too big either for individuals or for a private irrigation company to undertake. Outside aid, specifically that of the federal government, was needed. The project was authorized in late 1905, and ten years later the first irrigation water was turned into the tunnel. As the project neared completion the Strawberry Valley Water Users Association, the clientele of the federal government, put out a pamphlet stating, "In this valley where formerly stretched the gray desolation of sagebrush the rose is blooming, and all that has been accomplished in the way of putting Nature's bloom on these regions where the moisture has been insufficient, is small compared with what is under way and will come into realization. The Reclamation hose will soon be turned on." The dam on the Strawberry River was completed in 1917, just in time to have guards posted on it to keep away any potential World War I saboteurs, should any have found the remote site and deemed it of importance. Although the original contract between the water users association and the federal government called for a repayment schedule over ten years, fifty years after completion of the dam the project had still not been paid off by revenues from power and grazing fees. But Utah Valley prospered and was to get additional supplies of water from the Central Utah Project that was authorized in 1956. Instead of one alfalfa crop, farmers could get three cuttings a season with the additional amounts of water.

The Strawberry Valley Project was one of a number of early, noncontroversial projects the bureau built. They were small undertakings, compared with what was to come, but larger than the locals could afford to finance. Although the concept of federal Reclamation was somewhat experimental in the early years, and there were some mistakes, the bureau built up a loyal clientele by responding quickly to demands. The early projects in the Colorado River basin, all initiated around the same time, included Strawberry, the Gunnison Tunnel in western Colorado, Roosevelt Dam on the Salt River in Arizona, and Laguna Dam on the lower Colorado River straddling Arizona and California. Financial difficulties, poor management, surplus crops, and depressed farm prices stalled the Reclamation program through the 1920s. With the authorization of the Boulder Canyon Project in 1928, which would result in the building of Hoover Dam through the Depression years, the bureau embarked on its great dam-building era that was to end with the completion of Glen Canyon Dam in 1963. The third phase in the conquest of the Colorado River consisted of large aqueducts, like the central Utah and central Arizona projects, where dams would be secondary to conveyance facilities. During this phase, due to last through most of the 1980s, smaller units tacked onto projects authorized in earlier years, such as Savery–Pot Hook, would probably be completed. The fourth phase, launched in 1974, was designed to clean up the mess caused by the preceding phases. Specifically, a bill was passed to build projects (again, primarily the plumbing-system approach) that would lower the increasing amounts of salinity in the river. The last phase, although neither politically nor economically feasible in the 1970s, would be to import more water into the basin from the Northwest, Canada, or Alaska. Although the plumbing system that would be required to accomplish this augmentation of the Colorado River system boggled the mind and proposals made openly in the 1960s had gone underground in the 1970s, such plans most probably would, like the city of Phoenix, rise from the ashes at some future time.

Utah is the best place to get some historical perspective on the development of the Colorado River. Nowhere else is there such a sense of continuum. In no other state has there been such a unified drive toward the acquisition of water, although this unison among white Mormons and gentiles would result in others, like the Indians, being left out of what mattered most in this arid land. It persisted into the 1970s, when a few hairline cracks appeared in the otherwise monolithic structure. Certainly the state's elected officials presented a solid front. "In my own state of Utah," wrote Senator Frank E. Moss in 1967, "water is always a political but seldom a partisan issue. Ultraconservatives, who deplore expenditure of federal money on welfare, readily campaign for more water projects. Without such developments as the Central Utah Project, which will utilize the water from the Colorado River system, the state would stagnate."

A water project is a whole concept, a totality for a region. A unit is one portion of a project that itself can consist of a number of individual dams, canals, power plants, visitor centers, transmission lines, reservoir sites, boat-launching ramps, and picnic tables. The Central Utah Project has six units. The second to be built would be the Bonneville Unit which, among other things, would inundate the original Strawberry Dam under the water of an enlarged reservoir, and furnish additional amounts of water to the urban populations in the Great Basin. The last unit scheduled for construction was the one to benefit the Ute Indians in the Uinta Valley of eastern Utah. Avoidance and delay in satisfying Indian water rights that were assured in concept, if not specific quantity, by a 1908 Supreme Court ruling would be one of the recurring themes in the West. Politically, the Colorado was a white man's river, although historically the Indians had been there first and had their own thoughts on the law of prior appropriation. The Supreme Court ruling, known as the Winters Doctrine, assured the Indians they could use as much water as they needed. This was later refined by the court in 1963 to mean as much water as necessary to serve all "the practicably irrigable acres" on a reservation—which could mean a vast amount of water, indeed. For five small tribes along the lower Colorado it would mean, according to the 1963 ruling, one million acre-feet. Other tribes in the basin did not have the amount quantified. The whites trembled at what an extension of the ruling might mean to their planned and already completed waterworks, and hastened into advantageous agreements with the Indians in the 1960s that were to be challenged by a more militant Indian leadership in the 1970s. From Strawberry to Bonneville, the pattern had been consistent.

When the Mormons arrived in Salt Lake Valley, their first encounter with the Ute Indians was ominous. Measles spread from the Mormon children to the Indians, and thirty-eight of the latter died. A short time after the founding of Provo, to the south of Salt Lake City, the Indians killed some Mormon cattle and horses. The Mormons retaliated in 1850 by killing twenty-seven Indians. By 1861, following the pattern established elsewhere, a federal Indian agent and the Mormons were searching for some suitable place to send the Indians. Brigham Young dispatched an expedition to the Uinta Valley; it reported back that the area to the east was worthless and "even wondered why God had created the area unless it was to hold the other parts of the world together." The Utes were given the valley for a reservation. But it was not entirely valueless.

As the years went by, miners and farmers began to trespass on the reservation, and its lands were chipped away by the whites. Mormon cattle overran the reservation and heavily overgrazed it without the compensation of any fees being paid the Indians. One Indian agent reported, "They have pastured their cattle for years on the reservation, and swindled these Indians at every opportunity." The problem was particularly acute in the Strawberry Valley, where there was water. As early as 1879 white ranchers had been illegally diverting water from streams in the valley. Ditches were dug and canal companies formed and stock issued to illegally divert this water, some of which went over the divide. A bill was introduced into Congress to legalize these actions, but it failed to pass. But the water continued to be diverted illegally, the whites hoping legislation would eventually be passed removing the land from Ute control. This, of course, was accomplished in time to build the Strawberry Valley Project, not by legislative but by administrative means. After completion of a survey for a reservoir and dam site in the valley, the Reclamation Service offered to buy 56,000 acres for $1.25 an acre. The Utes refused. The land was then condemned by the service and the Utes were paid $71,000. And so the Strawberry Valley Project was made possible. The federal Indian Office did little to protect Indian water rights under the Winters Doctrine, stating that "The state laws of irrigation would be followed in order to avoid criticism by white land owners."

The avoidance of satisfying the water rights of the Ute Indians remained the shaky foundation on which the billion-dollar-plus Central Utah Project would be built. The Utes, like the Navajos, would find themselves caught between the proverbial rock and a hard place. There was the Winters Doctrine, of course, which seemingly promised great quantities of water. Except for those five tribes along the lower Colorado, however, the Supreme Court had not spelled out how much water this actually meant for each tribe. Even if it should do so after long and costly litigation, this would not come close to assuring the Indians any water. There had to be a project to store and divert it, and this took an act of Congress. Not only one act, but two acts plus two presidential signatures were needed—the first act to authorize the project and the second to appropriate funds. Even if by some outside chance it should get to the second stage, funds could be appropriated very slowly, indeed, if Congress so chose, even after making a binding commitment. Congress in such matters acted on the advice of its almost invisible subcommittees dealing with water and Indian affairs, and they were dominated by white westerners. To shortcut this labyrinthine process with little hope of success at the end, the Utes decided to negotiate an agreement with the whites who desperately wanted to build the Central Utah Project. The Secretary of Interior, in this case Stewart L. Udall, had to approve the agreement, which he did as trustee for the Indian interests. For the whites, the incentive to negotiate an agreement rested on the fact that the Indians had first call upon the water. Under the white man's doctrine of prior appropriation, the Utes' priority dated back to 1861, while the earliest white rights were acquired in 1906. A deferral agreement was signed in September, 1965.

Although not a reality yet—it would become so with completion of the Central Utah Project—most of the state's remaining share of Colorado River water had already been appropriated, and additional filings greatly exceeded the supply. That there was any controversy in the 1970s, or even a difference of opinion on something so basic as the use of water, was a rarity in monolithic Utah and a testimony to how tight things were getting. Again, the Indians were central to the differences, but there were other elements—environmentalists, Uinta Valley residents, agricultural and energy interests. The consortium that wanted to build the Kaiparowits power plant had located the coal they needed in 1962, and then went after the water. All water is interconnected in the West, either physically or on paper, so water for Kaiparowits in southern Utah would affect the vast water project in the eastern and central portions of the state, although the power plant would obtain its water from Lake Powell and the water project would intercept streams tumbling down from the Uinta Mountains to the north—it was all water from the same Colorado River system and committed by compact to Utah.

The utility companies, whose presence was much desired in Utah, asked for 200,000 acre-feet a year from Lake Powell, but were restricted to 102,000 acre-feet because the larger amount would have cut into the needs of the Central Utah Project. In May, 1965, the utilities entered into an agreement with the Central Utah Conservancy District, the bureau's client for the project, recognizing that even with the lower figure, there could be a shortage in dry years. After the year 2010, the utility companies would begin surrendering some of their water so the Ute Indians could then use it. A few months later the deferral agreement was signed, assuring the Utes of their full Winters Doctrine rights and promising that the water project to make those rights a reality would be pursued "in good faith." The whites, in turn, got the Bonneville Unit. No water project to fulfill these promises to the Indians appeared. A suit filed by some Utes claimed: "As a result, after ten years of water deferral, the Ute Tribe has received only a small reservoir and duck ponds." The Ute Tribal Council would complain a number of times about the delays in satisfying the water needs of the Indians, but when the Carter Administration in 1977 challenged the Bonneville Unit on environmental and economic grounds, the Utes rushed to its defense, still seeing the deferral agreement as the only way they could obtain water. Thus, Mormon whites and Ute Indians were united on this issue, and the Administration backed down. As Interior Secretary Andrus wrote in a memo to the President in April, 1977, "The Bonneville unit is admittedly one of the weaker of those projects remaining, but, as with most, there are some complex issues involved."

The Central Utah Project had been a gleam in Utah's eye since 1903, when State Engineer A. F. Doremus drew a map showing a "proposed intercepting channel" along the southern foothills of the Uinta Mountains leading to Strawberry Reservoir, the tunnel, and eventually a canal that would traverse the Wasatch Front to a point north of Salt Lake City. Originally Echo Park Dam in Dinosaur National Monument was part of the project, but conservationists managed to get that dam moved to Glen Canyon by the time the Colorado River Storage Project Act was passed in 1956. The projects distributed throughout the upper basin by that act mostly benefited Utah, a fact that did not escape Wayne Aspinall, who credited the Mormons with domination of the Bureau of Reclamation's upper Colorado regional office in Salt Lake City. Twelve years later Aspinall was to see to it that Colorado got its share. A National Science Foundation–financed study stated, in reference to the 1956 act, "As an experience in distributive politics, the act has some interesting features. In terms of money spent and project size, Utah clearly has received the best of it." The Vernal Unit was completed in 1962, and Soldier Creek Dam and the enlargement of Strawberry Reservoir were finished ten years later. Other portions of the Bonneville Unit were still under construction in 1977 and the other four units were in the planning stages when the Carter Administration took another look at it.

Such long-held dreams do not die easily, especially when they begin to emerge as reality. Environmentalists within Utah, somewhat of an endangered species themselves, objected to the greatly reduced stream flows below the intercepting ditches of the Central Utah Project, as did the Forest Service and the Environmental Protection Agency. The growth-inducing impact the project would have on the metropolitan Salt Lake City area, already beset by smog, was also questioned. The Carter Administration had its objections, as had certain officials in the Department of Interior during the Nixon Administration. An assistant secretary in those years said it was a "costly, unnecessary project and that it should be stopped." He and the undersecretary were overruled by Interior Secretary Rogers C. B. Morton after a visit by Utah Senator Wallace Bennett, who also made his pitch to the White House. The Salt Lake County Attorney opposed portions of the costly project, which he termed "an unnecessary exploitation of our resources." The whites who lived in the Uinta Valley were not particularly happy with the project, because it took water away from agricultural and energy use in eastern Utah and gave it to the cities to the west, an argument less divisive in tone but reminiscent in content to the differences between east and west slopes in Colorado.

But these arguments advanced against the Central Utah Project, or certain aspects of it, amounted to mere hairline cracks in the overall structure. In contrast to the Central Arizona Project, the arguments would be low-keyed and confined to isolated utterances within the state by ineffective interests. The Arizona project would gain more notoriety in the 1970s, a function not only of its greater size and more far-reaching effects, but also because of the more open society in which it was discussed. Utah could still act with near unison when it came to water. At the 1972 public hearing in Orem on the draft environmental-impact statement for the Bonneville Unit, there were high school bands, signs, and banners promoting the project. The atmosphere was patriotic, the statements were of the cheerleader variety, and only a few "radical conservationists," as they had been branded in a newspaper editorial, showed up. The Mormon ethic of putting natural resources to productive use was best enunciated by the Utah Farm Bureau: "It would be extremely short-sighted for us to leave our mountain streams to run wild, to rush down into the valleys only to stand stagnant, producing swamps when the prudent installation of dams at strategic places can regulate the flow of water, conserve it for culinary use and irrigation, and contribute to the beneficial development of our soil resources, thus resulting in the establishment of homes, and communities, the evolvement of our culture and an opportunity for our people to live rich and useful lives."

Farming had been the traditionally approved Mormon occupation in Utah; search for and production of minerals was historically discouraged. Water followed socio-religious preferences, with 90 percent of it diverted to agriculture. But in the last ten to fifteen years, mineral production—particularly if it was energy-related—has gained an accepted, even preferred, status in the Beehive State. Employment was a principal reason. No longer did agriculture supply jobs in sufficient numbers, so water was to be directed more and more toward municipal and industrial use. Agriculture would remain stable, but energy uses of water would expand, or so it was hoped. A 1967 policy statement of the Utah Board of Water Resources, reiterated in 1975, was revealing: "The Board considers the use of water for the irrigation of new lands to be relatively less advantageous to the State under conditions now existing than some possible alternative uses." Energy, for instance. The state's large reserves of coal, oil, oil shale, tar sands, and uranium were located in the Colorado River basin, precisely in those remote regions where the population was dropping most alarmingly and that basic Mormon unit, the family, was disintegrating. The problem was to get the water there, back to its basin of origin, since most of it was already committed to urban areas west of the divide.

As the Colorado River leaves Wyoming and Colorado and starts to flow through the desolate but scenically spectacular regions of eastern Utah, the claims on its use begin to multiply, as they continue to do the further south the river travels. In Utah, there is that first sense of urgency, spurred by the fear that there eventually might not be enough water to go around. A 1975 report by the Bureau of Reclamation, *Critical Water Problems Facing the Eleven Western States,* commented on Utah's predicament: "Rather, the water supply problem is associated with the cumulative effects of all development in the region. Utah does not have enough water remaining in its Colorado River entitlement to meet present and projected long-range water requirements for a significant level of electric power generation, oil shale development, projected transregional diversions, and potential regional agricultural development."

This was what worried the Ute Indians. Like others, they had plans for agricultural and energy use of water, but there would not be enough water for all their desires to be sated even with the guarantee of the Winters Doctrine. From 1960 to 1970 there was an 18.8 percent increase in population along the Wasatch Front. There were also fish and wildlife to consider. The squeeze was on, and nowhere was that more evident than in the case of coal, Kaiparowits, and other Southwest power plants.

Coal and Utah coalesced in the mid-1960s, as did the energy needs of southern California and the policies of the federal government toward natural resources and Indian tribes. In Utah the courtship was serene, but when it came time to consummate the union things got a bit turbulent. Elsewhere in the Southwest, objections would arise, mostly after the fact. Power plants and coal mines would be built and, like the river that made them possible, the energy would flow toward southern California—the ultimate consumer of the West. More than ten million people lived in southern California, more than twenty million in the whole state—a number equaled in the West only by the combined populations of the eighteen other most western states, minus Texas. To attend a meeting of Hopi elders in the centuries-old village of Shungopovi, to hear the governor address the Utah legislature, to see a Navajo Indian on horseback herding sheep by a coal strip mine in northwestern New Mexico, to follow from the air the undulations of a coal-slurry line 274 miles from Black Mesa in northeastern Arizona to the Mohave power plant in southern Nevada, to watch the tankers pulling into Los Angeles and Long Beach harbors with crude oil from Alaska and more distant points, to trace on a map the natural-gas and crude-oil pipelines and high-voltage electrical transmission lines converging from all parts of the West on the Los

Angeles Basin, like spaghetti being drawn into the mouth, and to know water was being sucked into that same vast, insatiable maw from as far away as the Wind River Range in Wyoming, the Feather River in northern California, and the streams falling into Mono Lake and the Owens Valley from the eastern side of the Sierra Nevada, was to know intuitively that something was drastically out of whack.

But that was not all. There were other plans in the works, such as liquefied natural gas from Alaska and Indonesia, natural gas via pipeline from the north slope of Alaska and perhaps the Mackenzie River delta in northern Canada, and Utah coal by train to power plants in California. Simply put, southern California should never have been allowed to be populated to the extent that it was. The land slips, burns, trembles. It is a desert with no natural harbor, San Diego excepted, nor any measurable

Power from the Four Corners Plant in northwestern New Mexico is transported to southern California, among other places. In the foreground is the holding pond for cooling water pumped from the San Juan River.

natural resources of its own, save sunshine. Most has to be brought to it. It is a drain on the rest of the West. As Raymond F. Dasmann wrote in *The Destruction of California*, "We insist that water must be shipped to the places where people and industry have located. We could equally well insist that people and industry locate in the areas where water is available."

Through the 1960s, southern California experienced dramatic increases in population and demands for electricity. The demand kept increasing yearly at rates averaging 7 and 8 percent, which meant a doubling of a utility company's generating facilities every ten years or so. The utilities were overly reliant on oil-burning power plants in the Los Angeles Basin, where air-pollution requirements were being tightened. They wanted to diversify to other, less costly fuels. Coal and water, the two main ingredients needed to start planning a power plant for the interior West, were available; and with the increasing interest in protecting California's coastline from such unsightly developments as power plants, the rush toward the interior regions was on. There the utility companies would find, at first, compliant local and state agencies and citizens, all of whom would be greatly impressed by the monetary rewards of locating such a plant in their bailiwicks. This was an argument that had lost its appeal on the West Coast, where some of the disadvantages—such as smog—had already become evident, but it had tremendous effect in the interior regions. Then the entities in these regions, too, began to doubt. For Utahans and Indians, these plants would come to be seen as a drain on such valuable resources as water, and the compensation would be viewed as unequal to the risks or value of the resource. For the West Coast utilities, coal plants were cheaper to build than nuclear plants, the principal alternative at that time for large blocks of power, and could be put into operation more quickly. In addition to all these favorable factors, there was a compliant federal policy of cheaply leasing coal-bearing federal lands during the 1960s. It would be a boon to destitute Indian tribes, return some revenues to the federal treasury (to be earmarked for dam construction), and, in the case of the Navajo plant, be the necessary alternative to dams in the Grand Canyon that would make the Central Arizona Project possible.

So Southern California Edison Company bought into the Four Corners Power Plant, to be operated by Arizona Public Service Company near Farmington, New Mexico, and began to build the Mohave plant in 1967. The Mohave Project, Edison executives pointed out in Las Vegas, would increase the tax base of surrounding Clark County by 10 percent—the greatest single economic impact ever felt in that money-conscious area. For Utah, the Kaiparowits plant would have provided 8,500 jobs and a 15 percent boost in the assessed valuation for the whole state. Clark

County would later become worried about the Mohave plant's air pollution and tighten its regulations, while Utah would require greater in-state participation in such projects. It was the Mohave plant, located next to the river in southern Nevada, that would be supplied by coal mixed with water and shipped through a slurry pipeline from Black Mesa. Other western states later would not allow valuable water to be shipped out of their boundaries in this manner. Coal from the Navajo-Hopi Indian reservations on Black Mesa would also be shipped seventy-eight miles by railroad to the Navajo Project to be built on Indian land near Page, Arizona. Like the plans for Kaiparowits, the Navajo plant took its water from nearby Lake Powell. The Los Angeles Department of Water and Power would get a major share of the Navajo plant's output, as would the Bureau of Reclamation to pump water through the Central Arizona Project, a chore that had been previously assigned to dams within the Grand Canyon. The two southern California utilities would share in the output of other southwestern power plants, both coal and nuclear, and would attempt to build their own giant facilities, such as Kaiparowits and the Intermountain Power Project, also in Utah. These plants were the result of a regional planning effort by some two dozen utility companies named Western Energy Supply and Transmission Associates (WEST), whose formation was blessed by Interior Secretary Stewart Udall. WEST had ambitious plans. By 1985 it foresaw producing three times more power in the Southwest than the Tennessee Valley Authority, and seventeen times as much as was being generated by the Aswan Dam in Egypt.

Then something went wrong. The dream of WEST, Udall, Indian tribal councils, southern California utilities, and Utah dimmed as the air quality deteriorated with the immense amounts of contaminants being poured into the air by the Four Corners Plant, the first to go on line. The plant became the harbinger of doom for clear skies in the Southwest. These were the years of Earth Days, and environmental concern loomed large in the Four Corners region, where within a two-hundred-mile radius of the common point shared by Utah, Colorado, New Mexico, and Arizona there were six national parks, twenty-eight national monuments, three national recreation areas, several national forests, thirty-three national historic landmarks, and sixteen million visitors a year recorded by the Park Service. It was a region unrivaled in the nation by such a complex of officially recognized natural beauty spots—recognized by the same entities, Congress and the Department of Interior, that would also allow the power plants. The question was, were they compatible? In June, 1973, Interior Secretary Rogers C. B. Morton decided that Edison could not build Kaiparowits at the site it wanted, but did not rule out the possibility of a new location not too many miles away. The basis for the decision was that Kaiparowits, in conjunction with the nearby Navajo plant, would degrade air quality in the Lake Powell and Grand Canyon areas. Morton's action was unprecedented. The federal government had said no, or at least not in that place, to a coal-burning power plant.

Edison and Utah were stunned. Governor Calvin Rampton said of Morton's announcement, "That is the most arrogant statement I have ever heard anyone make." The new site was fourteen miles distant, and the utility and federal government quickly commenced the process heading toward its expected approval. But before the secretary's ruling, Edison announced it was dropping all plans for a power plant on the Kaiparowits Plateau. The process had dragged on for too long. The economics had changed. In the wake of the first Arab oil embargo, the annual increase in demand for electricity had declined in Edison's service area to 0.5 percent in 1975 and forecasts for the future would put it at 3.8 percent, less than half of what it had been ten years previously. Various plans emerged later to use the coal on the plateau for another Utah power plant, a coal-gasification plant, or to have it mined and shipped by railroad outside the state. With the coming of the second energy crisis in the spring of 1979 and the renewed emphasis placed on domestic energy production and synthetic fuels, the large-scale production of western coal once again seemed a possibility. After all, it had been around for a long time and had a history of booms and busts.

To watch the Green River disappear at the southern end of Brown's Park into the Gates of Lodore in Dinosaur National Monument, Colorado, and to see it reappear near Jensen, Utah, is to experience the plunge from lush tranquillity into desolate wastes. For me, the Gates of Lodore mark the point where the change begins. It is completed once the river flows into the Kingdom of Zion, the state of Utah, that separate nation whose accessibility by river is, appropriately enough, through a gate. The gate's precipitous canyon walls, viewed from the entrance, rise in smooth, massive blocks from the flat terrain of the park. They glow blood-red in the morning. The blocks on either side of the river, it seems, would interlock in a decisive closing should they ever be joined together like the fingers of both hands. The tranquil river, sweeping around a bend, disappears into this fastness, a disappearance made more ominous by its placid surface. The rapids are within the Canyon of Lodore, hidden from outside view.

I once camped at the entrance, only to be kept awake most of the night by a Memorial Day rafting party that was overactive with nervous anticipation. Other times and at other seasons I have been there alone, and slept no better. That late May weekend I climbed to a point above the river near where John Wesley Powell had observed a sunset more than a hundred years previously. Powell wrote, "The vermilion gleams and rosy hues, the green and gray tints, are changing to sombre brown above, and black shadows below. Now 'tis a black portal to a region of gloom. And that is the gateway through which we enter [on] our voyage of exploration tomorrow—and what shall we find?" Emerging on the south side of the national monument, the impression is one of dusty, white light—intense yet vapid. The light is reflected off the heaps of sandstone, whose disarray is made more evident by the lack of vegetation away from the river. Suddenly, it is a scratchy land.

This is mineral country, a place where riches can possibly flourish only underground, locked in ancient geologic formations that indicate this land, too, was once well watered. But that was long ago, and too far removed from the present dusty reality to imagine with any clarity. Here there are oil shale, crude oil, tar sands, uranium, and coal, and isolated exploration camps and clusters of tin-roofed buildings around producing oil wells. Eastern Utah is a place where only the dispossessed and the resolute—Indians, Mormons, and energy workers—can make a go of it. The landscape is neither true desert, high country, nor canyonlands, and its impreciseness has always bothered me. For me, this region is the Sargasso Sea of the interior West, and I am repelled by its stagnant desolation. Only a few streams, not really deserving of the appellation "river" but retaining it nonetheless, flow sluggishly into the Green and Colorado. The White, Duchesne, Price, and San Rafael rivers feed into the Green; the Dolores joins the Colorado above Moab, Utah. Their meager flows will be lessened, halted entirely at certain times of the year, with the completion of water projects upstream. They will become non-rivers, like the lower ends of the Gila and the Colorado.

2

Although they too had been discriminated against and had withdrawn, like the Indians, to their separate place, the Mormons were very much a part of the institutional structures that divided up the Colorado River and authorized the projects to put its water to use upon the land. Utah participated in the negotiations that led up to the 1922 Compact dividing the waters of the river system between the upper and lower basins and the 1948 Compact that apportioned the water between the four upper-basin states. The Indians participated in neither the negotiations nor the signing of these two decisive interstate compacts. Nor did the federal government, supposedly the trustee in such matters, assert Indian water interests in an aggressive manner. A perfunctory and ambiguous clause was inserted in both compacts stating that they would not affect the obligation of the United States toward the Indians. That was all. Supposedly, this kept the states from infringing on Indian water rights, but it remained for the federal government to assert its obligation—something it would shy away from doing at the request of the states. Utah had its share of congressmen and senators on key committees to watch out for its water interests. The Indians had no such protectors. True, a smattering of legislators, mostly from the East, would make occasional noises about Indians sharing in the West's water; but they would be outside the mainstream of those powerful western institutions that decided where water projects went and how fast they were completed. The Mormons had the additional advantage of being unified within their own ranks and cohesive in the pursuit of outside goals, something the Indians lacked.

When it has come to distributing water in the West, it has been the politically strong and aggressive who get it. To be tenacious and knowledgeable helps. Moral rights, historical priority, and legal merit count for less. And the Indian tribes, like backward nations, have been historically weak in this arena. There are some recent signs of change.

Not the right of the Indians to more, probably much more, water from the Colorado River (that has already been determined by the Winters Doctrine), but the question of how much water, is the sword of Damocles that hangs over the West. It threatens, like nothing else, to sever the complex web of laws, agreements, regulations, quiet understandings, and court decisions that, collectively known as "the law of the river," constitute the major determinant in the growth of the West—the white man's West, that is, since the Colorado is essentially a white man's river. More than that, it is an Anglo river. The Mexicans, also, have only grudgingly been given water. Even within this last concentric racial circle, the river has been divided up and diverted by only a very small segment of Caucasian society. Others do benefit, but only the few distribute the water. This small group of decision-makers and technicians has been invariably white, Anglo, male, and extremely conscious of how they wanted the economic benefits of water dispersed. Few others have taken the time or interest to understand or deal with the esoteric and arcane world of water law and politics.

In the last ten years the Indians have begun to take more interest, their somewhat blunted wedge being their assertion of Winters Doctrine rights on their own behalf. The white man covets Indian water and the vast energy resources on Indian reservations that it might unlock, as the white man in the previous century coveted Indian lands. The Indians, like the oil-rich developing nations of the Middle East, have perceived that they might get more by banding together and alternately threatening to withhold their coal and uranium and actively pursuing quantification of their water rights. Such a strategy has to be pursued with extreme care, since there can be massive retaliation—a feint here, a feint there, like the guerrilla warfare the Indians waged with the white settlers and troops in the nineteenth century.

The Mexicans, across the border where the river begins to end, can exert pressure only through world opinion by projecting the image of a poor nation being oppressed by its rich, powerful neighbor. This does not mean much to the Colorado River basin states, but it does attract the attention of the President and the Department of State. One gets the distinct feeling that, if the seven basin states had their way—if they had not already coerced the federal government into giving them the least costly solution for furnishing Mexico with water of suitable quality—little if any water of such quality and significant quantity would flow south across the international boundary. Water flows toward the powerful and rich. Paradoxically, for such a vital element in an arid land, it costs those who benefit from it very little. Where once it was the law of the democratic six-gun—people have been shot over water disputes in the West— it is now the more polite and discriminatory forums of the water bureaucracy, Congress, and the courts that hand out water. There is no better example of the racial inequality of water distribution in the West than a little-noted incident that involved the Supreme Court.

The governors from the western states were gathered at their annual conference in Albuquerque on November 2, 1953, listening to Governor Howard Pyle of Arizona give the keynote address. A Republican administration held office back in Washington, D.C., after twenty years of Democratic rule, and Pyle, himself a Republican, was making a pitch to it. He dwelt on the importance of the West to the nation, its contributions, its "virtually untouched" resources, and the need for an expanded Reclamation program. "The question facing us today," Governor Pyle continued, "is how we of the West can lend a helping hand rather than asking for a helping hand." Before the day was over the West, which prides itself on its sense of independence, would once again be asking the federal government for help.

On that same day in Washington the federal government filed a petition with the Supreme Court to intervene in the lawsuit initiated the year before between Arizona and California. It would take twelve years to resolve; by 1964 the court had divided the waters of the Colorado between the three lower-basin states and quantified the amount under the Winters Doctrine for five small Indian tribes along the lower river. The basis for the 1953 intervention of the United States in the suit was, according to the petition filed with the Clerk of the Supreme Court by the Department of Justice, the "prior and superior" rights of the Indians to the water in the river system. It was a strong statement, but not without precedent, since the 1908 *Winters* decision, involving the Milk River on the Fort Belknap Indian Reservation in Montana, held that the Indians had prior and paramount rights to the water that traversed their reservation. Those rights dated in priority from when the reservation was formed by treaty, generally at a time for most tribes in the West when Anglos had not yet filed for water rights.

But the phrase "prior and superior" ignited the anger of the western governors and their water experts when they were informed of the substance of the petition in Albuquerque. The author of the phrase was William H. Veeder, the only attorney at that time knowledgeable about Indian water rights within the Department of Justice. Veeder, a native of Montana who had practiced law in Colorado before coming to Washington during the Roosevelt Administration, was known as a New Dealer within the Republican Administration, but was kept on because of his knowledge of western water law. In the early 1950s he had met with Arizona representatives who had asked him to "play down" Indian water rights in the upcoming legal struggle. At the direction of Attorney General Herbert Brownell and working with Assistant Attorney General J. Lee Rankin, Veeder began to research and prepare the petition for intervention. There were meetings within the Justice and Interior departments. Veeder later wrote, "At the first meeting after it was agreed the suit would be initiated, this strong caveat was given: Play down the Indian rights! To that caveat the response was made that there probably would not be a justiciable issue in the case if the Indian rights were played down." The "Petition of Intervention on Behalf of the United States of America" was filed in *Arizona v. California.* The United States, as trustee for the Indians, asserted "that the rights to the use of water claimed on behalf of the Indians and Indian Tribes as set forth in this Petition are prior and superior to the rights to the use of water claimed by the parties to this cause in the Colorado River and its tributaries in the Lower Basin of that stream."

Immediately upon learning of the content of the petition at the governors' conference, western water interests notified Brownell of their extreme displeasure. Veeder was told by Assistant Attorney General Rankin to withdraw the petition on November 6. "I was advised," he later testified, "that the political pressure was so great in opposition to the Indian claims as set forth in the petition it would have to be withdrawn from the Office of the Clerk of the Supreme Court where it had been filed and the offending provision revised. . . ." Veeder went to the clerk's office in the Supreme Court Building and was handed the original petition and some of the extra copies required for filing. A few copies had already been served on parties to the suit. Someone in the attorney general's office had called ahead of Veeder and asked the clerk's office to hand over the petition. Within a month a delegation of westerners showed up at the Department of Justice for a stormy meeting with Brownell. They included Governor Pyle; Governor Edwin L. Mecham of New Mexico; attorney Northcutt Ely, representing California; and a Colorado attorney by the name of Jean S. Breitenstein.

Pyle led the attack. Brownell later recalled that the Arizona governor —a former radio personality and ardent supporter of the Central Arizona Project—asked why a Republican Administration kept a "damn New Dealer" on its payroll, to which Brownell replied that Veeder was the most knowledgeable person available on western water rights. It was very clear, the Attorney General added, that it was the role of the federal government to be the guardian of Indian water rights. Pyle objected to such a strong assertion of Winters Doctrine rights in the petition. Such wording, he said, would hinder the Central Arizona Project, which had not yet been authorized; it could result in its never being built. The governor of the state containing the nation's largest Indian reservation and the largest Indian population added, "Indian claims [are] a nuisance and excess baggage." Pyle then launched into the political aspects of the situation, an argument that was not lost on Brownell, since he had run President Dwight D. Eisenhower's successful campaign for office. Testifying later at a Supreme Court hearing, Ely recalled, "There was prolonged discussion of it. The argument presented to the attorney general was primarily political, and I told him on behalf of California I would not join in any political consideration of the matter, and that, so far as California was concerned, we would not join in any request that there be deleted from the government's petition any language whatever." Not that California felt any moral commitment to the Indians—such a position was just part of its legal strategy to oppose Arizona in the court case.

On December 8, without the Supreme Court having authorized any amending action, a revised petition of intervention was submitted by the Attorney General's Office. The revised petition was the same as the original, except for two pages, which had simply been removed, reworded, retyped, and inserted back in the original petition. The Attorney General's Office did its work rather sloppily. The original petition was narrowly spaced. The revised pages stuck out glaringly in more widely spaced type. Needless to say, the phrase "prior and superior" no longer appeared on page 23. The offending language was replaced by the following: "The United States of America, as trustee for the Indians and Indian Tribes, claims in the aggregate on their behalfs rights to the use of water from the Colorado River and its tributaries in the Lower Basin of that stream. . . ."

It was a much weaker pleading, and the Indians did not like it. In a later petition to the court, they declared that "This extraordinary procedural lapse . . . permitted the United States to make a radical shift in position without the embarrassment of setting forth the reasons for the change as part of an application for leave to amend." The attorney for the Navajos, Norman M. Littell, added in an oral argument before the special master appointed by the Supreme Court to hear the case:

December 8—this was withdrawn in the Supreme Court of the United States by the simple expedient of drawing inked lines through the record, through the entry in the records of the Supreme Court of the United States. The Attorney General prevailed on the Clerk of the Court to break his own rules. Rule I, subparagraph 2, of the Supreme Court provides: "Original or filed copies of pleadings and papers or briefs may not be withdrawn by the litigant." This was withdrawn.

THE MASTER: You mean the paper itself?

LITTELL: The paper itself was withdrawn. It was physically withdrawn without the permission of the Court and in violation of the Court's own rules. The Clerk himself, when he discovered this, was somewhat astonished. Ink lines through the entry.

The western water interests had prevailed, once again. Now they went after Veeder: they did not want the government's case argued by an expert in western water law who was sympathetic to the Indians' cause. Again they were successful, in a situation not unlike the removal of Gary McVicker from the Kingman, Arizona, office of the Bureau of Land Management when grazing interests felt threatened by a decisive federal presence. At the time Veeder and Brownell did not believe that the revised petition was harmful to Indian water rights. Later Veeder would change his mind, stating that such actions by Anglos interested in building water projects without regard for Indian water rights was "Corrupt. Totally, completely corrupt." Indians, and those white lawyers who represented them aggressively, would come to view the incident as symbolic of the two-faced policy of the federal government. Among the small cadre of those knowledgeable about western water law and Indian water claims, it would become a *cause célèbre*. References to it would appear in learned, though obscure, papers. It rankled quietly.

Veeder would remain a hard-liner on Indian water rights within the Department of Justice, where Brownell sheltered him, and later within the Bureau of Indian Affairs. But his active role would be diminished, and he would not argue the federal government's case before the Supreme Court. In the late 1970s, Veeder was working as a consultant to the Bureau of Indian Affairs. He had survived because of protectors within the Senate, and because he was too visible a symbol to be eliminated entirely. Indeed, he had briefly reemerged into favor when a wave of reform swept through the moribund bureau in 1970. But when the backlash to reform developed in mid-1971, he was transferred to Phoenix. The transfer precipitated tribal protests and a march by Indian militants on the bureau's Washington headquarters in a vain attempt to place top officials under citizen's arrest. The transfer was rescinded. Throughout the years Veeder has remained an intransigent, zealous protector of Indian water rights, as dedicated in his pursuit of this one goal as Ben Yellen was in his prolonged attempts to make the 160-acre-limitation law apply to the Imperial Valley of California. Because of these two men, one barely within and the other outside the federal bureaucracy, these issues would remain irritants to western water users, who would undertake periodic, complex maneuvers to circumvent them. Brownell's method of sheltering Veeder within the Justice Department was to enlarge the section dealing with Indian matters. In this way he would not be such a red flag to western water interests. In January 1954, Governor Pyle was appointed by President Eisenhower to be his special assistant on federal-state relationships. In this White House role, Pyle asked the President to remove Veeder from the Supreme Court case.

The Justice Department attorney who took Veeder's place was ill prepared, and faced an imposing array of extremely knowledgeable lawyers representing the interests of western states. Water law is a legal specialty, a narrow field; western water law is a subdivision of it; Colorado River water law is a subdivision of the subdivision. And each basin state and both regions have their separate versions of the law. The attorney for the Indian tribes, Littell, claimed that the new federal attorney had to ask where the Navajo Reservation was located, did not properly cross-examine witnesses, was not aggressive in presenting his arguments, and had been employed by the Department of Justice only six months prior to being assigned to the most important and intricate water case in the West's history.

Whatever the reason, the final Supreme Court ruling giving the five small tribes a surprisingly large amount of water was a lot more circumscribed than if the court had ruled on the quantity of water due all the tribes in the lower basin. A legal authority on Indian trust matters, Reid Peyton Chambers, would later review the legal conflicts of interest within the federal government and conclude, "The shortchanging of Indians by the white men—thought by some to be an historical phenomenon—is a present-day occurrence, abetted by the government itself."

Years later, Brownell was again to confront western water interests. He was appointed by President Nixon in 1972 to negotiate an agreement with Mexico on the quality of Colorado River water. The basin states would again come out on top, this time in an international arena.

For the government to be so cavalier about Indian water rights was not unusual. The white man's policy toward Indians has vacillated between assimilation—meaning the dissolution of the reservations and the absorption of Indians into white society—and self-determination, the preservation of Indian culture and their separate identity while seeking ways for them to survive economically on the reservation. During the 1950s the policy of the federal government was one of assimilation, also known as termination, and its leading congressional proponent was Senator Arthur Watkins of Utah, who was also an ardent backer of the Central Utah Project. A number of bills advocating termination policies, such as cutting off federal services to Indians and permitting long-term leasing of Indian lands (a few reservations were actually extinguished by executive action), were passed by Congress during these years. Senator Watkins proclaimed the policy of termination thus: "Following in the footsteps of the Emancipation Proclamation . . . I see the following words emblazoned in letters of fire above the heads of Indians—THESE PEOPLE SHALL BE FREE." And some of the Indians were freed of their land and water, usually with disastrous results for the tribes and with economic gains for the white men who took them over. An earlier advocate of the policy of assimilation was Senator William Stewart of Nevada, the author of the mining act. Stewart proclaimed in the early 1870s, "I regard all those treaties as a sham," adding, "The idea of thirty or forty thousand men owning in common what will furnish houses for five or ten millions of American citizens, cannot be tolerated."

In 1887, when the policy of termination was first formally launched in the West, there were almost 2 billion acres of land under Indian control. By 1924 this had shrunk to 150 million acres. During the first administration of Franklin D. Roosevelt a policy of self-determination was launched and, except for the Eisenhower years, was continued through the administration of President Carter. But still, the Indian land base kept diminishing; in 1975 it amounted to 50 million acres. Most Indian lands are in the West, and many reservations, including the largest, are within the Colorado River basin. Some of the lost lands went toward reservoir sites in the western states. The American Indian Policy Review Commission, established by Congress, concluded in its 1977 report, "However, historical experience has shown that it is less politically sensitive and less expensive to take Indian lands for federal water projects than non-Indian lands." The report pointed out that the loss, amounting to about 13,000 acres a year, was "during a period when the official policy of the United States Government was to assist tribes in consolidating their land base and seeking economic self-sufficiency."

Up to 1955 the vast bulk of Indian lands that had been reduced were outright losses to the tribal land base. In that year Congress passed a bill permitting long-term leasing of Indian lands for the first time, and the trend quickly changed. Previously lands were sold, frequently at prices below market value or at prices quickly exceeded by the market value once the Anglos made a few improvements, such as diverting irrigation water or digging a mine. Now, the Indians would retain ownership, but the mineral resources and agricultural land would be leased at prices that were below market value for equivalent lands off the reservation, or exceeded by market value once production of coal or uranium was begun. The 1955 act, backed by such western congressmen as Representative Stewart L. Udall of Arizona, was designed to provide the Indian tribes with income and at the same time open up their lands for use by white men. The Indians formerly could lease their lands only for short periods of time, but the 1955 and subsequent acts permitted them to lease lands, with the approval of the Secretary of the Interior, for up to one hundred years for a variety of uses, including coal and uranium mines, farming, industrial-plant sites and real-estate developments. In the early 1970s about 15 of the 50 million acres of Indian lands were leased by non-Indians for mining or agricultural purposes. These were most of the choicest Indian lands.

Two legal authorities have argued that the leasing act, by limiting secretarial approval to financial review only, "is functionally analagous to a sort of termination," since the secretary's role is circumscribed and the leases are long-term. Monroe E. Price and Reid Peyton Chambers wrote in the *Stanford Law Review,* "It is not far-fetched to view the kinds of current residential development on the reservation—made possible by the 1955 act—as a sort of termination, or a 20th century equivalent to 19th century negotiated reductions in Indian land. For value received, the tribe surrenders a significant portion of its land, irrevocably in effect, to the non-Indian community." Indian lands leased for potential coal development in the 1960s, with no escalation clause for increased royalties, would become vastly undervalued in the 1970s when power plants began sprouting up in the Southwest. Eventually these lands would be handed back to the Indians in a depleted and exhausted condition. Such leases, undertaken when Udall became Secretary of Interior and ostensibly inaugurated a policy of self-determination, would amount to a virtual termination in the effective use of that land. There would be a value received, but it would be underpriced. There would be jobs provided, but they would be mostly menial. The policy of self-determination would be continued through the Carter Administration, through one of those periodic, emotional upsurges of white interest in Indian matters during the early 1970s that led to the mistaken belief that Indians were careful caretakers of the land (there was, after all, the vivid example of Indian overpopulation of Chaco Canyon) and through two energy crises when Indian lands were once again avidly coveted for use by white men.

Yet another issue sensitive to the West which the Carter Administration chose to interject itself into during that watershed year was Indian water rights. As part of the wider National Water Policy Review initiated by the Administration, an Indian water policy was drafted in 1977. It stated, "Any review of the situation regarding the Indian water resource development leads invariably to the conclusion that the need to meet present and future Indian water requirements has been sadly neglected. To change the momentum of that historical trend will be a difficult task." It would be difficult for many reasons: the federal government itself could not be entirely trusted to press Indian claims diligently; Indians would be hard-pressed to pay for the costs of extensive litigation; such cases were increasingly being heard in unfriendly state courts; and the states, along with their congressional allies, could be expected to protect vigorously non-Indian interests that had made extensive capital investments in water projects and water-dependent developments. These interests felt threatened by Indian water claims. The Navajo tribe alone talked about claiming an amount of water equal to two-thirds of the average annual flow of the entire Colorado River system.

The assertion of such rights, if too vigorous, could arouse anti-Indian hostility in the western states. The Indians had to tread a careful line. Forrest J. Gerard, assistant Secretary of Interior for Indian Affairs, warned American Indian leaders in October of 1977 that "This situation is a political fact of life which represents the most serious threat to the tribes' ability to secure the water supply to which they are legally entitled. While legally these considerations are irrelevant to the tribes' rights, as a practical matter these competing interests cannot be ignored. They represent the most formidable obstacle against any effort to apply Indian water to beneficial uses for the Indian owners."

That the Indians had been wronged, were continuing to be wronged in terms of a proper determination of their water rights, has been adequately recognized on paper. "With few exceptions," the National Water Commission concluded in 1973, "the [water] projects were planned and built by the Federal Government without any attempt to define, let alone protect, prior rights that Indian tribes might have had in the waters used for the projects." But this and similarly brave statements by the Carter Administration and certain segments of Congress would do little to alter the fact that the Indians shared very minimally in the western water pie. About 1 percent of all Indian agricultural lands were irrigated, compared to 5.1 percent of all agricultural lands in the seventeen most westerly states. Serving the approximate 50 million acres of Indian lands were 123 irrigation works, for the most part simple diversion structures supplying water to a few parcels of land. Only 20 to 22 such works could be classified as genuine water projects, and of this number only 7 or 8 served the full project area as defined in the original studies, and these projects were in need of extensive repairs. The exception was the Navajo Indian Irrigation Project in northwestern New Mexico, which had its own inconsistencies.

Three rivers border or cross the sixteen-million-acre Navajo Reservation, one reason why the Navajos believe they are owed a lot of water under the provisions of the Winters Doctrine. Beginning at the 11,500-foot level of the wilderness area near the top of Mount Baldy in the Apache-Sitgreaves National Forest of eastern Arizona, the Little Colorado River drops down through successive stands of spruce, ponderosa pine, piñon-juniper, and gamma grass. Along the way there are a series of small dams and reservoirs constructed mostly between 1890 and 1920, when typical high-country irrigation agriculture and dry farming peaked in the region. Because of overgrazing, livestock declined after then. Native Merriams elk were exterminated, to be replaced by Rocky Mountain elk imported into the area. Most of the national forest has been heavily logged. A Forest Service report stated, "This watershed has in general had a history of exploitation with resource development. As each resource has developed it went through a period of accelerated use." Now the area gets heavy recreational use in the summer months, it being higher and cooler than the densely populated lowlands to the west in Arizona. The same report, a watershed plan for the forest, added, "Water draws these users to the area with a magnetic effect." It is a phenomenon occurring throughout the West, from the crest of the Sierra Nevada in California to Escalante Canyon and beyond; campers invariably cluster around water, sometimes to the point of fouling it. One prime example is raft trips through the Grand Canyon. Another is the Easter-week crowds along the lower Colorado. Recreation is one more heavy demand on the limited water resources of the West, one more element in the crowding of the West that becomes more noticeable as the river flows south.

But in this case, the Little Colorado flows northwest, down through narrow, deep canyons at its higher elevations until it begins to widen and meander through the open country across the southern and eastern edges of the Navajo Reservation to its junction with the Colorado River within the depths of Grand Canyon National Park. Its lower course is usually dry, except when there is a large runoff or an intense local storm. Such flows, typical of a desert river, can be extreme. In 1938 there was no water in the river for 213 days at Grand Falls in the Navajo Reservation. Then in a few days there was a flood of mocha-colored water measuring 38,000 cubic feet per second (1 cubic foot per second being equivalent to about 450 gallons per minute). It is a thirsty, sun-baked region. About 98 percent of the rain falling in the 25,000-square-mile watershed of the Little Colorado either is consumed by plants, evaporates, or percolates underground. What little remains to flow on the surface has long ago been appropriated, mostly by white ranchers on the upper reaches of the river. In western New Mexico, the Zuñi Indians make use of some of the water from one of the tributaries to the Little Colorado, named the Zuni River. The Bureau of Reclamation in 1977 launched a study into the possibility of building a 118-foot-high dam eight miles above the Zuni Pueblo (which was a flourishing community when first visited by Coronado in 1540) to lessen the flood danger to the pueblo. The older, smaller dams had filled with sediment and could no longer contain flood waters. Like the Gila River to the south and the main river at its mouth, the Little Colorado has been depleted by upstream use. As early as 1946 the Bureau of Reclamation noted that "As a result, virtually all of the waters of this area have been apportioned. Only during extreme floods does any water escape the region."

Along the northwestern border of the Navajo Reservation, the main stem of the Colorado, joined above by three of its four major tributaries, flows through Marble and Grand Canyons. Alternately running within the reservation or forming its northern boundary, flows the San Juan River —the most Indian of all the streams in the basin. Four tribes—the Navajo, Jicarilla Apache, Ute Mountain Ute, and Southern Ute—draw on the San Juan or its tributaries, and a fifth, the Hopi, live nearby. It is along this river and its tributaries that most of the evidence of the ancient Indian civilization of the Anasazi, including Chaco Canyon, can be found. This is the Four Corners Region, the potential Ruhr Valley of the West in terms of energy development. In Colorado the San Juan is essentially an Anglo river, but in New Mexico it is mostly Indian and it is here the two cultures have come into conflict over water.

The San Juan River drains nearly 16 million acres in New Mexico, Colorado, Utah, and Arizona. About 40 percent of the land area is in New Mexico, the remaining three states having about 20 percent of the land area apiece. But Colorado contributes most of the river's flow. Less than 20 percent of the basin area—the San Juan Mountains in southwestern Colorado—contributes more than 90 percent of its water supply. From elevations of 14,000 feet along the Continental Divide, the river drops to 3,200 feet above sea level where it joins the Colorado River at the upper end of Lake Powell. From fifty to sixty inches of precipitation fall annually at the higher elevations compared to ten inches and less in its lower desert portions: this is the story of the source of water in the West. From the West Fork of the San Juan, where I traced its beginnings during the rainstorm that broke the drought, comes an average annual amount of 64,100 acre-feet of water from a 41-square-mile area. Beaver Creek joins the West Fork above Borns Lake. Below, the West Fork of Wolf Creek and the East Fork of the San Juan come tumbling down from the divide to slow and unite in the broadened ranch-dotted valley where the San Juan takes on a form recognizable as a small river bordering State Highway 160—to the east an asphalt ribbon stretching over Wolf Creek Pass into the watershed of the Rio Grande River.

Not too far down the road from where the San Juan forms is the San Juan River Resort—home sites by the river, all utilities provided, and fresh air. The latter attraction is emphasized. BREATHE, advises one of the Burma Shave–type signs advertising the resort. It is duplicated by a more sophisticated recreational subdivision to the west of Pagosa Springs, named Pagosa, which boasts a small lake, golf course, tennis courts, stables, homesites, and one of those large motel complexes that have come to serve as community centers in the West: a place for service-club luncheons, wedding receptions, small conventions, sales meetings, cocktails, dinner, and, only as an afterthought, sleep. As in the lower reaches of the river basin, and outside it where Colorado River water is imported to southern California, a small lake is a necessary selling point for a desert subdivision. Together with the inevitable green lawn around the sales pavilion or golf course, it creates the impression of easy livability in what is essentially a hostile, arid land. Air conditioners do the rest.

From Pagosa Springs, Colorado, the river leaves the highway and drops toward the southwest and drier country before reaching the state line and its first, total entrapment in Navajo Reservoir. And behind every dam in the West there is not only a reservoir but also an extensive history. In 1863, after a series of Indian raids on settlements along the Rio Grande River, the federal government decided to subdue the Navajos once and for all. The scout Kit Carson was dispatched with about seven hundred local troops and some Indian allies. They laid waste to the Navajo lands, destroying crops and driving off the livestock, and killed about fifty Indians during a six-month campaign, before convincing most of those remaining to surrender. The Navajos' resistance was broken when Carson pursued them into their most hidden sanctuary, the Canyon de Chelly in eastern Arizona, and marched up and down that canyon and the adjoining Canyon del Muerto cutting down the peach trees. Eventually about eight thousand Navajos were sent on the "Long Walk" of three hundred miles to Fort Sumner in southeastern New Mexico, an alien land where they suffered greatly. The Navajos promised not to cause any more trouble, and on June 1, 1868, a peace treaty was signed, designating a reservation on part of their homeland to which they could return. During the negotiations that led up to the signing of the treaty, one of the Navajo chiefs, Barboncito, stated: "I thought at one time the whole world was the same as my own country but I got fooled in it, outside my own country we cannot raise a crop, but in it we can raise a crop almost anywhere, our families and stock there increase, here they decrease, we know this land does not like us, neither does the water." The water, as it turned out, was highly saline.

Article V of the treaty granted to the head of each family a 160-acre tract of land to farm on the new reservation, while single adults got 80 acres. It was this promise of farmland, land that would be useless without irrigation water, that prompted the Bureau of Indian Affairs to look into the feasibility of an irrigation project on the south side of the San Juan River in northwestern New Mexico in 1920—slightly more than fifty years after the treaty was signed. A private irrigation project had been proposed for that area as early as 1901, but it failed to interest any potential investors. Up to the 1940s an Indian project was not deemed economically feasible. With World War II Indian veterans returning to an impoverished reservation and the Department of Interior finishing its survey of the water needs of the entire basin and the signing of the upper-basin compact in 1948, the Bureau of Indian Affairs moved in 1949 to begin the planning process for the Shiprock Indian Irrigation Project, whose name was later changed to the Navajo Indian Irrigation Project. In 1950 the formal recommendation was made to build it. The federal government's representative to the 1948 compact negotiations estimated a possible 192,000 acres of irrigable Indian lands and 787,000 acre-feet of water available for such a project. This was just too much precious water to go solely to Indians. So non-Indian interests along the more heavily populated, water-short Rio Grande valley and the Albuquerque area also put in a bid for part of the water from New Mexico's share of the Colorado River system in the upper basin. In 1951 state officials said they would not support the Indian project unless "competing projects" were authorized concurrently. Since no project would pass Congress without the state's support, the Bureau of Reclamation quickly amended its recommendation to include both projects. Thus was born the hermaphroditic concept of the Navajo Indian Irrigation Project and the San Juan–Chama Project existing under a single umbrella. Each sheltered the other, but the latter definitely rode to authorization on the coattails of the former. The Navajos' share of the water in the San Juan River would shrink from this point on.

The strategy adopted by officials representing New Mexico to get their project through Congress was to laud the benefits of the Navajo project for the Indians. Such an interest in Indians gaining water was unprecedented in the state, and was rarely to be heard again. Of all the western states, New Mexico has been the most dogged and aggressive in protecting the water interests of its non-Indian citizens. One congressman remarked that the bill authorizing the two projects would never pass without the "glamorizing" effect of the Navajo Project. Senator Clinton P. Anderson foresaw the first crops being raised by the Navajos in 1966—a date, as it turned out, fully ten years before the Indians received the first water from the project. Grossly inflated figures of the numbers of Indians who would benefit from the project—one estimate put it at 20 percent of the tribe—were inserted into the hearing record by Interior Department officials. Governor Mecham said the project would alleviate the "severe and chronic economic distress" of the Indians. A New Mexico congressman, citing Article V of the peace treaty, said there was an obligation to make the reservation "a fruitful land."

The sales pitch continued for the Rio Grande basin. Senator Anderson referred to the "urgent need" to rehabilitate the Navajos, then launched into the benefits for the Rio Grande valley: "Presently the Rio Grande waters are fully appropriated, which makes it increasingly necessary that we make use of the surplus San Juan water by transporting it across the mountains to the Rio Grande." New Mexico State Engineer S. E. Reynolds weighed in with population projections for greatly increased growth in the Albuquerque area. Water was also needed for the small communities around Santa Fe and Taos, for Pueblo Indians, for larger irrigated lands further south along the Rio Grande, for urban development and recreational subdivisions, for the defense industries that were locating around Albuquerque, and for the nuclear community at Los Alamos and the White Sands Missile Range. It was the time of the Cold War, and national defense was a selling point. It was also hinted that Colorado River water in the Rio Grande would be used to satisfy New Mexico's commitment to its downstream neighbor, Texas.

With all this water flowing east over the Continental Divide to the Rio Grande, the Navajo project taking a large chunk, the claims of three other Indian tribes in the basin yet to be considered, and some of the water going to white agricultural, industrial, and municipal interests in the San Juan basin, what about shortages? It was a question that preoccupied Congressman Wayne Aspinall, who was knowledgeable about water matters and exceedingly protective of western Colorado's interests. One of the projects being considered in the San Juan basin was the Animas–La Plata, which would benefit both Colorado and New Mexico water users. (It was later challenged by the Carter Administration, one of the objections being a possible shortage of water.) In order for non-Indians also to benefit, the Indians had to be made to surrender their Winters Doctrine right to San Juan water, agree to a set amount, and to share in any shortages that might occur in dry cycles—in other words, to give up their prior rights dating back to 1868. This the Navajos did under a great deal of pressure. They, too, saw the irrigation project as an economic benefit for the tribe. Within another ten years, the leasing of Indian lands for coal and uranium developments would be presented in the same light. In the late 1950s the average income for a Navajo, including relief payments, was $450, compared to the national average of nearly $2,000. Paul Jones, chairman of the Navajo Tribal Council, testified at a hearing on the project: "Obviously, if the treaty promise was meant in good faith, it contemplated irrigation." Jones assured the House Subcommittee on Irrigation and Reclamation at a May, 1960, hearing that only Indians would farm on the project, then conceived of as benefitting Navajo families on small plots. The tribal chairman was questioned closely by Aspinall, who was chairman of the subcommittee:

ASPINALL: Do you know of any further demands by the Navajo Tribe on the waters of the Colorado River within the state of New Mexico?
JONES: [No,] other than for municipal use as I read in the report.
ASPINALL: In other words, at the present time you think the tribe will be satisfied as far as reclamation and irrigation development within this particular area?
JONES: Yes.

The Colorado congressman, a very thorough person, persisted one year later. Aspinall asked Governor Mecham, "And that Indian rights as such in New Mexico will be limited to the amount of water provided for in the Navajo part of the project?"

To which the New Mexico governor replied an emphatic, "Yes, sir."

J. Maurice McCabe, executive secretary of the Navajo Tribal Council, was a direct descendant of Barboncito, one of the Navajo chiefs who signed the 1868 peace treaty. He gave the subcommittee members the language they wanted: "All water users from Navajo Dam would have equal priority. The Navajo tribe has consented to this, and relinquished its rights under the Winters Doctrine for the water necessary to irrigate the Navajo Indian Irrigation Project, in order to provide a practicable plan for comprehensive development of the resources and industrial potential of the San Juan Basin of New Mexico. We have taken this important and far-reaching step because such development is necessary for our very survival." Aspinall still wanted more, so McCabe wrote him a letter to the same effect when he returned to tribal headquarters at Window Rock, Arizona, and assured the congressman that the necessary resolutions had been passed by the tribal council. "They are getting some value for the value they forego. As far as any water rights that they had which are undetermined, they have made their agreement," concluded Aspinall.

There was one other assurance that the subcommittee needed, besides those from the state and the tribe, and that was from the federal government, the guardian of Indian interests. Interior Secretary Stewart L. Udall was asked if the Navajos would be deprived of any water rights they then possessed. No, they would not, said Udall; in fact, the irrigation project would allow them "to use and establish a water right, and it will enable them to develop a whole agricultural resource that they would not be able to develop in any other way. . . ." Some 1,120 Navajo farms would be established on irrigated lands and the project would benefit between eighteen and twenty thousand Indians, said Udall. The concept of the irrigation project would eventually change to an Indian corporate venture, and at least up to 1980, nowhere near that number of Indians would benefit from it.

Both projects containing the split of water between the two watersheds and two cultures were given final authorization in 1962, ninety-four years after the peace treaty was signed and one year after the Navajos had made all the necessary pledges. The bill provided for an Indian irrigation project of 110,630 acres served by 508,000 acre-feet of water. But the project had to be built first. Before it became a reality, it would fall victim to the two-step authorization and yearly appropriation process. The first hint that all might not be completely equitable in the distribution of construction funds for the two projects was contained in an early Biennial Report of the New Mexico State Engineer, which stated that, although there had been a need for joint authorization, "the two projects may not be able to proceed simultaneously with construction." And, indeed, they did not. By 1970 the Navajo project was only 17 percent completed and the San Juan–Chama Project was 65 percent finished. The appropriations would be speeded up by Congress for the Navajo project during the 1970s, after the inequity was pointed out. The first water was delivered in 1976, and the project was scheduled to be completed ten years later. Water was diverted through the San Juan–Chama Project in 1970 and two years later, for the first time since 1942, New Mexico delivered excess water to Texas via the Rio Grande.

Writing on Indian water rights, or the lack of them, in the Colorado River basin and the Navajo irrigation project in particular, two authorities on western water law, Monroe E. Price and Gary D. Weatherford, noted: "The stark truth of the matter is that, beginning at the turn of the century, the offices and powers of national government were marshalled to plan, construct and finance non-Indian agricultural development in the West, and nothing comparable was done for the Native American."

Not too long after both water projects got under way, the leasing of Indian lands for coal mining, along with energy production, began to replace irrigated agriculture as the preferred government policy to lift the Indians out of poverty and into the affluence of the modern world. As plans for additional energy demands on water were announced in the Four Corners area and some developments became an actual reality, little water remained and the Indians' share would shrink even further—from an original estimate of 787,000 acre-feet to a request by the Navajo tribe for 610,000 in 1952, to 560,000 the next year, to 508,000 in the 1962 bill, and to perhaps 370,000 with installation of a sprinkler instead of a gravity-flow irrigation system for the project. There was no doubt the more modern, sophisticated sprinkler system would save water, and would be more desirable in a region where non-Indian irrigators had the amazingly high inefficiency rate of 20 percent of water diverted actually being consumed by crops. An opinion by the Solicitor's Office of the Department of Interior in 1974 stated that the Indians were entitled only to the amount of water necessary to irrigate the 110,630 acres—in other words, 370,000 acre-feet. The Navajos insisted they had a right to the full 508,000 acre-feet promised in the legislation. New Mexico then filed suit for a determination of San Juan River water rights, and the case wound up in a state, instead of a federal, court; there would be certain legal advantages to the state if the suit were to be appealed from the state court system to the Supreme Court rather than from a lower federal court. This touched off a number of races to the courthouse in the 1970s to obtain the jurisdiction of the court system likely to be most favorable to a cause. And water was inevitably the issue.

As the water began to be delivered in the late 1970s to the Navajo Indian Irrigation Project, which was being opened in segments, it was decided to operate it on a corporate, not an individual, basis. The Navajos formed a corporate structure, the Navajo Agricultural Products Industry, because of the need to attract capital to what was the largest such irrigation project in the nation. Also, there had to be coordination in such a large venture, such as was not possible with a lot of individual farmers. Each year a new 10,000-acre block of land had to be readied for the initial delivery of water. Such a task would have taxed any of the large private agribusiness corporations so prevalent elsewhere in the Southwest. The Navajos, who had never irrigated on a large scale before, found it very difficult to handle. Further, alfalfa was the dominant crop in the first years and this caused ill feelings among white farmers in the surrounding area who felt the Indians were depressing the market. Large amounts of irrigated pasturelands could also lower the cost-benefit ratio of the project and attract the criticism of Congress. Navajos on the west side of the reservation did not like the attention and money the irrigation project was getting on the east side. They, too, wanted their share of new paved roads. Navajo workers who were given agricultural training soon left for higher-paying jobs. Even with the corporate structure, loans could not be obtained and, like Lockheed and Chrysler, the Navajos sought loan guarantees from the federal government. A vice-president of the First National Bank of Chicago was quoted as stating: "In view of the social aspects of the project and its importance to the Navajo Nation, it would seem doubtful to me that their objectives could be attained and meet the economic needs to return the proposed indebtedness." In 1976, there was a loss of $3.7 million. In 1977, the loss would be higher. Managers and top employees came and went with disturbing frequency. There were charges of equipment thefts, and the Federal Bureau of Investigation looked into the matter. The thought occurred that perhaps large doses of Colorado River water were not an instant panacea for long-standing Indian problems.

Emerging at the same time was another solution that would possibly deal with the perennial economic problems of the Indians and at the same time satisfy white interests. This was mining and the production of energy on Indian lands using Indian water, with the income from leases and sales of such commodities going into tribal coffers and some of the jobs going to Navajo laborers. It sounded like a good idea. Again, it depended on water, more particularly a use of water that was, like large-scale irrigation, outside the past experience of the Indians, that was solidly a part of the white man's world and thus a difficult concept for the Navajos to deal with. It required a system of bartering that contained strange computations and an alien expertise. The Indians had to depend on the federal government, with its divided loyalties. As matters turned out, the Indians would get only the peripheral benefits, and in the process become part of the industrialization of the interior West.

At a Senate subcommittee hearing in 1975, New Mexico State Engineer S. E. Reynolds was asked if there was enough water in the San Juan for the Navajo Indians and all the energy projects envisioned for the northwestern corner of New Mexico. "In the ultimate," Reynolds replied, "there is always a way under New Mexico law to get the water." Reynolds was asked by whom, and he answered, "By the person who needs it." The state engineer should have added, or perhaps that is what he implied, that it also took a person who knew the system and was within it.

Against the cultural cohesion and technical expertise of the whites, the

Indians were fractionalized and unknowledgeable about the ways of western water. Historically, they had been herdsmen in an arid land that happened to be bounded by rivers but contained only a few perennial streams running through it. And that land had not been treated very well. The huge reservation had been badly overgrazed. Before Glen Canyon Dam was completed and trapped the silt further upstream, the Navajo country contributed only 2.5 percent of the water but 37.5 percent of the silt load to Lake Mead behind Hoover Dam. Eighty percent of the livestock range used jointly by the Navajo and Hopi Indians was judged to be in poor condition in 1964. Ten years later it had deteriorated further. Crowding on the 16-million-acre reservation was becoming a major problem. There were less than 10,000 Navajos when the reservation was established in 1868. By 1940 there were about 50,000. There was a 50 percent increase in the population of all Indian tribes in the decade that ended in 1970; the Navajos were no exception. By 1974 it was estimated that they numbered 140,102. By the year 2000, it was thought, they might number 348,000. As the crowding worsened, the Navajo and their neighbors, the Hopi, had a bitter dispute over land. About 40 percent of the Navajo tribe was on welfare, and income was far below the national average. The Navajos had not learned the lesson of Chaco Canyon. Nor had the whites, who moved onto the reservation with promises of vast energy projects and income in return for water and land.

It seemed a fair trade at first, but when they realized the extent to which they had been duped, the Navajos began making noises in the 1970s about asserting their full Winters Doctrine rights. Navajo Tribal Chairman Peter MacDonald, addressing a group of Colorado River water users, stated: "The Federal Government has, over the last 70 years, aggressively subsidized and promoted the interests of virtually every water user in this audience. Only the Indians have failed to benefit. . . . As I see it, the Indian's exclusion from participation in basin development has resulted from a silent conspiracy between the western states and the Federal Government. . . . In fact, the Federal Government has breached its trust, not only by silence, but by actively subsidizing projects in contravention of Indian rights." MacDonald then went on to cite the example of the Navajo plant near Lake Powell: "That experience has also left many Navajo leaders determined never again to discuss water rights outside the courthouse." It was an effective threat to the established order that was not lost on that particular audience. The Navajos ostensibly had a greater claim on Colorado River water than any other tribe or possibly combination of tribes in the Southwest. They felt cheated by the federal government and its water clientele over the deal to furnish the Navajo plant with water. After all, the Department of Interior as trustee for the Indians was supposed to look after their interests.

Shortly after the Kennedy Administration took office, the previous administration's policy of termination was supposedly reversed and one of self-determination was instituted. Interior Secretary Udall formed a task force in 1961 to come up with recommendations on how to deal with the Indian problem. It was headed by a part-Indian who was also an executive for Phillips Petroleum Company. The task force's conclusion was to speed up industrial and commercial development on Indian lands. Udall held a series of meetings with corporate executives to promote this goal.

There were some limited successes, but nothing really outstanding until four years later. By 1965 a consortium of western utility companies had formed a planning organization called Western Energy Supply and Transmission (WEST). The same year the Salt River Project, a Phoenix-area utility, filed for Lake Powell water to serve a coal-burning power plant to be built near Page, Arizona. Udall, a native of Arizona, played a decisive role in getting the utility company into WEST and had encouraged the formation of the regional energy group. In April, 1965, the general manager of the Salt River Project wrote Udall, wondering about the antitrust aspects of WEST, but adding that water from Lake Powell and coal from the Navajo and Hopi reservations "would offer a real basis for our Project's participation in WEST." In September of 1965, Interior Secretary Udall, acting as trustee for the Indians, signed a letter of intent that would lead to the Peabody Coal Company strip-mining Indian lands on Black Mesa, a massive highland on the Hopi and Navajo reservations that rises to a 8,110-foot height at its steep northern rim and drops gently to the Little Colorado River to the south. Udall stated that the pending industrial developments would mean "new jobs, large tax benefits and tremendous economic advantages not only in royalties and jobs for the two Indian tribes but also for all of the entire Southwest."

One year later the importance of Black Mesa and the Navajo plant had increased greatly. When dams in the Grand Canyon became politically infeasible because of opposition by conservationists, the Navajo plant was selected as the source of power to pump Colorado River water through the tunnels and aqueducts of the Central Arizona Project. As trustee for Indian interests and as the prime promoter of the Arizona project within the Johnson Administration, Stewart Udall played the key role in this decision. The coal would go by railroad from Black Mesa to the Navajo plant and by slurry line to the Mohave plant in southern Nevada. Southern California would get a large share of power from both power plants. Ten years later Black Mesa had already come and gone as one of those periodic, emotional symbols of white concern over injustice to the Indians. It would be generally agreed that the Indian coal lands had been greatly undervalued and environmentally degraded. And the Indians' experience at Black Mesa, in the Four Corners area, and elsewhere would lead to the formation in 1975 of the Council of Energy Resource Tribes (CERT), an intertribal organization that has been likened to the Organization of Petroleum Exporting Countries. The Indians, not unlike developing Third World nations, were looking for a better deal for their energy resources, which were considerable.

After the coal, the water had to be obtained for the Navajo plant. This was accomplished in 1968 when the Navajos waived their claim to 50,000 acre-feet of water that had been apportioned to Arizona from the Upper Colorado River Basin Compact of 1948. Most of Arizona is in the lower basin, so this amount represented only a minuscule portion of its total entitlement to Colorado River water. The Indians were not a party to the compact nor had they taken part in its negotiations. The 50,000 acre-feet represented a bone that was thrown to them, since it was far less than

New Navajo housing project for energy workers at Black Mesa coal strip mine. Black Mesa is in background.

what they could probably obtain in court under the Winters Doctrine.

In February of 1967 Interior Secretary Udall unveiled the Johnson Administration's new plans for the Central Arizona Project. They included no dams in the Grand Canyon, but authorized the federal government to become a 24.3 percent partner in the Navajo Generating Station, the third large power plant to be built in the Southwest under the regional planning concept of WEST. The legislation authorizing the Central Arizona Project was passed by Congress in 1968. Now the Indians had to be talked out of their water. As with the leasing of Indian coal lands and the irrigation project, the overriding consideration was for jobs and tribal revenues. On December 11, 1968, the Navajo Tribal Council passed the necessary resolution. The draft resolution was handed to the tribal council by a Department of Interior representative who had been dispatched from Washington, D.C., by Udall. A Navajo legal-service organization later claimed the Salt River Project also had a hand in drafting the resolution, as did the executive director of the Upper Colorado River Commission, an agency representing the four upper-basin states. The waiver of claims—some thought it pertained not only to the 50,000 acre-feet but also to any Winters Doctrine rights above this amount—was demanded by Congressman Aspinall, ever the careful protector of upper basin and western Colorado interests.

The information available to the tribal council to aid it in arriving at a decision came from the Salt River Project, the Upper Colorado River Commission, and the Bureau of Reclamation. All had a decided stake in the outcome that was not necessarily synonymous with the best interests of the Navajos. There was little discussion or information on Winters Doctrine rights. The economic benefits were emphasized, especially the chance for Indians to get jobs on the reservation. Within the Department of Interior, the Bureau of Reclamation—which would share in the output of the power plant and build the Central Arizona Project—played the dominant role in advising the Indians. A Bureau of Indian Affairs official said if he had not followed the orders of the Secretary of Interior to favor the waiver, he would have had to resign. Also, the Bureau of Reclamation had the expertise in water matters that the Bureau of Indian Affairs lacked, and Reclamation was traditionally tied to western water and energy interests who did not look too favorably upon Winters Doctrine rights. Udall was heavily committed to the power plant and Arizona's interests. Price and Weatherford, the latter at one time a special assistant to the solicitor under Udall, later wrote:

The deep and certain conflict of interest prevented the Secretary from performing the traditional trust role. The Bureau of Reclamation had too great a stake in the outcome. Nor did the Secretary perform the second role [that of independent, objective advisor] adequately, largely because of the lack of expertise and forcefulness of the Bureau of Indian Affairs. Here the United States was purporting to act as trustee and the Navajo Tribe was relying on that representation. But nothing in the period leading up to the December 1968 resolution suggested that the role could be performed adequately by the department.

It was not until a group of young activist lawyers, outside the hierarchy of the Department of Interior but inside the antipoverty program, pointed out the poor deal the Navajos had gotten that the tribal council had second thoughts. A paper prepared by Dinebeiina Nahiilna Be Agaditahe (DNA), the legal-services organization, stated: "In essence, then, the relationship between the southwestern power complex and the water rights of the Navajo Tribe is this: Navajo coal mined from Black Mesa will be burned at the Navajo Generating Station on the Navajo Reservation in order to pump Navajo water from the Colorado River to the Central Arizona Project."

There was one final irony. The Navajo Construction Workers Association was formed in the spring of 1971 by Navajo workers who felt they were being discriminated against in hiring and firing practices at the Navajo Generating Station. The Salt River Project had agreed in the lease for the plant site to give hiring preference to Navajos who were "qualified," a term that was never defined. Construction began in 1970, and during the early months of activity about 150 Navajos were hired by the Bechtel Corporation, contractors for the project. Soon Bechtel workers from the Mohave plant in southern Nevada, which was nearing completion, began arriving in Page looking for further employment. They got it. In early 1971, the level of Navajo employees began to drop markedly. At the same time Navajo employees were being fired from the Four Corners plant in New Mexico. Those who remained on the job felt their fellow tribesmen had been discriminated against, but did not want to jeopardize their own jobs by complaining. Finally a few Navajo workers sought help from the Bureau of Indian Affairs, who referred them to the young DNA lawyers. The antipoverty lawyers helped the Indians form the association and, with some help from the Office of Navajo Labor Relations, there was a gradual improvement in the hiring of Indians. The number of Navajos employed in energy-related jobs in 1977 outnumbered non-Indians. Among their fellow Indians, however, they were a prosperous elite representing only 2.7 percent of the total Navajo work force. In terms of economic benefits, the state of Arizona received $10.5

million in taxes from the Navajo plant, an amount equal to two-thirds of the total income the tribe received from all coal and coal-related developments on the reservation.

When entering the Navajo reservation, you experience a sense of separateness from the remainder of the West. There is, of course, the visible reminder of something different about to be encountered: "Welcome to

Indian employe at Navajo Generating Station, on reservation near Page, Arizona.

the Navajo Nation,'' the signs proclaim, a clear declaration of sove-reignty. The trappings of white society in Flagstaff, Page, Farmington, and Gallup—nearby urban centers that impinge upon the vast reserva-tion and live off it much as civilian towns live off adjacent military reserva-tions—quickly drop away. Not all these non-Indian artifacts disappear, however, because the Navajos, too, have acquired many of the trappings of a modern affluent society. They just tend to diminish in the greater sense of space.

One feature of Indian life is immediately noticeable, besides the many new pickups, and that is the sky-blue color of the roofing material on many of the traditional hogans as well as some modern structures. The roofs are the reflections of the color of the sky, or at least the color that used to predominate for most of the year until the power plants were built. Willa Cather described it in *Death Comes for the Archbishop:*

> The sky was as full of motion and change as the desert beneath it was monotonous and still,—and there was so much sky, more than at sea, more than anywhere else in the world. The plain was there, under one's feet, but what one saw when one looked about was that brilliant blue world of stinging air and moving cloud. Even the mountains were ant-hills under it. Elsewhere the sky is the roof of the world; but here the earth was the floor of the sky. The landscape one longed for when one was far away, the thing all about one, the world one actually lived in, was the sky, the sky!

The sky becomes less dominant when one enters the Canyon de Chelly, what Cather referred to as ''an inviolate place, the very heart and centre of their life.'' The walls of red sandstone, cut from massive windblown dunes frozen into place millions of years ago, constrict the sky and direct the visitor's attention toward the floor of the canyon, narrow and aston-ishingly fertile in this overused land. I once went there to live a few days with a Navajo family in a hogan whose roof was sky blue.

Chauncey Neboyia took me on a tour of his domain one day. Across the shallow riverbed, then dry, were about a dozen small peach trees. They were sheltered from the wind by rags draped over a wooden fence. Some older trees were individually fenced off to protect them from sheep and goats. Chauncey burned leaves in the orchard to keep the porcupines away, and occasionally trapped one or two. Peach trees were brought into the canyon in the seventeenth century by Hopi Indians, to whom they had been introduced by the Spaniards. Some survived the depredations of Kit Carson and his troops.

From the orchard, Chauncey and I recrossed the riverbed and climbed partway up a gentle incline sloping down from the abrupt wall of the sandstone cliff. There his wife, Dorothy, was tending the sheep with the help of a toddling granddaughter. A hawk wheeled above in the violent updrafts of the canyon on that windy spring day. The leaves of the sun-splashed cottonwoods along the river were a light, fluttering green against the shaded canyon walls.

We climbed a little higher and came upon an Anasazi ruin. It lacked the sweeping, elegant grace of White House Ruin further down the canyon, being a stolid, crude structure, more a storage pit than a habitation. It probably dated back some 2,000 years. Lying scattered about were a few ears of corn, their kernels gone as if gnawed by ancient teeth. Chauncey warned me not to take away any artifacts. Something bad might happen, but I do not remember the exact form of retribution he described. I certainly had no desire to make off with those ears of corn. They seemed too personal, even after all those years, even to touch.

The tour that day ended at Chauncey's proudest possession, what set him apart from the other Navajo families living in the canyon. It was his well. "It cost a lot to put it in, but we didn't have to go down far. Only twelve feet," he explained, adding that no one else in the canyon had such a well. The old couple had another home up above on the mesa, but they had moved into the canyon earlier than usual this dry year because of the dependable water supply.

CHAPTER FIVE

Canyon Country:
The Ultimate Ditch

1

The route I chose to scramble down from the canyon rim above the confluence of the Green and upper Colorado rivers was the one John Wesley Powell laboriously climbed on his epic voyage down the river in 1869. I had gotten to the rim in Canyonlands National Park by means of a four-wheel-drive vehicle on Valentine's Day of that drought year. The jeep road was a legacy from the post–World War II uranium-mining boom, which saw prospectors fan out all over the canyon country to search for the mineral that was so much in demand to build nuclear weapons. With the boom subsequently going bust, until becoming revitalized in the 1970s, it was possible to establish a national park in 1964 containing about two hundred miles of usable jeep trails. Before mining, there had been cattle grazing, which reached its peak in the area at the end of the last century.

Those whites who first saw the region judged it to be worthless. Captain John N. Macomb, who made the first scientific survey of the area in 1859, reported, "I cannot conceive of a more worthless and impracticable region than the area we now find ourselves in." Macomb echoed the humid-area sentiments of Lt. Joseph C. Ives, another topographical engineer, who had gazed down into the depths of the Grand Canyon shortly before and reported, "The region last explored is, of course, altogether valueless. . . . Ours has been the first, and will undoubtedly be the last, party of whites to visit the locality. It seems intended by nature that the Colorado River, along the greater portion of its lonely and majestic way, shall be forever unvisited and undisturbed." More than a hundred years later, controls would be placed on visitors to both national parks. The Park Service's plan for roads in Canyonlands National Park stated, "Virtually unknown by Americans before the 1960s, Canyonlands is now attracting so many visitors that current demands for use must be managed through planning to ensure protection of resources."

It has been estimated that thousands of cubic miles of rock have been washed from Canyonlands National Park into the river, leaving the standing rock for which the area is famed. The color from the minerals in the rocks averages out to a reddish brown or ocher cast, which is what gave the river its name. They said of the Colorado, "Too thick to drink, too thin to plow."

Before the gates on Hoover Dam were closed in 1935, the Colorado River carried an average annual load of 180 million tons of silt past Yuma, Arizona. With the closure of the gates and the construction of two other dams on the lower river, this amount was cut to 13 million tons. The trouble was, the silt then began to accumulate in Lake Mead, the reservoir formed by Hoover Dam. The huge mud flats that were formed at the upper end of the reservoir reduced the storage capacity of Lake Mead by 137,000 acre-feet a year. Combined with a yearly evaporation loss of 800,000 acre-feet from Lake Mead, the effectiveness of that dam could be seen to have its limits. To trap the silt elsewhere was one reason for building Glen Canyon Dam. The waters behind Glen Canyon in Lake Powell, when that reservoir is full, come to within one mile of the southern boundary of Canyonlands National Park.

Although John Wesley Powell has never been accused of lacking for hyperbole, no one has better described the canyon country. On reaching the rim above the confluence on July 19, 1869, he exclaimed, as he wrote later in *The Exploration of the Colorado River and Its Canyons:*

> And what a world of grandeur is spread before us! Below is the canyon through which the Colorado runs. We can trace its course for miles, and at points catch glimpses of the river. From the northwest comes the Green in a narrow winding gorge. From the northeast comes the Grand, through a canyon that seems bottomless from where we stand. . . . Wherever we look there is but a wilderness of rocks—deep gorges where the rivers are lost below cliffs and towers and pinnacles, and ten thousand strangely carved forms in every direction, and beyond them mountains blending with the clouds.

The climb down took a little less than three hours, including the time it took me to backtrack when I lost the way a few times and the extra care I took on the steep slope because of being alone on a winter day in one of the wilder portions of a national park where I had seen no others. The way was first marked by rock cairns, but then became an educated guessing game. It led down a dry watercourse, from one layer of slickrock to another, and then to a temporary respite in a small amphitheater. The rationale of the way down dawned on me. It led along the tops of the different rock formations until they crumbled. Then there was a steep pitch and a scramble until the next ledge was gained. The process then began over again until the last talus slope eased onto the dark red silt of the riverbank opposite the confluence of the two rivers.

At the bottom, scattered amongst the tamarisk, were some faded yellow boards which, when mentally pieced together, spelled out "Upper Colorado." The boards, I surmised, had probably been placed there to guide power boats on the 183-mile Memorial Day race from the town of Green River, Utah, to Moab. The race has now been turned into a more leisurely two- to three-day cruise. Between 400 and 500 boats participate, stopping overnight at Anderson Bottom to use the pit toilets and barbecue facilities. Should the boats have turned downriver instead of up at the confluence, within a few miles they would have found themselves in the turbulent waters of Cataract Canyon. I guessed this was the reason why the sign had been placed at this critical junction.

I ate my lunch on the bank of the river, as close to the water as I could get. This meant forcing a way through the dense growth of tamarisk, also known as salt cedar, which bordered the river. That these shrubs should have found their way one thousand miles upriver was a tribute to their hardiness and the foolishness of man. Eight species were imported into the country in the early 1900s by the Department of Agriculture. They came from the eastern Mediterranean, which has a similar climate to southern California, where they were planted to stabilize riverbanks and serve as windbreaks. The tamarisk spread like wildfire, to which they were periodically exposed but are relatively immune, and quickly became the vegetative pest of the West. Native vegetation has been forced out by the hardy exotic, and the tamarisk has become the dominant species of a new vegetative zone within the Grand Canyon that was formed when Glen Canyon Dam smoothed out the river's flows. Like cottonwoods, the tamarisk in time came to be viewed as competitors for water. By the late 1940s, the Bureau of Reclamation decided to wage war on these plants and initiated a multi-million-dollar river-management program along the lower Colorado to rid the riverbanks and backwaters of such water-sucking plantlife, termed phreatophytes. Chemical and mechanical means were used to eradicate these plants, which were thought to be consuming too much water from a water-short river.

I finished off my lunch with a cup of river water, whether from the Green or upper Colorado it was hard to say. The winter sun was hazy, and low in the south. The air was cool along the river, but warm on the slopes above. Every once in a while there would be a muted whoosh, as a small piece of riverbank gave way and fell into the water. The ripples of reddish-brown silt, exposed by the declining flows in this drought year, were tinged with white saline deposits. I walked south a few miles to Red Lake Canyon, beyond which Cataract Canyon and Lake Powell lie, and returned through the "strangely carved forms" Powell had seen from above.

There was one other stop to make before a float trip through the Grand Canyon could be understood within the context of what the Colorado River had become—that was Glen Canyon Dam. Any experience on the river or close to it in the Grand Canyon is dependent on what is happening, and what has happened since 1963, at Glen Canyon Dam. The flow of water released from the dam to satisfy water-storage and hydroelectric needs elsewhere affects where one camps in the evening, the ferocity of rapids, toilet habits, and when one awakens in the morning. The flows have changed the physical aspects of the river and its immediate surroundings. The Colorado, which has probably run for about ten million years through the Grand Canyon, has become a new river since the gates closed on the dam in 1963. In the summer, when most float trips leave from Lee's Ferry, fifteen miles downstream from the dam, the daily fluctuations of the river—now more a cold, clear tidal flow—are mostly determined by the diurnal energy needs of such cities as Phoenix. A large block of this electricity is gobbled up by air conditioners in the summer, especially when the temperature rises to above one hundred degrees, which it often does. If what was going to happen to me on the river was to be determined to a large extent by who was hot in central Arizona, then I was interested in tracing that cause-and-effect relationship.

There are few man-made things more physically impressive than a dam, and Glen Canyon Dam, a gleaming white arch pressed tautly against the red Navajo sandstone and holding back the azure waters of Lake Powell, stands near the top of the list. Nature itself cooperated by providing a backdrop of canyon country that simply cannot be improved upon. Fortunately, the engineers who designed the dam let the cleanliness of its utilitarian shape speak for itself rather than consciously trying to make it monumental with the Depression-era filigree of a Hoover or Parker Dam. The reason Glen Canyon is not right at the top of the list is because

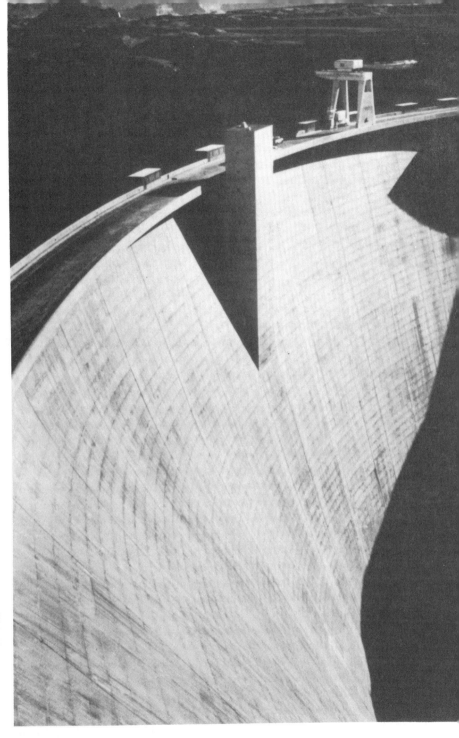

Glen Canyon Dam.

it lacks a certain lightness and grace. It has a decidedly pedestrian quality, best understood when one realizes that the reasons why it was located there were entirely practical.

Glen Canyon was a compromise site; and as such, nobody was entirely happy with it. For the conservationists, Glen Canyon was the virtually unknown trade-off for no dam in Dinosaur National Monument. For the Bureau of Reclamation it was a site about which there were initial reservations. For all, including the water interests in the seven basin states, the decision-making process that placed Glen Canyon where it was marked a new era in western water politics. It was the first time outsiders, in this case conservationists, influenced a major decision concerning water in the West. Previously, the water interests had feuded among themselves, frequently with a fervor that came close to matching their feelings about meddling outsiders. But no organized outsiders had interfered in any meaningful way or even showed any sustained interest in such matters hitherto. This was also the birth of the modern conservation movement and witnessed the invention and honing of that movement's techniques and styles that would take it through such later controversies as dams in the Grand Canyon and the exploding environmental concerns of the 1970s. While the Bureau of Reclamation and the interests it served were to be consistent throughout the controversies—they just wanted more dams and aqueducts—the conservationists were to show a disturbing lack of consistency. True, their opposition was total, but the alternatives they suggested, seemingly in good faith, changed with the expediency of the moment. But it was to be a better river because of their efforts.

Much of the political history of who got what amounts of water from the Colorado can be best understood in terms of gargantuan California, and the reactions of the other six basin states to its huge liquid appetite. From 1922 to 1974, the use of the Colorado River system's water has been determined by fifteen major agreements and a host of minor ones, collectively known as the law of the river. When Glen Canyon Dam was authorized by Congress in 1956, it represented the consummation of legal and political efforts extending back to the first attempts to divert significant amounts of water from the main stem of the river around the turn of the century.

In 1896 the California Development Company was formed to bring water by canal from the Colorado River to the Imperial Valley, lying just north of the Mexican border. With the promise of water and its actual delivery in 1901, thousands of settlers poured into the searing desert where only three inches of rain fell each year, on the average, and summer temperatures hit 120 degrees. By 1909 the population of the valley was 15,000, and 160,000 acres were being irrigated, making it by far the largest such development in the basin. Problems, not the least of which was the failure of the canal's headgate in Mexico and the resulting diversion of the whole river into the valley, led Imperial Valley interests to push for construction of a dam upstream to control the river and a canal that did not cross the border into Mexico, later to become known as the All-American Canal. Coastal southern California originally wanted the dam for its power supply, and only later for its water. In 1902 the federal Reclamation Service, the forerunner of the Bureau of Reclamation, saw the need for a dam—eventually to emerge as Hoover Dam—as part of a comprehensive plan to develop the river. Thus, southern California and federal interests became allied, to the distress of the upper-basin states, which saw the water originating in their mountains disappearing into the insatiable maw of California. The realities of the political situation were such that the Imperial Valley interests realized they could get no dam—needed for storage, for flood and sediment control, as well as for power revenues—without the support of the upper-basin states, whose congressmen were on key committees. The slower-growing upper-basin states wanted an assured water supply for future years. So the seven basin states sat down in 1922 to draw up an interstate compact dividing the waters. Secretary of Commerce Herbert Hoover presided over the sometimes acrimonious proceedings.

After tossing various alternatives around, including apportioning water to each state, the delegates finally decided to split the waters somewhat equally on a basin basis. The river was arbitrarily divided into upper and lower basins, and the line of demarcation was drawn through Lee's Ferry, a rather natural division point in the no-man's-land of canyon country. If a state's waters drained into the river above Lee's Ferry, it was an upper-basin state; if below, then it was a lower-basin state. Wyoming, Colorado, Utah, and New Mexico wound up in the upper basin, and Arizona, Nevada, and California in the lower basin. Small portions of New Mexico and Utah streams flow into the Colorado below Lee's Ferry, while a bit of Arizona water flows into the river above the division. There is nothing to mark the actual spot of demarcation on the river, other than a gauging station; but it has been indelibly fixed in the politics of western water, and the Bureau of Reclamation has divided its administration of the river into an upper and lower region, which compete with each other for water projects.

According to the terms of the compact, each of the basins got 7.5 million acre-feet a year, with the lower basin obtaining an additional 1 million acre-feet, ostensibly for Arizona to use, from the Gila River which empties into the Colorado far below Lee's Ferry. To compensate for the vagaries of wet and dry cycles, the upper basin was to allow 75 million acre-feet to flow into the lower basin over ten-year periods. The problem and real fear, voiced from the compact negotiations to this day, is that the river does not contain the estimated amount of flow on which the compact was based, and somebody, someday is going to suffer.

At the time the compact was negotiated the Reclamation Service estimated the average annual flow to be 16.8 million acre-feet at Lee's Ferry, based on records extending from 1896 to 1921. The compact apportioned virtually all that amount plus what it was thought the tributaries, namely the Gila River, contributed below Lee's Ferry. But the measurements were made during a wet cycle with faulty techniques. A stream gauging station was not installed at Lee's Ferry until 1921, and estimates for the amount of water flowing past this point before then were made at Laguna Dam near Yuma, a long way downriver. As early as 1905 an engineer for the U.S. Geological Survey, which is responsible for such stations, noted the process needed "improved methods."

It was a very costly error for the West. From 1922, when the compact was signed, to 1976, the estimated flow past the division point has averaged 13.9 million acre-feet, which approaches the 13.5 million acre-foot figure a study of tree-ring growth in the basin has shown to be the average flow over the last 400 years. From 1931 to 1940, a period of minimum flows, there was only an average of 11.8 million acre-feet. The Bureau of Reclamation looks on the Colorado as a "deficit" river, as if nature were to blame for the faulty estimates. But the basic law of the river was established on the faulty measurements of man.

The fear there might not be enough water for all plans and authorized projects on the Colorado River has haunted decision-makers in recent years, but it has never been firmly confronted. In 1946 the Department of Interior issued what was referred to as the "blue book," the first truly comprehensive report on the river and the guide for all subsequent water projects. Of the 134 potential projects identified in the report, Interior Secretary Oscar L. Chapman warned, "There is not enough water available in the Colorado River system for full expansion of existing and authorized projects and for development of all potential projects outlined in the report, including those possibilities for exporting water to adjacent watersheds." Twelve years later Congressman Wayne Aspinall warned Interior Secretary Stewart Udall, who was pushing hard for the Central Arizona Project in his home state:

Anyone who knows the river and the river's history knows that there is not sufficient water for the Central Arizona Project, as far as its continuing efficiency is concerned, unless one of two developments takes place. The first is that the Central Arizona Project takes water that rightfully and legally belongs to other basin states. The second is that the supply of the river basin is increased by some means or other.

Aspinall then attached five water projects in his home state of Colorado to the bill. Besides being based on faulty measurements, the 1922 compact failed to deal in any meaningful way with water quality and Indian water rights—issues that were to haunt the basin states in the years to come. The compact was historic in the sense that it was the first major interstate agreement on water in this country and established a precedent for other states to follow.

After the delegates from the seven states signed the compact at Sante Fe, Arizona began having second thoughts. The water had been divided between the two basins but not among the states, and while the upper basin was now protected against California's voraciousness, Arizona was not. Arizona then got a bit paranoid about protecting its interests. First, during lobbying against authorization of Hoover Dam, there were references to the "yellow peril" below the border. Large numbers of Chinese laborers had been imported to work on American-owned farms in the Mexicali Valley. Hoover Dam, it was argued, would help establish "a great Asiatic city and state" in the Mexican delta area and result in a war with Asia, claimed a zealot who had the ear of Arizona Governor George W. P. Hunt, a fiery orator in his own right. During the 1930s Arizona three times filed Supreme Court suits directed at California's interests in the Colorado River, only to have them thrown out each time.

The situation got to the point where Governor Benjamin B. Moer called out the Arizona National Guard in 1934 to halt construction of Parker Dam, which was designed to help transport water 242 miles through the Colorado River Aqueduct to the Los Angeles metropolitan area. The calling out of the troops has been treated as a humorous incident, but actually it was a good lesson in water politics. Feeling that any construction activity would be an infringement on the state's sovereignty, Governor Moer ordered one major, a sergeant, a cook, and three privates to the dam site, 18 miles over a tortuous route from Parker, Arizona. They were to report to him "on any attempt on the part of any person to place any structure on Arizona soil either within the bed of said river or on the shore." The reconnaissance party traveled to the site first by ferry boat—quickly dubbed "The Arizona Navy"—and then by horse before deciding to use cars, which bogged down in the sand. They found cables stretched across the river to which barges were attached. Reports were made daily by radio to Phoenix. It was 120 degrees during the day; at night the temperature dropped to 105 degrees. There were poisonous snakes and sandstorms. One private caught pneumonia and was sent home, where he died. More than a hundred national guardsmen, referred to as machine-gunners and infantrymen in press accounts, arrived in November when the government contractors began building a trestle bridge toward the Arizona shore. Moer issued a proclamation entitled

"To Repel an Invasion" and declared martial law. Construction halted and the workers were laid off in that depression year.

Privately, Governor Moer was playing another game. A Democrat, he was an ardent admirer of President Franklin D. Roosevelt. Moer offered to drop his objections to the dam if the Roosevelt Administration approved the Gila Project, an irrigation scheme in southwestern Arizona. The Administration moved quickly and the troops went home. Interior Secretary Harold L. Ickes called a hearing on the Gila Project for December 17 in his office. The situation was being referred to as "a state of war" by Department of the Interior officials. By next May, the Bureau of Reclamation had recommended to Congress that the Gila Project be included in that year's public-works bill. The project included the Welton-Mohawk Project, which was later to strain U.S.-Mexican relations. The Arizona legislature held out against ratification of the Colorado River Compact until 1944, when it decided to get what interstate protection it could shortly before a treaty with Mexico was signed. With Hoover and Parker Dams completed and the All-American Canal carrying water by 1942, California had gotten its share of the pie. Counting projects already built or authorized, California had the capability to divert as much as 5.6 million acre-feet from the main stem of the Colorado. Arizona was not to get its share until after the Supreme Court ruled on the division of the waters between California and Arizona in 1964 and Congress authorized the Central Arizona Project in 1968. In the meantime, it was the turn of the upper-basin states.

The Department of the Interior's "blue book" provided the catalyst for apportioning the water among the upper-basin states and getting the projects authorized to accomplish this task. The report's recommendations were hardly any surprise to the upper-basin states, since they had pushed for such a comprehensive review of the whole river and carefully monitored the results in draft form. Subtitled "A Natural Menace Becomes a National Resource," the study pointed out that an average annual flow of 20.2 million acre-feet was needed to justify construction of all 134 proposed projects. The message was clear. The states were to divide the waters between themselves and then pick and choose from this master shopping list. This the upper-basin states accomplished with relative ease and the Upper Colorado River Basin Compact was approved by Congress in 1949. The four states had learned since signing the first compact not to apportion water by amount, which might vary, but by percent of whatever was available. Thus, Colorado got 51.75 percent, Utah 23 percent, Wyoming 14 percent, and New Mexico 11.25 percent. For its small contribution to upper-basin flows, Arizona obtained 50,000 acre-feet a year. The upper-basin states acted quickly because they lacked the historical divisiveness of the lower basin and realized that unless they got their projects into place right away and found uses for the water, it might disappear downstream regardless of the compact's supposed guarantees.

With the river producing less than the amount apportioned in 1922 and the upper basin states obligated to deliver 7.5 million acre-feet a year or 75 million acre-feet over a 10-year period to the lower basin, the four states might find themselves, for instance, with only 4.3 million acre-feet, using the minimum flows for the 1931–1940 period as an example. That would be far less than the 7.5 million acre-feet seemingly promised the upper basin by the compact. Also to come out of this amount was one-half the water promised to Mexico. So far there had been no problem, since the upper basin was using only 2.1 million acre-feet a year—but it had great expectations for growth. To realize these expectations, a storage place was needed in the upper basin where water could be held for delivery to the lower basin during dry periods. In this way it was hoped that no shortages would occur in the upper-basin supply. Following ratification of the upper-basin compact, bills were introduced in Congress. These bills included plans for two dams in Dinosaur National Monument, which straddled the northern Utah-Colorado border.

The National Park Service, traditionally overshadowed within the Department of Interior hierarchy by the Bureau of Reclamation, fired the first rather muted broadside at the dams. Designed as a companion study to the bureau's "blue book," a Park Service report, with the noted landscape architect Frederick Law Olmsted, Jr., serving as consultant, merely said of the two dams, "Construction of dams at these sites would adversely alter the dominant geological and wilderness qualities and the relatively minor archeological and wildlife values of the Canyon Unit so that it would no longer possess national monument qualifications." It was left to Bernard DeVoto to focus national attention on this relatively obscure national park property. Writing in a mid-1950 issue of the *Saturday Evening Post,* DeVoto, a member of the National Parks Advisory Board, pointed out the politics of the situation: "By the time a project is laid before Congress it has already been decided upon, the local interests have been organized and the Western senators and representatives—one of the most powerful blocs in Congress—have been lined up. Within the West there is severe infighting for the allocation of projects, but when it comes to projects to be allocated, there are neither state nor party lines: there is only a solid West." DeVoto saw the entire national park system as being "in danger of being subverted by engineering construction." The article received an even wider circulation when it was reprinted in the *Reader's Digest.* After western water interests talked to the editor of the *Post,* DeVoto was never again published by that magazine. But the theme he had struck—the establishment of a precedent for the national park system—was taken up by organized conservation groups, which set out to crack the western water bloc.

In their fight to get the dams out of the national monument, the conservationists developed all those techniques that were to be perfected during the Grand Canyon dam battles ten years later and the environmental movement of the late 1960s and early 1970s. The only tool that was not used was the lawsuit, which was to come into full flower after passage of the National Environmental Protection Act of 1969. The Wilderness Society and the Sierra Club led the battle, but it was the latter with its eloquent executive director, David Brower, and its history of being founded in California during John Muir's vain attempt to keep a dam out of Yosemite National Park, that was out front most of the time. The strategy was to whip up enough interest outside the arid West to overcome that near-solid western bloc. Since easterners did not understand what living in an arid land meant, they could best be attracted by an emotional appeal. But most westerners did know, and this meant a rational appeal to those congressmen knowledgeable about water projects. It also meant presenting an alternative, rather than just saying no dams and losing credibility.

However, conservationists did have an ally within the bloc. That was southern California, which had an interest in seeing that no dams were built upstream. Southern California was worried about lower flows, which would mean less power being generated by Hoover Dam. The chief engineer for the Los Angeles Department of Water and Power even went so far as to testify before a Senate subcommittee that the interest subsidy alone for the Colorado River Storage Project, of which the dams in Dinosaur were a part, would amount to $4.4 million for each $1 million of non-interest-bearing loans advanced by the federal government. In later years, such breaking of the ranks would be unthinkable.

Dinosaur National Monument, straddling the two states, was a relatively unknown portion of the national park system before the dam controversy arose. Established by President Woodrow Wilson in 1915 to protect some fossilized dinosaur bones, the original 80 acres was expanded into a 210,000-acre national monument by President Roosevelt in 1938. At the height of the controversy, 900 persons floated through the monument on either the Green or the Yampa river and 70,000 visited it as more conventional car-bound tourists. But to the conservationists it symbolized the threat to all national parks, as witness this rhetorical question in a pamphlet put out by the dams' opponents: "Will you DAM the Scenic Wild Canyons of Our National Park System?" The pamphlets, newspaper stories, editorials, advertisements, magazine articles, books, and a movie ground out by conservationists or their sympathizers put the monument on the map. Some thirty conservation organizations joined together under one common heading, with Brower telling congressmen "a dam would be the tragedy of our generation." (It was the same generation that had lived through World War II and the start of the Korean conflict.) With such tactics, congressmen's mail started running 89 to 1 against the dams. It was the intense focusing of this aroused interest on the political process that differentiated this conservation issue from previous ones.

On the rational level, Brower, with a piece of chalk and a blackboard, became his own expert witness. In the later Grand Canyon controversy, the conservationists would rely on academic experts, but now they were only developing this technique. Brower told the House subcommittee it "would be making a great mistake to rely upon the figures presented by the Bureau of Reclamation when they cannot add, subtract, multiply, or divide." An incredulous Congressman Aspinall asked Brower, "And you are a layman and you are making that charge against the engineers of the Bureau of Reclamation?" Brower replied, "I am a man who has gone through the ninth grade and learned his arithmetic. I do not know engineering. I have only taken Mr. Tudor's own figures which he used and calculated an error to justify invading Dinosaur National Monument." Undersecretary Tudor of the Interior Department later admitted the error.

Brower, who made it through ninth grade only to drop out in his sophomore year of college, was to derive his reputation as chief voice of the conservation movement in the 1960s from the Colorado River. There would be no dam in Dinosaur National Monument nor in the next decade would any dams be authorized by Congress for the Grand Canyon, mainly because of Brower's innovative tactics. Other conservationists would contribute to these two major achievements, but it was Brower who was visible most of the time. Less well known was the fact that the alternatives to these dams, perhaps equally pernicious to conservationists, would later appear with Brower's prior blessing. There were actually going to be no clearcut conservation victories on the Colorado. The water was too much in demand. Brower, from the California university city of Berkeley, was not of the Colorado River basin, although he made occasional forays into it, being the first in 1939 to climb Shiprock, a volcanic plug resembling a schooner on the desert of the Navajo Reservation that, for the Indians, had religious meaning. He also floated down the Colorado through Glen Canyon after the dam was authorized but before the water was impounded, and came to regard that compromise as the greatest failure of his life. Prodded by Brower, Interior Secretary Udall in 1966 set up a task force to study coal- and nuclear-fired power plants as an alternative to dams in the Grand Canyon. A coal plant was finally selected. Brower went on to serve the Sierra Club as its executive director for seventeen years; after being ousted in 1969 by its board in a philosophical and organizational dispute, he founded his own conservation organization, Friends of the Earth.

The point Brower was making was that the evaporation rate would be less for a reservoir at Glen Canyon than at Dinosaur and, hence, Glen Canyon would be a preferable site for a dam. As the conservationists' chief spokesman, Brower stated, "I know, and I will bet Reclamation knows, that if the river disappeared in its course through Dinosaur, or was somehow unavailable, a sound upper Colorado storage project could be developed elsewhere." That elsewhere, although Brower hedged somewhat, was Glen Canyon, as long as Rainbow National Monument would be protected from the reservoir's waters. The Sierra Club would, in time, come to violently oppose their prior advocacy of nuclear and coal-burning power plants.

If the conservationists were later to change their positions on the desirability of these energy sources, the Bureau of Reclamation was also somewhat expedient in its position during the authorization process. The bureau went from opposition to Glen Canyon to lukewarm acquiescence as the political winds changed. Besides citing a greater evaporation loss at Glen Canyon, the bureau was initially worried about the safety of the site. The Glen Canyon area was first identified as a possible dam site in 1916 by E. C. La Rue, the hydrologist who made the first survey of the entire river system for the U.S. Geological Survey. Not too long after, doubts arose about the safety of a dam built into the bedrock of Navajo sandstone. This dramatic formation is the predominant feature of many national parklands in canyon country; its beauty is matched only by its porosity. The dark red rock is actually solidified sand dunes and is rated moderately porous and highly absorptive. It extends above and below the 710-foot-high Glen Canyon Dam and serves as the leaky, crumbly frame for the gleaming white structure. A panel of three engineers and a geologist said of the canyon walls in 1922 that "there is a tendency for the rocks to fall off in blocks." One of the engineers later wrote: "It does not seem feasible to build any type of masonry dam of the necessary height for effective storage on the soft sandstone at Glen Canyon. . . ." The chief construction engineer for Southern California Edison Company thought otherwise and stated in 1924 that he had "no hesitation" in recommending the site for a dam. There the matter rested until the authorization hearings in the mid-1950s.

In 1954 the bureau was opposing the Glen Canyon site and pushing for Dinosaur. Commissioner of Reclamation William A. Dexheimer wrote Brower in October: "Our design specialists are quite concerned whether or not the foundation characteristics of Glen Canyon and Gray Canyon sites are capable of safely supporting high dams. . . ." The following month the Secretary of Interior also wrote Brower, stating that "The poorly cemented and relative weak condition" of the rock at the dam site has given the engineers "some concern as to the competency of the foundations to support any structure higher than 700 feet." As the conservationists kept up their pressure, the Eisenhower Administration began to change its position. Finally, in the spring of 1955, a Department of Interior geologist said Glen Canyon would make an excellent site for a dam. In February of the following year, as the final debate took place in the House of Representatives, an eastern congressman dropped a piece of shale gathered from the reservoir site in a glass of water and watched it dissolve. Stewart Udall, then a supporter of the project in Congress, countered that gesture by dropping a different specimen of shale into a glass of water and announced he would drink it upon completion of his speech.

President Dwight D. Eisenhower signed the Colorado River Storage Project Act in April, 1956. It contained an amendment that no dam or reservoir authorized by the act would be built within any national park or monument. Measures were to be taken by the bureau to "preclude impairment" of Rainbow Bridge National Monument. The act authorized construction of Glen Canyon Dam and three other major dams and paved the way for such other water works as the Central Utah Project, the San Juan–Chama and Navajo Indian Irrigation Projects. It set up an upper-basin fund into which revenues from power generation—mostly from Glen Canyon Dam—would finance construction of other projects in the basin. Most importantly, though, it provided for a large reservoir behind Glen Canyon Dam to provide storage space so the upper basin could use its entitlement to Colorado River water while releasing those flows specified in the 1922 compact.

Conservationists came to rue their compromise. Writing in the foreword to the Sierra Club book *The Place No One Knew*, Brower lamented, "Glen Canyon died in 1963 [the year the water started to be impounded behind the dam] and I was partly responsible for its needless death. So were you. Neither you nor I, nor anyone else, knew it well enough to insist that at all costs it should endure." This mistake would add greater fervor to conservationists' upcoming fight against dams in the Grand Canyon, where everything, including political battles, would be on a much larger scale.

The first construction contract was awarded later in 1956 for building a portion of the road to the remote site, 75 roadless miles from the nearest community of Kanab, Utah, and 135 miles from the railhead at Flagstaff, Arizona. A town, named Page for a former commissioner of reclamation, was built on a mesa two miles from the dam site where nothing had stood before. The town quickly boomed, as it was again to do nearly twenty years later with construction of the nearby coal-fired Navajo power plant. When work started on Glen Canyon Dam, the first large-scale permanent habitation of the canyon country had begun. With the exception of Moab, Utah, near the start, all other towns in this region are perched back along the rims; use of the heart of the canyon country by ranchers, miners, and tourists has been temporary, or on an almost hermitic basis, at best.

Glen Canyon Dam is one of those public works that by its sheer audacity and immensity captures the imagination and stands as a tribute to the force of the natural resource it seeks to either bridge or contain. Two tunnels, measuring 41 feet in diameter, were built to divert the river 3,000 feet through the Navajo sandstone. When the river rose, it seeped through the porous sandstone into the partially completed tunnels and had to be pumped out. All through construction, there were problems with falling rock slabs. The walls of the canyon had to be held secure by multiple lines of rock bolts inserted 40 to 80 feet into the sandstone and cemented into place. In both abutments adjacent to and just downstream from the dam, 514 such bolts were placed, and more were installed near Glen Canyon Bridge, 865 feet downstream from the dam and with a span of 1,028 feet over the Colorado River. The dam was built as a series of concrete blocks, the largest being 60 feet wide and 210 feet long. The first was poured in place on June 17, 1960, the last a little more than three years later and 710 feet higher. Storage began in Lake Powell on March 13, 1963, when the gates to the diversion tunnels were almost completely closed, allowing only a mere trickle of 1,000 cubic feet per second to run down the river through the spring months. The next May releases were again cut back to 1,000 cubic feet per second, ironically almost stranding a Sierra Club dory trip on the river. The finishing touches were put on the dam in January, 1967, when two anodized aluminum plaques, costing $7,000 and bearing the Great Seal of the United States and the lesser seal of the Department of Interior, were installed on the two elevator towers. Along the way 18 persons had died and 348 were seriously injured during construction. The total cost was $300 million, of which 96 percent was to be repaid by power sales. Glen Canyon, the cash register for other upper-basin water projects and savings depository for upstream users, was in place.

As the waters rose in Lake Powell during the late 1960s, it became apparent that no serious effort was going to be made to "take adequate protective measures to preclude impairment of Rainbow Bridge National Monument," as stated in the 1956 act. The reservoir's waters were going to spill into the 160-acre national monument and under the bridge's Navajo sandstone arch, despite the provision that "no dam or reservoir constructed under the authorization of this act shall be within any national park or monument." The compromise that allowed passage of the act with no dam in Dinosaur National Monument and seeming protection for Rainbow Bridge was based on these two sections. Despite urgings by Secretary Udall, both the House and Senate public works subcommittees refused to appropriate money for protective works estimated to cost $20 to $25 million. This money went into other water projects. A western congressman simply said of the refusal, "The law has been changed." Embittered conservationists, again led by Brower, filed suit in 1970. Federal District Court Judge Willis Ritter ruled in their favor, stating, "It was pretty sneaky of Congress to pass a law and then ignore it completely." The basin states appealed the ruling. A Utah congressman and senator promised to introduce legislation reversing the court's opinion. If upheld, the decision would have limited storage in Lake Powell to half its usable capacity. The whole carefully constructed edifice of revenues and water supply would come tumbling down for both lower- and upper-basin states and, according to one study, "trigger major conflicts and potential litigation involving all of the basin states" with time, energy, and money "spent in adversary type proceedings for many years in the future." In mind was the long-drawn-out Supreme Court suit between Arizona and California. The lower-court decision was not upheld by the Supreme Court, which stated that later congressional actions, or the lack of them in terms of nonappropriation of funds, took precedence over the provisions of the 1956 act.

The dam appears on the outside as a tiara—a glistening white arch set against the auburn hair of the canyon walls. Yet, as I walked through the wet passageways alive with the sound of running water within the bowels of the dam far off the regular tour route, I could not help thinking that it resembled a graceful woman unknowingly beset by the first stages of cancer in the prime of life. Most probably the disease would be arrested in time and not prove fatal, as doctors disguised as engineers assured me, but the very encounter proved to me the dam was a vulnerable, living entity and not the monolithic structure it first appears to be. As I was later to remark to a boatman, many miles downstream from the dam and within the walls of the Grand Canyon, any journey down the river should start with a tour of the dam. And there is no better guide than Grant Jones.

I first met Grant on a catwalk above the eight turbines inside the dam's powerhouse. I was being escorted by Harry Gilliland, the public-affairs officer of the bureau's Glen Canyon Division. Gilliland and I had walked to the powerhouse from the dam, where he pointed out the smooth carpet of grass planted between the two structures to demonstrate how water can make the desert bloom. Between the lush green carpet and the whiteness of the dam's concrete is a gutter that collects water pumped from the interior of the dam. "I refer to it as seepage, not leakage," said Gilliland, who then handed me over to Jones.

On first meeting him, I knew that Jones's job was different from those of the other bureau employees I had met in the dam or in the administration offices on the canyon rim. He is a small man, his size suggesting a benign underground creature. On this day, with the temperature hovering near 100 degrees outside, he was carrying a quilted jacket slung over his shoulder and a flashlight. It is Jones's job, and has been since the dam was completed in 1964, to prowl the interior galleries of that seeming monolith and take readings from various meters and instruments that will tell if the dam is safe or not. The readings are sent to the Bureau of Reclamation office in Denver. Jones has never learned what is done with them, but he faithfully continues to collect them. There was some nervousness among bureau employees working within Glen Canyon after the

Bureau of Reclamation employe Grant Jones, inside Glen Canyon Dam.

collapse of the Teton Dam in Idaho, but that has mostly disappeared with time. They did not find very amusing the suggestion of writer Edward Abbey that the dam be blown up.

When Glen Canyon Dam was built, 1,142 strain meters, 264 joint meters, 74 resistance thermometers, and 60 stress meters were imbedded within the concrete. The meters and thermometers are connected by electrical cables to 74 terminal boards and 116 outlet boxes located in about 60 reading stations scattered throughout the various corridors of the dam. The instruments measure stress and temperature changes within the 4.9 million cubic yards of concrete in the dam. To measure the dam's movement—and it does move—five plumblines were installed within the dam, each consisting of a 637-foot-long stainless-steel wire holding a 26-pound weight immersed in a barrel of oil so as to damp its movement. Grant uses an ohmmeter to obtain readings from the first set of instruments and a microscope mounted on a micrometer slide carriage for the plumblines. He also measures the amount of leakage, or seepage, by the volume of water passing over about 50 weirs placed in gutters running along the sides of galleries, and has seen it increase in recent years to a high of 2,500 gallons a minute.

Jones works alone most of the time. He used to have a steady helper but now only gets occasional company on his rounds. In prior years Jones used to leave, for safety reasons, a note saying where he was going within the dam, but he doesn't bother with that anymore. Nobody else seems to want the job, and Jones's superiors find his willingness to spend daylight hours in the dank galleries invaluable.

The quilted jacket—which tourists descending into the dam from the desert heat find disconcerting—and the flashlight are musts. It is chilly, with only a slight variation between seasons, in those concrete corridors beneath the lake's waters, and should the lights go off in that ultimate source of electricity it would be a difficult task to find the way out through the rabbit warren of galleries.

We are in the foundation gallery at the 3,480-foot level. In other words, 3,480 feet above sea level. But it feels as if we are beneath the ocean. When the dam filled, there were many leaks, but over the years the dissolved minerals in the saline waters of the Colorado River have plugged the cracks. As Lake Powell rose, the water pressure on the Navajo sandstone abutments increased and Jones's readings have shown a proportionate increase in seepage through the porous rock. When I visited the dam, the reservoir level was down from its record height and the seepage rate had accordingly declined to 1,900 gallons a minute. It was to rise again in the next few wet years.

The sandstone-and-shale banks of the 186-mile-long Lake Powell absorb about 1 million acre-feet of water a year. This is called bank storage.

When combined with evaporation losses of about 450,000 acre-feet a year and displacement by siltation of approximately 70,000 acre-feet, the efficiency of such large collecting ponds can be seen to lessen. The maximum capacity of the lake is 27 million acre-feet, an amount about to be stored for the first time and not all of it usable. The average annual evaporation losses for the major reservoirs in the entire river system was 2,244,000 acre-feet for the period 1971–1975. This does not count evaporation losses from small impoundments, such as the thousands of stock ponds that dot the basin.

Besides the seepage, the massive dam lives in other ways. In the summer, with the hot sun shining on the downstream side of the dam, the concrete expands and the structure bows out an additional three-quarters of an inch into the blue waters of Lake Powell. Because of pressure on the bottom of the dam, the top has moved as much as one and one-half inches upstream. Most noticeably in the power house, there is the constant throb of the river compressed into the eight penstocks leading to the turbines. To those who work inside the dam and are concerned about the generation of electricity, the river is measured by that throb and what it does to the equipment.

We have climbed down a circular steel stairwell to the bottom foundation tunnel at the 3,247.5-foot level in the east side of the dam. This is the rain tunnel. Where the concrete ends and the sandstone of the abutment begins, it is like being in a misty rain forest during a torrential downpour. Water is spilling everywhere, and I scamper through it to emerge on the other side in the misty, sodden tunnel. In the poor light the Navajo sandstone looks substantial. But it is not.

Jones tells me, "We had a bunch fall down a while ago right here, and it was a mess." Jones lifts aside a board and out spurts some water.

The drains are overflowing. The confined space magnifies the sound of running, falling water. I consider where we are, beneath the accumulated waters of the Colorado River and encased by the fragile sandstone, and have a moment of trepidation. I cannot help thinking, perhaps unfairly, "But these are the folks who brought you the Teton Dam disaster." I ask Jones if he ever worries about the dam giving way while he makes his rounds. He replies, "No. It'll be here for hundreds of years." And that is also what the engineers assure me above.

The heart of the dam, its nerve center, is the hospital-green, semicircular control room. It is kept immaculately clean, as is everything else in the L-shaped powerhouse. In fact, cleanliness is the dominant characteristic not only of the structure but of the employees. In contrast, the tourists, numbering 180,000 a year, look grimy and disheveled as they take the thirty-minute self-guided tour of the dam and power plant.

The control room, which comes close to being a science-fiction fantasy of the command deck of a space ship, contains the unit control boards, the generator and unit-transformer relay boards, the line relays and control boards, capacitator-potential device adjusting units, the main alternating- and direct-current distributor boards, and a host of other functions, including a long desk-console for the two operators. Compared to Jones's job, the life of an operator is extremely sedentary; but it is by far the most prestigious job in the dam. With the switches and buttons of the console, the instant microwave communications, the graphs and gauges arranged around the operators, there is a feeling—although deceptive—of being in command of a potent source. Actually the commands come from the bureau's power-operations center in Montrose, Colorado, where orders for power are placed by the various utility companies in the West and the scheduling is done by three different computer systems. What the operators are allowed to do is further circumscribed by the complex laws of the river.

Most of the power generated at Glen Canyon goes to central Arizona, although some is used by the small communities in western Colorado and southern Utah. In Arizona the Salt River Project is the largest consumer and wholesaler of Glen Canyon power. It serves the area in and around Phoenix, where air conditioners and evaporative coolers can account for as much as one-third of the energy consumed during the hot summer months. On the midsummer day I visited Montrose, the power being distributed by the operations center from all the hydroelectric facilities in the upper basin, including Glen Canyon, and some of the steam plants was allocated this way: 810 megawatts to the Phoenix area, 560 to the Denver metropolitan area, 260 to New Mexico, and 200 to Utah. From Glen Canyon alone, the percentage going to central Arizona would be greater.

Hydropower is used to meet peak energy requirements, since water flowing through a dam can be turned on and off quicker than steam made from oil, natural gas, coal, or nuclear fuels. So the largest releases of water are from 8:00 to 10:00 A.M. and from 4:00 to 8:00 P.M., when people are most concerned about cooling and cooking. The afternoon peak is usually greater than the morning peak. Other types of plants meet the base loads.

Historic flows on the Colorado River through the Grand Canyon before the construction of dams varied from almost nothing to more than 200,000 cubic feet per second. With the beginning of the impoundment of water behind Glen Canyon in 1963, summer flows have varied between 5,000 and 35,000 cubic feet per second. More water is released in summer months than winter to meet the greater demands of urban energy users and irrigators—but always in accordance with the 1922 compact. These higher, more consistent flows make for better river running in the popular summer months, a condition that did not prevail before the dams. Low flows can be disastrous, since the rocks in the rapids of what is termed the most challenging whitewater run in North America become exposed and are difficult to avoid. During the spring months of 1977, in the second year of a record-breaking drought in the West, the bureau cut its flows from Glen Canyon to minimums of 1,000 cubic feet per second when they normally would have been between 12,000 and 15,000. During the Easter weekend ninety persons on eight separate float trips were stranded in Marble Canyon. Food had to be flown in by helicopter to passengers of one trip who had been stranded for four days, and a 22-foot dory was airlifted out because it could not negotiate the rapids. Finally, enough additional water was released to flush the boats and rafts down to a point where the passengers could either hike or ride out of the Grand Canyon. In a dry year the bureau was saving all the water it could for release during the summer months. The National Park Service wanted higher releases for the popular float trips and to protect the canyon's changing ecosystem.

It was these varying flows of the river, dictated by demands far removed from any realities I encountered within the Grand Canyon, that were to be the dominant factors in our 280-mile trip from Lee's Ferry to Pierce Ferry on Lake Mead.

2

The preparations for a trip down the Colorado River seem interminable. First, there is the decision of what to take. We will be on a river whose water temperature varies between 45 and 50 degrees at the start, yet passes through a broiling desert. It can rain and be chilly, or the sun will shine and the temperature climb to 120 degrees. A hat and long pants are needed so one will not become parboiled, yet no one wants to be overdressed when there is a chance to go swimming or work on a tan. The brochure makes it clear: Bring as little as possible. Space is at a premium in the small wooden dories.

Having made our choices, my twelve-year-old son, Alex, and I drive from the San Francisco Bay area to the Grand Canyon Dories' boathouse in Hurricane, Utah. There we are instructed to repack our clothes into watertight bags and park the car. It is clear that our lives will be run by others for the next eighteen days. We are driven by van to a motel near Lee's Ferry, where we spend the night. The other twenty-four passengers, many of whom have come much further, gather at the motel that evening along with the seven boatmen and two women cooks. We are given a short lesson in how to pack the Navy-surplus bags. This first group experience shows we are a diverse lot. Early to bed was a pattern that was to repeat itself on the river. Alex is excited and it takes him a while to get to sleep, as it does me. The air conditioner is faulty and there are crickets in the room.

In the morning Martin Litton arrives, piloting his own small plane from the scores of places he has just been. The peripatetic Litton shepherds the launching of his dories not only on the Colorado, but also the Green, Snake, Salmon, and Owyhee rivers. He has just come from the south rim of the Grand Canyon, where he had flown to check the accuracy of the park rangers' lecture to tourists on the river. Not very accurate, he reports. A former director of the Sierra Club and one of the first two hundred persons to run the river, Litton is afflicted by the same disease most persons contract who spend some time on the Colorado—an almost zealous possessiveness, a sense of proprietorship. I have noticed this not only about the river, but desert areas in general.

With the complicated logistics of getting seven boats launched and thirty-five people into them with all they will need for two and a half weeks ahead of us that morning, we go tearing off in the van for Lee's Ferry. Litton gives us a running commentary about how the Bureau of Reclamation has misnamed Marble Canyon, the money he is losing on the trips, the names of different rock formations, a short discourse on the Mountain Meadow Massacre, why Navajos are selling necklaces with Japanese-made glass beads at the bridge, and so on. There is a brief stop at the paved launching ramp at Lee's Ferry, where Litton chats with the park ranger who is going over his checklist. Then the group splits, nine to hike into the canyon with Litton and the remainder electing to float down the river to Badger Creek Rapids.

I choose to go with Litton, since I have ridden that stretch of river on a previous trip and want to soak up some additional folklore. He doesn't tell any of the passengers, but by getting some of them to hike in, Litton saves user days from his annual allotment that can be applied to other trips. The way down is over the lip of Kaibab limestone, the topmost layer of which will eventually tower over us as we float deeper into the Grand Canyon. The day warms up as Litton regales us with further tales. Grand Canyon rattlesnakes are chicken, he assures one passenger. True enough, we see a number of rattlers on the trip but suffer only one scorpion bite.

The river is a cool blessing, and once we reach it we swim. Lunch is eaten and then Litton departs, hiking back up the canyon to fly off somewhere else that day. We rejoin those who have come by boat and gaze down at our first rapid. It is the practice of the boatmen to scout difficult rapids. But Badger Creek is only in the moderate class and I get the feeling the pause is mainly for the benefit of the passengers to become used to the idea of dropping off sharp inclines in small boats into roiling whitewater. I assure Alex it is a piece of cake, that we will just slide down that smooth V-shaped tongue and be carried through the standing waves in a few moments. I am not sure he buys that.

Our boat, rowed by Greg Williams, is next to last to push off from the bank. From the start, I am a bit apprehensive. I learn later Williams has been down the river before, but never rowing a wooden dory which is more touchy than the rubber rafts most people use. As we are pulled by the current toward the tongue, Williams stands a number of times to make sure the boat is positioned correctly. Too many times. He is unsure. We enter the rapid too far to the right, the boat dips into a large hole, teeters on its left side, and Williams is thrown out. Alex and I are sitting in the stern, facing backward. The violent angle makes me look back. I

Alex getting splashed in a rapid.

spot Williams in the water by the left gunwale. With my left hand I hold Alex down. His principal interest is to get out. With my right I lean down to give Williams a lift back into the boat, which has righted itself, minus one oar. It is over in an instant, shorter than I had promised Alex. We recover the oar, bail out, and make for the shore where we will camp that night.

"A piece of cake, huh?" is Alex's later comment.

There is a tendency to think of the Colorado River through the Grand Canyon as one continuous set of booming rapids. It is not. It is a series of long, slightly angled pools and short, steep rapids. The river descends about 2,200 feet within the canyon, and rapids account for only 10 percent of this drop. In the first 150 miles of the river there are 93 rapids averaging a distance of 1.6 miles between each other. Altogether there are about 160 rapids on the total river run, with many being barely noticeable but still technically qualifying for the designation. Badger Creek Rapid, through which we had just run, drops 14 feet in a horizontal distance of 860 feet, a not uncommon rate. Speeds of the river's current through the flat sections are about 4 mph, with the water at a few steep drops approaching 30 mph. A trip down the river is essentially a long, lazy float interspersed with very brief, sometimes violent descents. Like books about sailing cruises that tend to emphasize storms at the expense of the more prevalent calms, accounts of the river journey tend to overly dramatize the whitewater.

The justifiably famed rapids of this stretch of the Colorado River are formed in one of four ways. Most of them are the result of storms dumping large amounts of water in short periods of time higher up in the canyon-country tributaries to the river. The resulting flash floods deposit gravel, small rocks, and large boulders into the bed of the master stream. An underwater barrier is flung across the river and its flow is constricted by the debris collected at the mouth of the tributary. A pool is formed above the new rapid caused by the resulting rise in the riverbed, and turbulent water rushes through the constriction.

Crystal is the classic example of this type of rapid. Fourteen inches of rain fell over the drainage of Crystal Creek on December 4 and 5, 1966, and the resulting floods carried away Indian ruins that had stood since the twelfth century. Thousands of tons of gravel and rock, with large boulders mixed in, hurtled down at speeds of up to 50 mph. The mass of material pushed the Colorado River toward the opposite shore, and overnight the second most dreaded rapid on the river had been formed. A U.S. Geological Survey crew that had made the first detailed survey of the rapids and river depth the year before had noted nothing significant at Crystal Creek. This was to be one of the two rapids we were required to walk around while the boatmen ran them alone.

The granddaddy of them all, Lava Falls, is just that, a falls, the second form of rapids. It is formed by an outcrop of hard basaltic rock, the result of a number of lava flows that had plugged the river only to be worn down over a period of time by the persistent force of water. Other rapids are formed by huge slabs of rock that peel off from the canyon walls periodically, or by gravel bars extending partway into the channel.

It is not the waves formed by these obstructions that are feared so much as the holes found on the downstream side. These holes are the black spaces of the river. The average depth of the river is about 40 feet, but one measurement of 110 feet has been obtained. The boatmen's terminology for such phenomena is descriptive of the havoc they can wreak. They are known variably as souse holes, suck holes, stoppers, reversals, or keepers. With water pouring over a boulder or ledge, a void is formed below, while the towering wave just beyond it keeps trying to rush back into the hole. Some boats have been literally kept in such holes. Others are held, then spat out. While the water in the curling wave vainly seeks to fill the hole, other water dives deep to travel at high speeds along the river's bottom to emerge as a large boil, or upwelling, as far as one-quarter mile downstream. These boils form powerful eddies below the rapid. Here river currents may be going in opposite directions with only a thin shear line separating them. Boats are pulled one way, then another in these unpredictable currents. I found the holes and boils the most powerful forces in the river.

Glen Canyon Dam is changing the nature of the rapids, along with riverside beaches, vegetation, and wildlife in the Grand Canyon. It is, in effect, a new river with different forms of life dependent upon it. Before the dam was built no thought was given to what it would do to the downstream river and only a cursory examination was made of the effects of the reservoir on the Lake Powell area. These decisions came before the era of voluminous environmental-impact statements, which now attempt to predict such effects, and possible alternatives, with relative precision. What was not taken into account was the fact that the great flood flows of the past would exist no more, the average amount of silt passing Phantom Ranch was going to drop from 500,000 tons a day to 80,000, and the cold, clear water released from Glen Canyon was no longer going to fluctuate from near freezing to 80 degrees.

The rapids are in the process of changing because the flood flows no longer flush boulders downriver. Some researchers think this will make the rapids generally easier to run, since the remaining boulders and rocks will now tend to rearrange themselves within the rapid, resulting in a more even cross section and filling in some holes. There are exceptions to this thesis, and one is the possibility that rocks may pile up at the toe of a rapid, thus creating a shallow area. With less rock being swept away, the common theory among boatmen is that the gradual accumulation of such obstacles will make the river more difficult to run. Another set of researchers has also come to this conclusion, stating the "rapids may become impassable to river traffic" in time.

To date, the change in beaches used extensively for camping in the summer months has been more dramatic than any alteration in the regimen of the rapids. Boatmen who have been on the river for a number of years can point to beaches where the volume of sand has decreased markedly. With most of the silt now being trapped behind Glen Canyon, this essential beach-building material no longer replenishes downstream areas. What has happened already in the fifteen-mile stretch between Glen Canyon Dam and Lee's Ferry is, in the words of one technical report, that beaches have "been degraded and stabilized with an armor bed of self-sorted riprap ranging from small cobbles to large boulders." Behind their new armored coating, some beaches have disappeared while a few are still being eroded. The report continues, "The lessons to be learned here should be applicable shortly to the river below Lee's Ferry." A comparison between aerial photographs taken in 1965 and 1973 shows the beach degradation process to be slow, perhaps taking a few decades to work its way downstream. But with less silt, it is sure.

The most noticeable change has been in the riverside vegetation. Prior to the completion of Glen Canyon there were three vegetation zones parallel to the river. Adjacent to the water was the ephemeral zone subjected to periodic flooding. This zone consisted mostly of such plants as seep willows, desert broom, and the true willows that would desperately try to put down roots before the next flood. Above the high-water line of these floods was the second zone, typified by such species as Apache plume, redbud, hackberry, mesquite, and acacia. On the dry talus slopes above, in a typical desert environment grew brittle brush, various cacti, creosote bush, and Mormon tea. With the dam came the establishment of a new riparian community at the lowest level. Here the ubiquitous tamarisk took hold, along with arrowweed, coyote willow, and other herbaceous plants. The new zone replaced the old ephemeral zone in many areas, and the plant life in the ephemeral zone then changed to such dominant species as red brome, tansy mustard, and fescue. Two exotic species, Russian thistle and camelthorn, found their way into the ephemeral zone.

With the increase in plant life along the river, the diversity and density of animal life have grown accordingly. During periods of blooming the lowest zone, dominated by tamarisk, contains nearly three times the number of insects as the next higher zone. Lizards have found an abundant source of food, and one species, the whiptail, feeds on small crustaceans when the daily "tide" goes out. More mice are found in this zone, and such birds as the yellow warbler and blue grosbeak have taken up residence along the river. With the change in water temperature and turbidity, the fish life has been altered. The humpback chub, an endangered species, has all but disappeared, while an excellent trout fishery has developed in the cold, clear water just below the dam.

With all these changes, it is hard to say the river and its immediate surroundings could be considered to be in their natural state any longer. But unless you whisper this repeatedly to yourself while floating down the river, it is difficult to realize that the Colorado within the Grand Canyon has become the ultimate ditch in the efficient transport of water from Lake Powell to Lake Mead.

3

There is a special river routine, and we get our first lesson in it from Bego, the wiry, bearded trip leader who once spent a hundred days in the canyon during the winter. Bego, whose single Indian name has replaced a more prosaic Anglo one, is a veteran of the technical rock-climbing scene at Yosemite National Park who drove out to the Grand Canyon one day and became ensnared by its geological charms. He is now one of those wandering young people—like ski patrolmen and climbing guides—who manage to eke out a living off the natural assets of the West. Bego is also quite possessive about his domain. He doesn't realize it but there is a strong tie between river boatmen and the Bureau of Reclamation employees back at the dam. They are the only people to make a living directly off the river.

A conch shell blown by Bego with a great deal of lung power summons us to meals and the first group meeting on the river. He tells us to put on life jackets when the boatmen do, and never to sit on them or the air pocket inside will burst. We are shown the proper position to assume in the water when we are thrown out of the boat. High-siding, whereby one leans into the waves to help keep the boat on an even keel, is explained —a bit too late for Alex and me to be of any help to Greg. Urinate in the wet sand, and the yellow rubber glove on the stick will show if the portable toilet is occupied or not. "Don't forget to dump the snow"—(lime) —"on the mountain," Bego reminds us. All human wastes and garbage are carried out on the one rubber raft accompanying the six dories. The

The Colorado River within the Grand Canyon, the ultimate ditch.

raft is nicknamed T.O.B., for "turds on board." Life is rather basic in the canyon, and dealing with those basics begins to bring the diverse group together.

Bego has a nice habit. Each evening he will read us those sections of Powell's book we will be experiencing the next day. Tonight:

> Riding down a short distance, a beautiful view is presented. The river turns sharply to the east and seems inclosed by a wall set with a million brilliant gems. What can it mean? Every eye is engaged, every one wonders. On coming nearer we find fountains bursting from the rock overhead, and the spray in the sunshine forms the gems which bedeck the wall. The rocks below the fountain are covered with mosses and ferns and many beautiful flowering plants. We name it Vasey's Paradise, in honor of the botanist who traveled with us last year.

Around the corner from Vasey's Paradise is Redwall Cavern, which Powell estimated would seat 50,000 persons. Next day we pull up just as another group in large rubber rafts, referred to disparagingly as "baloney boats" by us proud dory riders, departs from this major river attraction. The river is rather crowded at times.

There is a debate over which factor has had the greatest impact upon the river, the dam or people. Senator Frank Moss of Utah, heavily influenced by constituents who own raft companies, told his fellow lawmakers, "It is an undisputed fact that the greatest impact upon the canyon floor is the constant fluctuation in the Colorado River caused by the varying amount of water permitted to pass through Glen Canyon Dam." Moss was arguing against a river-management plan that owners of motor-driven raft trips thought would ruin them financially. His point was that the dam, not people on such rafts, had altered the river.

Yet what we were seeing at Redwall Cavern was human crowding and its attendant effects. The mere fact that we could clamber ashore and not have to fight our way through a dense stand of tamarisk was because the trampling of hordes of people before us at this popular stop had kept the growth down. We could not camp there because it was one of the prohibited spots on the river, having become too popular and overused. But the debate is meaningless. Nothing now can be done about the dam, but something can be done about the people. The National Park Service struggled mightily with that problem in the 1970s. It was heavily buffeted in the process by the intense competition between different types of users, all seeking a place on the longest, wildest stretch of whitewater running through a national park.

John Wesley Powell and his party were the first to be recorded as having floated through the Grand Canyon on the Colorado River, although I am convinced some Indians and a fur trapper or two must have made it through before but neglected to write or tell anybody about it. From Powell's first trip in 1869 until 1949, a period of eighty years, only 100 persons had run the river. By 1954, the first 200 had been down the river. The number rose steadily but undramatically until 1962, when 372 persons took the trip in a single year. Lake Powell was being filled during 1963 and 1964, and there was not enough water for a decent run. Then the numbers surged upward: 1,067 in 1966, 6,019 in 1969, and 16,432 in 1972. What had happened was that the releases from Glen Canyon Dam resulted in more reliable summer flows, the fight by conservationists against dams in the Grand Canyon in the mid-1960s had focused public attention on the river, and the environmental awareness blossoming at the end of that decade drew thousands of persons out of the cities and into the mountains and rivers in search of a wilderness experience. With the equally rapid growth in commercial outfitters available to take people down the Colorado, their advertising in turn promoted more customers until the river was nearly loved to death.

In 1972 the Park Service became concerned. The damage was not only visible but auditory and olfactory. Among indications of gross overuse that year was an outbreak of dysentery among river runners. In slightly more than two months, 132 out of 256 boatmen and passengers on thirteen separate raft trips were hit by the disease. And no wonder. More than twenty tons of fecal matter a year was accumulating on the beaches whose condition had become similar to a cat's uncleaned litter box. The Park Service rushed some interim sanitation guidelines into use for the 1973 season and clamped a lid on the number of people who could take the trip down the river. The number was adjusted to 14,000 a year, and the Park Service embarked on an extensive study to come up with some scientific backing for its position to ban motors from the river.

Beach space was at a premium. Somewhat less than 100 beaches receive 75 percent of the use during a summer season, with 30 to 40 persons camping almost every night at the more desirable spots. The constant trampling has denuded patches of 2,500 to 10,000 square feet. Charcoal is supposed to be burned in a pan, fires are not allowed in the sand, and there is the portable toilet–wet-sand arrangement for human wastes. Yet, according to one report, "Human debris (food particles, plastic, pop tops, etc.) is being incorporated into the sand/silt deposits at rates that exceed the purging capacities by natural means, causing beaches to look and smell like sandboxes found in heavily used public parks." The multiple trail system to popular hiking attractions added to the erosion potential of an already highly erodable area. Since the Park Service established rules on carrying out wastes and litter, conditions have improved, but some beaches still look and smell as if they had just been crossed by a herd of cows.

People, like the dam, have also changed the patterns of wildlife. Passengers on our trip, including Alex and me, were annoyed by the vicious sting of harvester ants and the large numbers of flesh and blow flies, whose numbers have gone up with the increased visitation levels. House sparrows, usually found in urban areas, have been spotted at some of the more heavily used camping sites, and virtually no overnight stop is complete without circling ravens waiting for a trip's departure in the morning before swooping down to pick up the refuse. Spotted skunks, ringtailed cats, rock squirrels, and mule deer live off these same remains.

Then there is the oars-versus-motors controversy. We happened to be by choice on one of the oar-powered trips, which constitute about 20 percent of all those departing from Lee's Ferry. Our run was the longest offered by any commercial operator. Some of the motor-powered trips can zip through the canyon in five or six days. Perhaps it was because we were late in the season, but I did not find the motors used by others too objectionable. A greater detraction from floating down the river in some semblance of peace is the large number of sightseeing aircraft. The Park Service announced it would phase out motor-driven rafts, since the agency found them incompatible with a wilderness experience. It met with political opposition. At stake was a portion of the $4.4 million annual business done by the twenty-one concessionaires operating on the river. Those with motor-driven craft felt threatened. So, using mailing lists drawn from past customers, they put out appeals to this effect: "We need your help in preserving the free public choice that has always existed— that has allowed a wide variety of river trips for different vacation schedules, vacation budgets, ages, and physical abilities." The customer was asked to send a letter to the Park Service, or Secretary of Interior, and their local congressman. August A. Busch III, chairman of the board and president of Anheuser-Busch Inc. of St. Louis, wrote Interior Secretary Cecil Andrus: "If the proposed plan is adopted executives will be unable to take five-day trips using motor-steered rafts as we have in the past." Nonetheless, the Park Service went ahead with its ban on motors, only to be taken to court.

4

A rhythm has set in on the river. At the end of the day's run I know exactly where to go to get my ration of one cool beer. Alex selects our campsite and I notice it becomes less romantically remote and nearer the kitchen and the boats as the days progress. We have our first layer or two of tan and do not have to cover up so frequently. I can even recognize some of the geologic layers the boatmen are constantly pointing out and through which we are slowly sinking. That is the predominant sensation, one of sinking into the earth. By now I have spent some time talking with most of my fellow passengers and the process of informally splitting into groups has begun, except on this trip there will be no firmly established cliques and relatively little acrimony. Altogether we are a rather bland group. I wonder if we will encounter an experience that will galvanize us in some way and take us out of our sun-drenched, wave-washed, full-belly torpor.

Lying atop our sleeping bags in the heat of the early evening, Alex asks if I had a wish what would I want. My mind is a blank. I finally say, "Nothing. I have everything." I ask him the same question. He answers, "Me too."

The next evening Bego goes over what we will be encountering the next two days. "We will have big, big water. There should be lots of turbulence. The boatmen will go over the safety procedures. Don't forget high-siding." Then his nightly admonition: "Don't forget we are in a desert. Drink lotsa water."

Bego has the slogan "Drink Lotsa Water" emblazoned on the inside of his boat's front hatch. I don't think I have ever drunk as much liquid in my life. I am constantly leaning over the side of the boat to dip my cup into the river. It reminds me of the other places along the river where I have also sipped the water, but never with such urgency as inside the furnace of the canyon.

That night Bego reads us another passage from Powell, which adds to the message of foreboding the boatman has already given us:

> We are three quarters of a mile in the depths of the earth, and the great river shrinks into insignificance as it dashes its angry waves against the walls and cliffs that rise to the world above; the waves are but puny ripples, and we but pigmies, running up and down the sands or lost among the boulders.

Powell may not have been first down the river, but that does not detract from his magnificent accomplishment. He had heard of none going before him, and the fear of the unknown is the greatest of all fears. Bego continues:

We have an unknown distance yet to run, an unknown river to explore. What falls there are, we know not; what rocks beset the channel, we know not; what walls rise over the river, we know not. Ah, well! we may conjecture many things. The men talk as cheerfully as ever; jests are bandied about freely this morning; but to me the cheer is somber and the jests are ghastly.

We are now in the black, dully polished granitic rocks of the Inner Gorge, perhaps two billion years old, or as we are told, half as old as the earth. When I tell Alex that these are probably the oldest things he will see in his lifetime, a passenger in our boat jokes, "Except for your dad."

We have sunk to our ultimate depth. The black rocks hold the heat, tossing it about in formless molten waves. The cold splashes of water as we run Hance, Sockdolager, Grapevine, and Eightythree Mile Rapids cut the heat with the immediacy of a razor slash. We stop for lunch just beyond the suspension bridge at Phantom Ranch, now being crossed by hikers and mules that have descended from the rim. It is the first and last evidence of present-day habitation we are to see in the Grand Canyon.

The history of man in the Grand Canyon and throughout the remainder of the canyon country is similar to what was experienced elsewhere in the Colorado River basin. Broadly speaking, it centered around two dominant activities—the search for transportation routes across or through the canyons and the extraction of resources to serve interests elsewhere. The Indians sought meat and a place to grow vegetables in the canyons, mostly to feed the main bodies of population on the rims. The Spanish were after riches, religious converts, and routes to link their missions. The fur trappers, who mainly worked the upper reaches of the river system but also made brief forays into the canyons, satisfied the fashion crazes of the cities. The Mormons populated the edges of the canyon country with small settlements and sought ways across and around it. The explorers, engineers, and scientists of the mid-nineteenth century satisfied the military and economic needs of the country for information on routes and mineral potential. The miners were out for a quick strike. In the present century, the dam builders satisfied compacts drawn up and ratified elsewhere for outside interests, the tourists took their film and memories home with them, and the power plants generated the electricity needed by far-off urban areas. Each in their time left their mark on the canyon country that was knit together and formed by the river.

The Anasazi were the first to use the Grand Canyon. The Indians' use of the canyon dates back 4,000 years. Within the national park there are 1,200 known ruins. Most likely the early Indians occupied the canyon country in numbers greater than any who followed and, as in Chaco Canyon, there is evidence they overused it. The Anasazi disappeared from the Grand Canyon during a climate change about the same time they left Chaco Wash. After they left, the Navajo, Hopi, Southern Paiute, Havasupai, and Hualapai appeared on the scene to use the canyon sparingly. But while the Anasazi were there, it was most likely a seasonal occupation. Most of their ruins are found where there are springs or permanent streams. During the spring and summer months beans, squash, and corn were grown in the canyon, and the small tributaries were diverted by check dams and guided onto the fields by stone walls

Rapids on the river within the inner gorge.

and ditches. Present-day trails follow many of the ancient Indian paths in the canyons. As we floated through Marble Canyon our boatman pointed out a gap in the limestone wall above us where an Indian bridge made of driftwood was still in place. Seeing that bridge was like finding the corncobs in Canyon de Chelly.

Next came the Spanish, marching in their heavy armor or religious robes across the deserts only to be confronted by those ultimate barriers, the canyon and the river. The myth that led the Spanish to the Colorado River had its roots in the Moorish invasion of Spain in the eighth century. Its basic form was that there was something fabulously rich to the west —which was not unlike the later concept of making the deserts bloom, except that the Spanish had come to believe in the more immediate and tangible metallic riches of the Seven Cities of Cibola. In 1540 Coronado did not give up easily. He sent search parties in different directions and finally one, led by García López de Cárdenas, arrived at the south rim of the Grand Canyon in September and gazed down at the river, then known as the Rio del Tison. Cárdenas and his party spent three days vainly trying to climb down to the river, whose width from above they estimated at six feet, but gave up the attempt and returned with some salt crystals but no gold. Within fifty years after Columbus discovered America, the Spanish had penetrated the basin and gazed down at its prime life-giving source. But it was going to be almost another four hundred years before much was done about it.

The priests came after the soldiers, looking for a route to the missions in California and some converts along the way. The intrepid missionaries, braving hostile Indians and waterless deserts, walked or rode in and around the Grand Canyon throughout the eighteenth century. None remained, nor did any of the representatives of His Holy Catholic Majesty linger for very long. Father Francisco Tomás Garcés, writing of his 1776 trip along the south rim of the canyon, was the first to refer consistently to the river as the "Rio Colorado." Garcés remarked, "I am astonished at the roughness of this country, and at the barrier which nature has fixed therein." As a transportation route, or a source of riches, the Grand Canyon failed the Spanish.

From 1820 to 1840 the mountain men flitted in and around the canyon country, leaving little in the way of a permanent record to document their search for beaver. The trappers were roamers by professional need and temperament, so perhaps some of them, using cottonwood dugouts or skin bullboats, floated all the way through the canyon during the low flows of summer without bothering to write about it. One such roamer, James White, inadvertently ran at least a portion of the river on a make-shift raft. But access to the river was difficult. James Ohio Pattie, a trapper who roamed much of the Southwest, skirted the canyon and noted, "We arrived where the river emerges from these horrid mountains, which so cage it up, as to deprive all human beings of the ability to descend to its banks, and make use of its waters." The river, lying at the bottom of an arid maze, was virtually useless to the trappers, as it was to be to others in the nineteenth century.

Even those most organized of desert colonizers, the Mormons, never managed to inhabit the heart of the canyon country in any sizable numbers. They located their agricultural settlements on the edges at Moab and Bluff, at the start of the canyon country and where it ended at the mouth of the Virgin River, now flooded by Lake Mead. Like the Spanish, the Mormons were also looking for a means of transportation to the sea. This was part of their dream for an expanded Mormon empire, as was the search for a suitable crossing of the river in its middle stretches to facilitate colonizing Arizona. The only place where the cliffs fell back far enough to allow relatively easy access to the river was just below Glen Canyon. Ordered by his Mormon superiors to establish a crossing at this point, John D. Lee arrived in 1871 with his seventeenth wife, Emma, to operate a ferry and give his name to what is close to being the geographical center of the river system and certainly its fulcrum in terms of how the waters were divided in 1922. This rancher and ferryboat operator was the great-grandfather of Stewart and Morris Udall, who were to play key roles in the further division of the river's waters in the 1960s. Lee did not operate the ferry for long, as he was in hiding most of the time. He was a fugitive from the Mountain Meadow Massacre, and after he was caught in 1874, became the only person tried and executed among the group of Mormons who had killed about 120 men, women, and children in a wagon train bound to California from Arkansas.

Explorers, engineers, and scientists dominated the scene from 1845 to 1880, tramping up and down the plateaus and canyons, taking measurements, and, in the case of John Wesley Powell, twice floating down the river to prove that there were no insurmountable waterfalls, that it did not disappear into the earth as feared. In the process, Powell proved the Colorado was no ideal highway for commerce. These men were organized, had the proper instruments and equipment, and unlike the mountain men and Indians, they voluminously recorded and lavishly published their findings with a view toward further congressional appropriations and posterity. Still they could find nothing commercially useful about the 630-mile stretch of river from Moab to where Hoover Dam was to be built. Like the Spanish and Mormons, the explorer-scientists were looking for transportation routes, and since most came from the humid East and Midwest, they thought the best route would be a river. The Colorado, they were to find, was a different river in a different land. Although they failed in their primary mission, these men accumulated enough information to take the "unexplored" designation off the maps and plant two seeds that were to take root in coming years.

The first was the mining boom. Like the urban explosion in the basin nearly a hundred years later, this boom was a backwash from California. After his second run through the Grand Canyon in 1871, Powell briefly employed two prospectors in a land survey. At the mouth of Kanab Creek, where the second river trip ended, they found some gold the consistency of flour. A strike was reported, and hundreds of prospectors descended upon the canyon. The resulting gold rush lasted four months.

Powell's writings were avidly read for clues as to the location of valuable minerals in this newly opened land. One college student whose imagination was fired by the report and who later hoped for as prominent a place in the river's history as Powell was Robert Brewster Stanton. His subsequent life on the river combined a search for a transportation route and gold with vivid personal encounters with the rapids that included death for others and hardship for himself. He typified the extractive interests; his life spanned both centuries plus the eras of the explorer-scientists and miners. What Stanton found feasible when he made a survey for a railroad through the Grand Canyon boggles the mind today. But with the railroad-building era about to end and the Reclamation era about to begin in 1890, there was unbridled optimism in the development potential of the West. In two different stages Stanton and his party floated down the Colorado River from Green River, Utah, to tidewater at the mouth of the Gulf of California. Stanton wanted a firsthand look at the potential route from Grand Junction, Colorado, to San Diego—the old dream of finding a corridor to the sea. It was a disastrous trip, with three persons, including the president of the railroad company, being drowned along the way. Incredibly, after traveling through those remote, convoluted canyons Stanton found the route feasible. He wrote, "The line as proposed is neither impossible nor impracticable, and as compared with some other transcontinental railroads, could be built for a reasonable cost." Of course, there would have to be twenty miles of tunnels. But on the plus side no winter weather would be encountered in the canyons and the tributaries could be dammed to provide hydropower to run the line.

The principal reason for such a railroad was to haul coal from Colorado to California, an earlier version of the current assumption that the basin can provide much of the energy needs of the West Coast. Stanton saw an abundance of riches along this corridor and dreamed on. He believed the line would compete successfully with other transcontinental routes for freight and passengers. The scenic attractions were certainly present, although there was a dearth of populated stops. The railroad could open up the canyon country to coal and other mining ventures, agriculture, grazing, and timber production in the forests on the north rim. Despite Stanton's enthusiasm, the money men would not buy the scheme. Stanton was first and foremost an engineer, which is one strong argument against letting experts in narrow fields make decisions about the use of natural resources. In later years, after his gold-mining venture in Glen Canyon failed, Stanton became quite possessive about the Grand Canyon and, returning periodically, was rhapsodic in its description. He wrote of the engineer's dilemma, "Every engineer—possibly there are exceptions —is possessed of two beings. With one he loves Nature for Nature's sake, loves it as God made it and gave it to us; with the other he follows Telford's definition of the profession of an engineer—'being the art of directing the great sources of power in Nature for the use and convenience of man.'"

After the miners, the tourists came around the turn of the century, but their arrival had been prepared for, at least in part, by the second seed planted by the explorer-scientists. This one took longer to germinate but its blossoms were more durable. One clear by-product of their work was the first appreciation of the canyon country simply for its beauty. Most notably in the reports of Ives and Powell, and of Powell's aide, Clarence E. Dutton, the use of clear, vivid writing and magnificent illustrations helped to sell their concepts to Congress and the public. Illustrated by the noted landscape artist Thomas Moran, such reports were the beginning of the tradition of fine illustrations of the canyon country that has led to the expenditure of millions of rolls of color film on the Grand Canyon and the awakening to the fact that the canyons represented more than mineral riches and transportation routes. Not until the 1970s was the federal government again to print as much material on the Grand Canyon area, and then it was the more prosaic wilderness, village, and river management plans and their accompanying environmental-impact statements. These reports were strictly utilitarian in character, written in a simple factual style with virtually no illustrations. By then the need was not to draw attention to a natural wonder no one knew about, but to control the crowds who were threatening to overrun it.

The different modes of transportation have been the greatest influence on the tourist experience at the Grand Canyon. Each brought its own type of visitor. The more hardy came first on horseback, it being a three-day ride from the railhead at Flagstaff. Later, four- and six-horse stages, to which a second trailer was hooked if there was an abundance of passengers, made the trip to the south rim in one day. An unsuccessful mining venture first built railroad tracks close to the rim; then the Atchison, Topeka and Santa Fe extended them to the rim in 1901. With the completion of the palatial El Tovar Hotel in 1905, tourists could visit the Grand Canyon in style and comfort. Built of pine logs from Oregon and styled after a Swiss chalet and Norwegian villa, the El Tovar, named for one of Coronado's aides, was not exactly an indigenous structure. But it had the necessary comforts to attract those well-off travelers who were the predominant national-park visitors in the early years.

President Theodore Roosevelt, who visited the canyon in 1903, proclaimed the area first a game preserve and then a national monument. His actions did not protect such predators as mountain lions, although it did preserve their prey, and after leaving office Roosevelt returned to hunt the large cats. His party gathered at the El Tovar, then departed for the hunt, which Roosevelt described thus:

> It was a wild sight. The maddened hounds bayed at the foot of the pine. Above them, in the lower branches, stood the horse-killing cat, the destroyer of the deer, the lord of stealthy murder, facing his doom with a heart both craven and cruel. Almost beneath him the vermillion cliffs fell sheer a thousand feet without a break. Behind him lay the Grand Canyon in its awful and desolate majesty.

A truncated national park of 645,000 acres, whose boundaries included only a small portion of the total geological feature known as the Grand Canyon, was created in 1919. Arizona's congressman, Carl Hayden, whose career in Congress would span fifty-six years and many Colorado River matters, was successful in deleting vast tracts of grazing land and existing mining claims from the park. It was a mistake that was not rectified until 1975, when the park was expanded to 1,227,850 acres to take in the whole river length of the canyon. During this time the automobile and a democratization in the awareness of what such natural scenery had to offer had changed the Grand Canyon from a rich man's playground into everyone's national park. There were 56,335 visitors to the south rim in 1920. By 1927 more visitors came by car than train. During the 1930s such facilities as auto camps and trailer courts, the forerunners of motels, appeared—small log, frame, or stucco buildings, or tents mounted over wooden floors. By the mid-1970s, more than 3 million visitors a year were driving into the park—a quarter million to hike into the canyon and a similar number to view it from sightseeing flights. An average of 14,000 persons a year floated down the river from Lee's Ferry.

As in Los Angeles, the Park Service could report "the major air pollution problem at the Grand Canyon is the automobile." By the mid-1970s, the Park Service was no longer trying to put additional tourist facilities in the park, but attempting to get some out or move them back from the rim. As a result of severe traffic congestion, shuttle buses replaced automobiles on some roads and were optional on others. There was talk of turning away car-bound tourists, as was being done with backpackers and river runners, after a quota had been reached. As the lodges on the rim with no historic value became obsolete, they would not be replaced. But the El Tovar would remain, although the campground jammed with recreational vehicles was a more relevant symbol of park use by this time.

5

The early 1970s saw an explosion in organized outdoor travel to exotic wilderness areas. People went off trekking in Nepal, riding rafts down rivers in Ethiopia and Alaska, and climbing glaciers in Tierra del Fuego. On such trips there is enough promise of controlled adventure to add excitement, but also enough guaranteed comfort so as not to tax the

Tourists at edge of the Grand Canyon.

traveler unduly. Our trip down the Colorado River was part of this phenomenon.

Taken as a whole, our group is rather ordinary. So far, in eleven days
on the river no one person or subgroup has stuck out in any noticeable
way. We are on time for meals and ready at the announced departure
times in the mornings. No one has gotten lost on hikes, nor have there
been any accidents. But in the course of almost two weeks on the river
certain groupings have emerged. There is the camera clique. They tend
to see the canyon through a viewfinder and think nothing of shooting
between twenty to thirty rolls of film, then go around bartering for additional supplies. One couple has even taken pictures of each other squatting on the toilet.

Then there is the who, what, when, where, why crowd. For them there
has to be a fact behind every rock, a name for each height, a mileage for
each day's run, and a menu for each night's dinner. There are the chow
hounds, lining up for seconds before everyone has received his first
portion. There are the silent ones who prefer being by themselves. Undoubtedly some of the singles have dreams of a romantic encounter,
especially with the muscular, tanned boatmen; but somehow the daily
closeness, heat, and scratchiness discourage this. There is the cocktail-
party crowd, sharing their precious stash of liquor before dinner in a
canyon adaptation of the suburban ritual.

Some are dragging, probably wondering why they took such a long
trip, while others, like Alex, cannot get enough out of each day and dislike
only the thought of the trip ending. Alex asks me, "How many more days,
dad?" I tell him and he looks glum.

An older married couple function well together. A younger couple's
marriage is visibly coming apart before us. A father and his seventeen-
year-old son are rediscovering each other, after the estrangement of the
adolescent years. Although undoubtedly many of these reactions are
inevitable, I like to think the river—that image used so frequently to
symbolize the passage of life—is responsible for at least part of our
behavior.

For all of us, after Phantom Ranch there is one dominant reality,
threatening in a building crescendo to be both a cataclysm and a crucible.
That is Lava Falls, perhaps "the most difficult stretch of runnable
whitewater in the West, maybe in the world," according to one river
authority, historian Roderick Nash. Whoever created the river run
through the Grand Canyon had a fine sense of dramatic timing, because
Lava is placed near enough to the end of the trip to serve as the perfect
climax. It seems our developing capabilities and energies have been
pointed toward this fourteenth day of the trip. We have surmounted all
the other rapids, learned to high-side with élan and bail with fervor, and

have been switched around enough times between the boats to learn the capabilities of all the boatmen, and for them to learn our liabilities. It will be boatman's choice through Lava Falls. Our choice is whether to walk or ride. It is their choice whom they want to take as passengers. Bego reads us the relevant passage out of Powell's account:

What a conflict of water and fire there must have been here! Just imagine a river of molten rock running down into a river of melted snow. What a seething and boiling of the waters: what clouds of steam rolled into the heavens!

The preparations for running Lava are a mixture of Sadie Hawkins Day and the initiation rites for a college fraternity or sorority. There is a sense of being chosen and then the mysticism of the ceremony. The ritual starts that morning. The boatmen inspect Oracle Rock at the National Canyon campsite and the judgment of the rock is "Not too bad," in the words of Bego, who apparently has a direct line to the source. Actually he is measuring the most critical ingredient of the day—the level of water. Lunch a few miles down the river is desultory.

A short distance above the rapid we pull off to the right-hand or north bank so the boatmen can scout the rapid and the passengers can make their choice. Alex decides to walk, which relieves me, and I elect to run the rapid, if taken. This makes Alex a bit apprehensive. The boatmen have been unusually clannish today, as if they know something we don't. They stand apart on an outcrop above the falls.

We watch some kayakers run Lava, nimbly darting around the large holes and standing waves and carrying their light, fiberglass craft back for a second and third run. Still our boatmen wait, apparently hoping for divine revelation or the right water level to attempt the run.

Suddenly there is movement. The boatmen split apart and move up to where the passengers who have elected to run the rapid wait to be chosen. Everything now becomes a blur, like a movie sequence speeded up. I am picked by Nels Niemi, who is looking mostly for weight to hold the raft down. I would have preferred going on one of the more romantic dories, those craft now on the river that most resemble Powell's rowboats. Powell portaged Lava Falls, so I tell myself there would be no historical continuity even if I went on a dory. Nels places Dave Rimer, a tall doctor from Santa Monica, and myself in the bow. Ed May, a chunky plumber, holds down the stern.

Three dories have already gone before us. We quickly pull out and Nels positions the raft for the run through the slot. It is a narrow V. The entrance is placid and smooth, almost lulling, with the gigantic turbulence to quickly follow. Others have commented on the sexual parallels of such an experience, and I agree. The first wave is truly awe-inspiring, and I wonder how we can mount that height and hold on through the resulting deluge. It hovers over us, and then we are under it, not in the broken water of previous rapids, but solid black sheets. We emerge, climb, drop, another deluge. My eyes open and close quickly. Visually the experience is stroboscopic. Those who watched from shore said we seemed submerged for long moments at a time and that I looked like I was enjoying myself, a feeling I was not aware of at the time. Dave starts to lean overboard and I grab his arm. How Nels hangs on in his exposed rowing position, I don't know. It is over in a timeless twenty or thirty seconds.

The first three dories make it through upright, the last three turn over. Those on shore say the boat rowed by Rudy Petschek twice failed to surmount the bottom hole, was then sucked back in and spat out the third time upside down. The water had risen a critical nine inches since our arrival at Lava Falls and by the time the last three boats ran the slot it was at a stage where they could not avoid crashing through the bottom hole, a combination of an extremely deep hole and a gigantic following wave rated the worst such combination on the river. One woman passenger is briefly trapped under an overturned dory. Two other women help right a heavy dory as it is being swept downstream. There are a few scratches and bruises, and four lost oars that are recovered downriver next day. Some wine, kept for the occasion, is brought out and the cooks outdo themselves in putting together a superb dinner that night. There is a celebration, an intense sharing of the one event that briefly unites us. We feel exceptionally close to each other. Our experiences that day are examined and reexamined in every detail. When we repeat ourselves there is no sense of repetition as a new detail is found to make a fresh tale. Some say it was the greatest experience of their lives.

We dealt that day with three variables: the level of the water, the ability of the boatmen, and the design of the craft. The water level, as it had been on the whole trip, was the most crucial element. Kayaks are more nimble and baloney boats more stable, flexible, and shallower of draft than the gaily painted dories—which are romantic, traditional, cumbersome, and rigid. Kayaks and the inflatable rafts, whether motor-driven or not, have a greater choice in how to run Lava. Generally speaking, for the dories the higher the water the further left the run. Because the water came up quicker than anyone expected, what started out as a good slot day, an entrance position to the right of center, ended up rather poorly for the last three boats. An ideal stage for running the slot is between 14,000 and 16,000 cubic feet per second. At 18,000 cfs a tremendous punching wave forms in the slot. It will throw a boat askew and not allow the boatman those few precious strokes needed to miss the center of the bottom hole. At 22,000 cfs, routes further to the left become more feasible. The boatmen judged they had between 17,000 and 19,000 cfs when they ran Lava that particular afternoon. This was about the worst possible stage for the slot run, but they saw no alternative.

The Bureau of Reclamation's estimates of the flows at Lava when we were there differed from the guesses of the veteran boatmen, a not unusual occurrence, since each group tends to view the river from a different perspective and with its own particular brands of proprietary fierceness. The bureau said we rode Lava Falls on water released 44 hours previously and 194 miles upstream at Glen Canyon Dam. At 6:00 A.M. of Lava Day this amounted to 15,000 cfs, 25,000 at noon, and 27,000–28,000 when we ran the rapids between 2:00 and 3:00 P.M. "The increasing flows during your time at Lava Falls correlates to increased power demands during the evening of September 4th, as lights were being turned on and home activities were increasing," said the bureau in answer to my inquiry. On this day the Phoenix area was receiving twice as much power from Glen Canyon as was being shipped to Colorado and Utah. The temperatures were hot, a maximum of 101 degrees in Phoenix.

In one sense, the history of Arizona in the present century could be viewed as a continuing effort to keep cool. As usual, the Indians began it, wrapping their blankets around porous clay jugs, known by the Spanish name of *ollas*. The moisture seeping through the clay would keep the blanket wet and the breeze would cool it. It was the same principle Alex and I discovered when we moved our sleeping bags next to the river. The Mexicans picked up this practice and soldiers in the latter half of the last century adapted it to their canvas-covered canteens. The early pioneers suspended wet burlap over a box to keep milk and butter fresh. Wet burlap was also spread over windows, and the breeze passing through it would cool a room. Sheets were dunked in water or kept on ice, then spread over a bed so a night could at least be started off under cool covers. During the 1920s making an evaporative cooler became a handyman's craze in Arizona. An electric fan drew the air through a pad wetted by a garden hose. These devices went by the names of desert cooler, window cooler, wet-air cooler, excelsior cooler, and swamp cooler. Around this time coolers began appearing in commercial establishments. Before World War I the Tucson Opera House was cooled by fans blowing across ice stacked in the basement. Any means possible was found to keep cool in the nation's hottest region.

By the mid-1930s the evaporative-cooling industry had gotten a start in the Phoenix area. In 1935 there were 1,500 such coolers in Phoenix. The following year there were 5,000 and enthusiasm for the new invention grew rapidly. *Desert Magazine* proclaimed in 1938, "Many believe that history will record air conditioning as the greatest achievement of the age and rightly so." The *Arizona Republic* declared in 1939, "The air conditioning apparatus has enabled Phoenix to meet and conquer the summer heat—long the bane of Southwestern existence." That year there were between 15,000 and 17,000 coolers in the metropolitan area, and Phoenix referred to itself as "the air conditioning capital of the world."

World War II gave the infant industry a big boost, as there was a great demand to keep servicemen cool in the deserts and tropics. After the war,

Large, air-conditioned shopping center mall in Phoenix.

air conditioning units, invented by Willis Carrier on the East Coast, became more popular because they could control both the heat and humidity. The older swamp coolers only added to the humidity, especially during the rains of the summer months. But evaporative coolers use less energy and during the 1970s they again became popular, with many Arizona homes opting for both types of units. In the Phoenix area 500,-000 households owned one or the other type of basic cooling unit and 70,000 households had a supplemental unit, according to a 1978 market survey by Phoenix Newspapers Inc.

There were divided opinions about the benefits of cool interiors, actually cold in some cases. Doctors warned about possible heart attacks resulting from the quick transition of hot to cold. But a utility company said that in times of fractured homes, a cooler would bring the family back together again. It cited a study that supposedly showed that a family that was cool would spend twice as much time together in the evening. There was no doubt, for whatever reason, that cooling had come to stay in Arizona and was the major reason for its phenomenal growth rate after World War II. In the ten years ending in 1976, Arizona was second only to Alaska in percent of population growth with a gain of 40.6 percent. An Arizona historian, Bert M. Fireman, wrote: "The age of refrigeration, beginning with the evaporative cooler, triggered both the urban era and progress. It brought a dramatically changed lifestyle to Arizona, one which included rapid urbanization, unequalled immigration, and a richer and more varied way of life. This was progress for 20th Century Arizona, and the age of refrigeration made it all possible."

6

For me, the sense of a running, vital river ends at Separation Rapid. From here to where the Colorado River ends just short of the Gulf of California the river is either a wide reservoir, a narrow lined channel, a trickle after being diverted, or a sluggish flow where the used water is returned.

Separation. The name sounds right, and there is even a bronze plaque commemorating the event. This is where three of Powell's men left the first trip, tired of the hardships and fearing what was to come, only to be killed above by Indians. Much feared in Powell's time, Separation Rapid is now a mere shadow of its former self; the still waters of Lake Mead have backed up this far into the Grand Canyon.

After Lava Falls, the remaining four days on the river, two of them on Lake Mead, are an anticlimax. Our thoughts turn to what kind of junk food we will voraciously consume when we get off the river. There are all kinds of liquid fantasies, having to do with hot showers and cold beer. Some begin to exchange addresses, and there is talk about the logistics of returning to the real world, of how many days to be spent in Las Vegas or at Disneyland before returning home. A sadness prevails as we run the last rapid, an experience that would have been fraught with trepidation something over two weeks ago. We are approaching the end of the trip.

Not too far above Separation Rapid we had floated past the remains of Bridge City, where Bureau of Reclamation crews had lived while drilling exploratory holes for a dam site. The proposed facility was first named Bridge and then Hualapai Dam, in honor of the local Indian tribe whose reservation lands would have been used. The gray drilling tailings spilled down from the holes in fan shapes, much as they did at the other dam site in the Grand Canyon we had passed on the third day of our trip. The name of that site was Marble Canyon.

For fifty years, dams in the Grand Canyon were a gleam in Arizona's eye. To this day there is still some hope for their construction. The problem Arizona faces is that the people are in the central portion of the state while the water flows along the northern and western boundaries. First Gila River water, then underground supplies were depleted. The only other source left that a burgeoning population could tap was the Colorado River and regardless where it was taken from, the water had to be dammed, pumped, then transported by aqueduct to the Phoenix-Tucson areas. An early version of the Central Arizona Project was the Arizona High Line Canal, a proposal that surfaced around 1920. One feature of this version was a dam at Glen Canyon to trap silt and regulate the river's flow and another dam at Bridge Canyon to produce electricity. There were other proposals. The copper-mining interests filed an application in 1922 to erect a dam at Diamond Creek, also in the lower reaches of the Grand Canyon.

That early river hydrologist, E. C. La Rue, saw the hydropower potential for the Colorado River within the Grand Canyon and published his findings in a Geological Survey paper. The upper reaches of the canyon within Marble Gorge could handle four dams, and La Rue said of the lower canyon: "The physical conditions are similar to those in the upper section of the Grand Canyon. To create a head for the development of power it would be necessary to construct high dams. Private parties are considering plans for the development of power on this section of the river by means of six 100-foot dams. . . ." La Rue had not seen this stretch of the river when he wrote his first report, so he made a trip in 1922 and found twenty-three dam sites, some admittedly poor, between the Little

Colorado River and Parker, Arizona. Meanwhile, Arizona's interests were being well protected in Congress by Congressman Carl Hayden, who inserted into the 1919 bill creating Grand Canyon National Park a provision reserving space for future dams "when consistent with the primary purpose of said park." The ambiguities of this provision were to cause problems of interpretation—as far as the Bureau of Reclamation was concerned, out of sight from the rim was out of mind.

What all of the proposals for dams in the Grand Canyon had in common was the fact that they were needed solely for the production of power and its revenues. As the years went on, the money became politically more important than the product, although the arguments would be couched in terms of needing more electricity for southern California and of pumping the water to Phoenix—which were certainly genuine, but secondary considerations. The money was to pay for the massive public works needed to augment the Colorado River's flow with water from the Columbia River in the Northwest. This was the single, unifying factor that held the seven basin states together during the tempestuous debates over the Central Arizona Project in the 1960s. And this augmentation was needed because the Colorado's waters had been overcommitted and its supply overestimated in 1922. Thus, the dams in the Grand Canyon came to be referred to unabashedly by bureaucrats and western lawmakers as "cash registers." Water interests hoped the dams would pay for the ultimate bail-out of the West.

The conservationists were led by the Sierra Club, and their position against the dams was equally a product of history. It was John Muir, a founder and first president of the Sierra Club, who had led the unsuccessful fight against damming Hetch Hetchy Valley in Yosemite National Park. Muir was deeply disappointed, and he died shortly after the dam was approved. That added pathos, and perhaps a certain amount of vengefulness, to the issue. Frederick Law Olmsted, Jr., was a consultant to a National Park Service study published in 1950 that equivocated only on what height the dams should be in the Grand Canyon. A high dam in Marble Canyon, the study noted, "might not be objectionable from the scenic and recreational viewpoint," and it recommended a low dam at Bridge Canyon. The Park Service never did take a strong stand against the dams or power plants, being overshadowed in the Department of Interior's hierarchy by its sister agency, the Bureau of Reclamation. When the controversy was beginning to emerge in 1963, Senator Hayden complained to Interior Secretary Stewart L. Udall that "We have got to get the bird watchers in line and, by the way, the very nature lover we want to get in line is your National Park Service director." To which Udall cryptically replied, "We are working on that."

Olmsted, a noted landscape architect, also influenced the directors of the Sierra Club in 1947 when they took a position favoring a high dam at Bridge Canyon, providing certain conditions were met. The club found this to be an embarrassment when it came to oppose the dam vigorously two decades later. David Brower, executive director of the club, testified before a hostile congressional subcommittee that pointed out this discrepancy in 1967, ". . . we do not believe on our side that because we were once wrong we have to stay wrong. We dug further for the facts, found them, reversed ourselves, and have been reassured of our wisdom, at least on that subject, even since that reversal." Brower said his original vote favoring the dam had been influenced by Olmsted's belief that ". . . the reservoir would be far down in a deep canyon, would enhance the view, flood nothing of consequence, and would make the canyon more accessible to tourists." Brower then told congressmen of the deep regret of conservationists over backing the Glen Canyon Dam site. From Hetch Hetchy, to Dinosaur, to Glen Canyon, and to the Grand Canyon, the fervor of the conservationists had now been aroused. It seemed at times they were acting more out of a sense of past disappointments than of present realities.

The full arsenal of modern-day conservationist techniques for influencing public opinion and Congress was deployed, perfected, and brought to bear on the issue of dams within the Grand Canyon. The arsenal included coffee-table books, pamphlets, bulletins, movies, letter-writing campaigns, expert witnesses, and a great deal of attention paid to the media, especially the more sympathetic eastern media. Manning the barricades were volunteers who fervently believed in the cause. Such volunteers, when properly aroused, were more than a match for the paid help of large public and private institutions.

Then with extraordinary ineptness, the federal government handed the conservationists what they needed for a victory: a *cause célèbre*. The Internal Revenue Service questioned the Sierra Club's tax-exempt status the day after the conservationist organization placed full-page advertisements opposing the dams in newspapers. A huge deluge of mail descended on congressmen. It soon became apparent that if the bill was going to pass, dams in the Grand Canyon had to be taken out. An analysis done for the upper-basin states showed the hopelessness of the situation: "The distorted and misleading national campaign to 'save the Grand Canyon' adversely influenced members of. Congress whose support would otherwise automatically have been available. The massive campaign of a few organizations like the Sierra Club, aided and abetted by such influential publications as *Life,* the *New York Times* and *Reader's Digest,* made the Colorado River Basin Project Bill one of the most controversial domestic national issues of recent years. Many eastern members of Congress reported that they had received more mail on this subject than on any pending matter facing them in the 89th Congress."

Although conservationists would on occasion hedge their statements, the overall impression they gave was one of preference for nuclear and coal-fired power plants, in lieu of dams in the Grand Canyon. Said Brower in 1967, "The real source of energy, if we look ahead, is going to be the atom." Charles H. Callison, executive director of the National Audubon Society, spoke glowingly of the future of western coal supplies: "Electric power can be produced more economically in the same region by utilizing fossil fuels now owned by the people through their federal government on the public lands." In time these organizations would come to oppose nuclear and coal-fired power plants with the same vehemence with which they had opposed the dams, leaving something to be desired in terms of credibility.

An alternative to the dams had to be found, since the seven basin states considered the bill a dire necessity. It contained, at the very least, a little of something for each state, with Arizona and Colorado reaping the most benefits. For all, there was the promise of augmentation, so such an alternative had to contain a provision for future augmentation. As a Department of Interior memo describing a meeting of representatives of the seven basin states pointed out, "The one unifying factor among the basin states is the promise of substantial assistance in meeting costs of augmenting the Colorado River water supply."

The federal government was slow to change directions. Revenues from the sale of power generated from dams were the traditional method for the Bureau of Reclamation to finance its projects. A fossil-fuel-burning plant built by southwestern utility companies in which the bureau would have a share of the output was a new concept. The utility companies, two in the Phoenix area and one in Los Angeles, supported the idea. They wrote Interior Secretary Udall in March, 1967, "We think that such a plan is feasible and we will cooperate in attempting to work out a satisfactory solution." A coal-burning power plant on the Navajo reservation using Navajo and Hopi coal would fit into Udall's Indian policy.

Udall would gain a reputation as an activist secretary and a conservationist, greatly expanding the national park system while in office for eight years, a term exceeded only by Harold Ickes, who served as Interior Secretary under Roosevelt and Truman. Udall's term came at a crucial time, those years just prior to and at the start of the great spurt in western growth, and the policies he evolved were to be among the most decisive factors in shaping the contemporary West. But Udall was first and foremost a man of the arid West, a Mormon and Arizonan whose roots went back more than one hundred years into western history. It was his great-grandfather, John D. Lee, who reported back to Brigham Young on the irrigation practices of the Mexicans he observed in New Mexico a year before the Mormon leader brought his people to the site of Salt Lake City. It was Udall in late 1962 who had requested Wayne Aspinall, as chairman of the relevant congressional committee, to ask him, as Secretary of the Interior, for a comprehensive report on the water needs of the Southwest, which would emerge a year later with the recommendation for two dams in the Grand Canyon. At first, Udall would strongly favor dams in the Grand Canyon, eventually running the gamut in policy options from favoring two dams, to one dam, to no dams and the Navajo plant and Black Mesa coal mine. He was to spend more time on seeing that the Central Arizona Project got authorized than on any other single issue while in office.

As a three-term congressman from Arizona, serving under the more senior Aspinall on the House Interior and Insular Affairs Committee, Udall had witnessed the conservationists' effectiveness on the dams-in-Dinosaur issue. On taking over the Interior Department, he warned his brother, Morris, who had succeeded to his congressional seat, "At the present time it is my judgment that our chance for success in getting the new CAP through the Congress will be good if we can avoid an Echo Park type pitched battle with the conservationists." At about the same time, before the dams in Grand Canyon became an issue, Udall wrote in his book, *The Quiet Crisis*, "We cannot afford an America where expedience tramples upon esthetics and development decisions are made with an eye only on the present." Shortly before leaving office, Udall took a trip down the Colorado River. He then wrote he had approached the dams issue "with a less-than-open mind." Udall concluded, "The burden of proof, I believe, rests on the dam builders. If they cannot make out a compelling case, the park should be enlarged and given permanent protection." The park was later to be enlarged, but space was reserved for a dam site.

For Udall, a conservationist at heart who was bitterly attacked by conservationists and an Arizonan by birth who was attacked with equal vehemence in his home state, it was a difficult time. Defending his brother against criticism from his home state, Morris Udall said, "There is no one in the country more anxious to see our Colorado River water plan a reality than he." The secretary was particularly close to his brother, Morris, who was leading the Arizona forces that were pushing hard for their project. Stewart Udall frequently gave Morris information that was privy to no one else outside the Administration; and Stewart, in turn, was kept closely apprised by his brother of what was happening on the congressional side—all to the greater benefit of Arizona's cause. At one time, objecting to the many demands being placed on him by his home state, Stewart Udall told Senator Hayden, "I must serve the President as Secretary of the Interior of the U.S. and not the Secretary of Interior for Arizona."

As the dams became more of a political liability for the whole legislative package, the pendulum gradually swung toward coal, specifically coal to be strip-mined on Black Mesa and shipped by train to the Navajo plant adjacent to Lake Powell. It was to be development of Indian lands instead of dams in the Grand Canyon. That key issue, augmentation of the water supply, was not abandoned. But sights were lowered in view of the political opposition of the Pacific Northwest and more thought was given to weather modification. In a little-noticed but key provision of the bill, augmentation was made a "national obligation," meaning that the nation's taxpayers would pay for any future new supplies of water for the Colorado River basin. In the 1970s the battleground was going to be coal mining and power plants in the Southwest. In May, 1971, conservationists ran full-page newspaper advertisements, reminiscent of the Grand Canyon campaign, stating that strip-mining on Black Mesa was "Like Ripping Apart St. Peter's in Order to Sell the Marble." The quandary was best put by Jeffrey Ingram, the Southwest representative of the Sierra Club, who stated at a subcommittee hearing in 1967:

> Suppose the dams are dropped from this legislation in favor of coal plants. Then we get air pollution. But if we give up coal plants for dams to save the air, we lose water through evaporation (which is one of the dam's hidden fuel costs—sedimentation is another). Yet if we save water by building coal plants instead of a dam, we use up the coal. But if we then argue that we must save coal, a non-replenishable resource, and therefore build dams, we lose the river and canyon bottom, which puts us back where we were.

As six coal-burning power plants on the fringes of canyon country went into operation and more were planned in the early 1970s, there were those who thought the quality of air in an area famed for its clarity was beginning to deteriorate. These people tended to be Park Service personnel, a few permanent residents, and visitors who had made a habit of constantly returning to the Southwest over the years. What the laymen instinctively knew to be happening in mid-decade, the scientists had yet to verify with their instruments. The problem was that the instruments and standards were developed for urban areas and could not measure the emotional impact of having a light gauze curtain pulled over a scene that was once known for its absolute clarity. At Bryce Canyon National Park, perched on the rim of the basin with an excellent view over the canyonlands, the superintendent said he frequently saw a yellowish brown layer squatting over the Lake Powell area. At Capitol Reef National Park, slightly off the beaten path, there was the suspicion that smog was creeping up the valley of the Water Pocket Fold from power plants to the south. The superintendent of Mesa Verde National Park, that bastion of an ancient Indian culture, reported that ten years previously it had not been uncommon to see Mount Taylor, 145 miles distant in New Mexico, but in the five years he had been at the monument he had not seen the mountain once. The Southwest photographer Laura Gilpin commented:

> It was the magnificent skies and the great variety of landscapes that drew me to New Mexico where I have made my home since 1945. As a photographer I was challenged continually with the problem of recording on film the changing light and the far distant horizon one saw on every hand. The quality of light which yielded luminous shadows and crisp contours is today becoming veiled with smog.

We literally sailed out of the Grand Canyon, a snappy following wind pushing our jury-rigged wooden boats past the Grand Wash Cliffs and out onto the broader reaches of Lake Mead. When Powell reached this point, his predominant feeling was one of relief, and he expressed it in visual terms:

> Now the danger is over, now the toil has ceased, now the gloom has disappeared, now the firmament is bounded only by the horizon, and what a vast expanse of constellations can be seen!

CHAPTER SIX Deserts:
The Politics of Water

1

Play is the most visible activity on and about the water flowing from Nevada to the Mexican border which marks the dividing line and common swimming hole for Arizona and California. Where the water flows into Mexico after being diverted for use elsewhere in that country, the Colorado ceases to exist, until some miles further on when the drainage water from agricultural fields in the Mexicali Valley returns to the river-bed to flow south to where, now utterly devoid of any vitality, it finally disappears before reaching the Gulf of California. But what remains of the lower Colorado, whether squeezed between artificial banks or let loose in a few remaining natural meanders, is essentially a play river—a tepid puddle for urban crowds.

True, the water is put to work here more intensely than anywhere else along the system—more intensely, perhaps, than along any other 342-mile stretch of desert and river in the world, and certainly in the Western Hemisphere. But the irrigation of fields, the transportation of water to the coast for domestic consumption, its infrequent industrial use, and its journey through the penstocks of dams to make electricity are uses less apparent, although far more important, than the play of the crowds in and alongside the lower Colorado. It is like a trip to Las Vegas, where one seldom considers that that tinsel playground is totally dependent for its existence on the river system. What river? the casual visitor is liable to ask while observing the slot machines and showgirls. Which is more real?

Along this stretch of river from Hoover Dam to San Luis in the Mexican state of Sonora, the Colorado becomes a plumbing system of varying efficiency. Sometimes it gets stopped up and has to be relieved by dredging. Through the Grand Canyon, the river is a very efficient carrier of water, easily traversing a series of rapids and pools. Although the gradient is much less, the same combination of pool and running water is found along the lower river. But here the pools are formed by a continuous series of dams, while the water runs, for the most part, between channelized banks whose rock-ribbed sides have been stripped of all water-sucking plants.

What the river has become—in effect, a graded ditch—is the inevitable price that had to be paid for the establishment of an extensive culture in the desert. Given the nation's policy of westward expansion and its corollary of growth, a policy that has gone unchallenged until recently, there was little else the lower river could have become. The problems arise when that civilization believes it can continue to grow until all the water from the river is used, and then grow some more when additional water is poured into the Colorado from rivers to the north. There is a point when all this growth, and its costs, become unacceptable; a point, which is fairly imminent, where the costs greatly exceed the benefits, and someone has to say, No more.

Who will say it with authority and make the pronouncement stick? President Carter tried and failed, although he subjected the policy to an unprecedented scrutiny. The result of no effective action coupled with a climatic change could be, like the disappearance of the Anasazi and Hohokam Indians, a general realization of declining water supplies followed by a gradual winding down and a selective migration away. This is a plausible scenario, delayed for a few more decades by interim technical solutions and readjustments in the priority of various types of water usage—such as agricultural partially giving way to municipal and industrial consumption. Any rational analysis of the distribution of western water comes to the inescapable conclusion that the huge amounts of water going toward agricultural use are disproportionate and wasteful. But agribusiness and its legislative allies, an unbeatable political alliance in the past, can be expected to resist tenaciously the loss of any water. Some will go elsewhere, but not much. It will take forces greater than those existing at present to make any significant changes in the law of the river—the chief prop that shores up western water interests, as they are now structured.

Desire has already exceeded present supplies; next it will be actual need. Even with more groundwater pumping and whatever augmentation schemes are found acceptable, the ultimate limits have to be reached someday, in terms of both quantity and quality. As with the last Sumerian Empire in ancient Mesopotamia, it will be hastened when the strong, centralized authority—the cohesive political structure that gets water projects authorized and built—becomes weakened or diverted by other more pressing needs. Through the late 1960s and 1970s, starting with budgetary challenges within the federal bureaucracy and environmental challenges from without, there were strong indications that Reclamation was losing its favored status in the West, had in fact lost that binding national unanimity (actually based more on the absence of interest outside the West) that was needed to get such key bills as the Colorado River Basin Project Act of 1968 and the Colorado River Basin Salinity Control Act of 1974 passed by Congress and signed by the President.

Just before the decade ended, in what could be considered a symbolic act, that most important of all western federal governmental institutions, the Bureau of Reclamation, changed its name to the Water and Power Resources Service. The emphasis now would be less on building and more on making what had been built run more efficiently. The word "Reclamation," actually a revered concept, had been in the agency's title since the first years of this century; it was the rallying call for the one political movement that had bound the arid West together for almost a hundred years. And now the word had been removed and the role of the federal government altered, at least on a letterhead basis. In making the announcement of the name change in November, 1979, Interior Secretary Andrus stated, "National needs and concerns now call for greater efficiency in the operation of existing structures and their integration in new programs for renewable resources and alternative energy." This wording was too voguish for immediate consumption in the interior West. It was Washington, D.C., come to Craig, Colorado, Escalante, Utah, or El Centro, California. But undoubtedly policy would—at least to some degree—soon catch up with announced intent.

Over a period of seventy-seven years the Bureau of Reclamation and its predecessor, the Reclamation Service, operated 333 reservoirs, 345 diversion dams, 14,590 miles of canals, 35,160 miles of smaller laterals, 990 miles of pipelines, 230 miles of tunnels, 15,160 miles of project drains, 188 pumping plants, and 50 power plants. More than 16 million people in the West benefited wholly or in part from bureau-supplied water, which also was spread upon 9 million acres of farmland. With the help of $8 billion in capital outlays, which does not include privately financed projects or those funded solely with local district and state funds, the West had indeed bloomed and prospered. Nowhere else could the benefits and costs be better documented in political terms than in the scorching deserts of the three lower basin states and Mexico.

Hoover Dam, athwart Black Canyon, is everything a dam should be—a monument to man's apparent infallibility and triumph over a mighty river. It is a tall, elegant plug in a spectacular setting—white concrete set against black canyon walls and the depression-era WPA architecture of the 1930s enhancing its monumental qualities. The dam, besides being a plug, is a major tourist attraction with daily tours being bussed out from Las Vegas. By the drought year of 1977, eighteen million visitors had taken a tour of the dam.

A tour is interesting, but I find that the movie depicting its construction shown at the small theater in nearby Boulder City better captures its essence in time. The movie documents in newsreel quality the epic con-

The tour taken by millions within Hoover Dam.

struction feat of the depression years. Within the darkened, cool theater there is a sense of immense pride in such an achievement that is lost upon emerging into the white glare of summer heat in southern Nevada. Then it just becomes a matter of enduring. At the construction site in the bottom of the canyon, temperatures as high as 152 degrees were recorded. Thirteen construction deaths were attributed to heat prostration. The convoluted black rocks of the canyon trapped the heat and set it bouncing from wall to wall. It was all that moved in the depths, except the incessant construction activity. Frank Waters described a midnight construction scene in *The Colorado:*

> The vast chasm seemed a slit through earth and time alike. The rank smell of Mesozoic ooze and primeval muck filled the air. Thousands of pale lights, like newly lit stars, shone in the heights of the cliffs. Down below grunted and growled prehistoric monsters—great brute dinosaurs with massive bellies, with long necks like the brontosaurus, and with armored hides thick as those of the stegosaurus. They were steam shovels and cranes feeding on the muck, a ton at a gulp.

Hoover Dam was the biggest. It would involve the best of American technological know-how available at the time to overcome what seemed like insuperable obstacles, much like the space program of the 1960s. It was America. It was moving mountains or rivers to achieve growth. It was what people during those depression years hoped would portend this nation's future—a tangible promise of prosperity rising Phoenix-like from the muck of a black canyon. The firm that built Hetch Hetchy Dam in Yosemite National Park and such bold contractors as Henry J. Kaiser and W. A. "Dad" Bechtel pushed the dam to completion, and President Franklin D. Roosevelt dedicated it before 12,000 spectators on September 30, 1935. Never mind that a total of 110 men had died during its construction. It had been done. The river had been tamed, and now the water could be safely distributed below.

Behind Hoover Dam, Lake Mead extends 108 miles to Separation Rapid. At the Grand Wash Cliffs the river, now a reservoir, leaves the high province of the Colorado Plateau, through which it has been winding generally west through canyon country, and emerges into the more open north-south-trending Basin and Range Province. Underneath the waters of Lake Mead is the point where John Wesley Powell ended his historic journey in 1869 and the Mormon settlements of Rioville and Callville, which were the head of steamboat navigation on the river in the late 1800s. The reservoir bends south to fit the new contours of the land near Saddle Island, where the Colorado's water is first tapped in the lower basin, to be gaudily displayed in the multicolored fountains of the casinos and hotels of nearby Las Vegas and spread upon the golf courses and suburban lawns of that desert oasis.

When the Spanish first visited southern Nevada in the late 1700s they found some springs; the name Las Vegas, "the meadows," memorializes the patches of grass surrounding the seeps. The Mormons settled the area first, and by 1900 a few settlers in the valley were raising alfalfa and grazing cattle. Surprisingly enough, there still is a sizable Mormon population in Las Vegas leading quiet and prosperous lives behind all the glitter. The construction of Hoover Dam spurred growth in the sleepy desert town, as did World War II industrial developments and postwar military facilities, such as Nellis Air Force Base and the Atomic Energy Commission's Nevada Test Site north of the city. Then came the gambling casinos and hotels; in the decade between 1960 and 1970 Las Vegas led the nation's metropolitan areas with an astounding 115 percent growth rate. Like other western cities, Las Vegas now has a smog problem.

The first water was pumped from Lake Mead to the Las Vegas area in World War II to serve an industrial complex just south of the city in Henderson. The initial domestic deliveries of river water to the city were made in 1971. Presently about half the water consumed in the area is pumped from Lake Mead and the other half comes from underground sources. Because of continued underground pumping, the land has subsided by as much as three feet, causing fissures to open up in the earth and streets and sidewalks to crack. Future growth will be totally dependent on Lake Mead water, to be supplied by the second stage of the bureau's Southern Nevada Water Project. Water use in the Las Vegas area is high. Not unexpectedly, the casino-hotels and private residences are the largest consumers. Golf courses also require a lot of water in that dun-colored desert landscape where creosote and burro bushes are the rule and those startling green swatches the exception. The squeeze will come after completion of the bureau's project and the delivery of the state's share of 300,000 acre-feet per year of Colorado River water sometime in the mid-1980s.

Below Hoover Dam the river is a long, thin lake for sixty-six miles of its length, since Davis Dam backs the water up in Lake Mohave to the foot of the larger dam. The lower Colorado is much like a set of liquid steps, a slow, drawn-out descent into the dry sandbox of the delta. The mountains surrounding Lake Mohave are not high, but they are forbidding on this late June day; molten heaps, scoops of marbled chocolate, off which the heat shimmers. On the west shore of the lake, about halfway down its length, are a launching ramp, a few scattered house trailers, and a cafe —the essential social institutions of the lower river. It is a small fishing camp, one of a number scattered along the backwaters of the lower river, each with its own special clientele and sense of separateness from the world. River folk are different. Wizened by the sun and water, cooked by the heat, it takes a certain type to live along the lower Colorado. There is that twang that speaks of native lands to the east, perhaps Texas, Oklahoma, or Arkansas: stragglers that were on their way somewhere else but found a forgotten niche between the metropolitan areas of southern California and central Arizona. In a cafe that sells water dogs and frozen

The fountain at the MGM Grand Hotel, Las Vegas.

chicken livers for bait, the rules for a bass-fishing contest are posted on a wall underneath a giant, stuffed sailfish that obviously came from more exotic climes.

When Hoover Dam was built, the fishes that were native to the widely fluctuating, muddy waters of the Colorado began to fade away. Exotic game fishes, those that would be sought by people driving all the way from Los Angeles for a weekend, were imported and took over the river whose flows are now precisely regulated and much clearer. There was much less current and no need for the muscled hump that became so well developed in the humpback and boneytail chubs, allowing them to hold their position or make progress in the strong currents that would sweep down the natural river. Eight dams on the lower Colorado and an aggressive program of dredging and redirecting its course have contributed to the extinction or endangered status of eight species of fish known to have once been native to it. (There was a certain irony in all of this. As more water was diverted and more dams and other river works were constructed, thus changing the regimen of the Colorado, native fishes became rarer or endangered, thus threatening to block further water projects under provisions of the Endangered Species Act.)

On the other hand, introduced species, such as rainbow trout and channel catfish, have prospered, the trout finding a suitable habitat up against the downstream face of dams, where cold water is released from the depths of reservoirs. One of the more exotic species of fish introduced into the lower river is the plant-eating Mossambique mouthbrooder (*Tilapia mossambica*), imported from Africa to eat the weeds in irrigation ditches. On another day I was to see thousands of these fish lying belly up in irrigation ditches bordering the Salton Sea. It was winter, and *Tilapia* does not do too well in cold water. That is its drawback, so another fish was sought to do its work. The Colorado River Wildlife Council denied the request of the Palo Verde Irrigation District to import *T. zilli,* fearing it would compete too vigorously with the introduced game species.

Every river has its local character, an old-timer who lends a note of authenticity to its history, which otherwise exists only in a more removed sense on a printed page or in a museum. The lower Colorado has Murl Emery, a suspender-clad, grizzled ex-miner, ex-trapper, ex-boatman, and ex-fishing-camp owner. Murl lived up the road from the lakeside cafe in the small settlement of Nelson, Nevada. Ivy, a luxury and sign of longevity in this desert climate, climbed the posts of his home to the tin roof, which reflected the dazzling heat. Nelson was a quiet settlement, a few retirees and empty dirt streets. Around Murl's yard lay those scattered, rusting remains of mining machinery that never seem to decay beyond a certain point in that dry air and with a little coaxing can be improbably fired up again, as Murl did with the twelve-horsepower engine built in 1903 that drove his gold stamp mill. It chugged slowly back to life again, then hit its rhythmic, clanking stride.

The mining camps along the lower Colorado and back in the parched hills were served by steamboats. Murl was too late for that era, but he trapped the last of the beaver below Black and Boulder canyons, the route Jedediah Smith and his fellow trappers traveled nearly a hundred years previously looking for untouched beaver grounds and finding instead a new interior route to the Pacific Ocean.

Murl's father operated the Searchlight Ferry and the son started out on the Colorado in 1921 in one of the first gasoline-powered boats on the river, carrying at the age of twenty such people as Stetson, of hat, and Gillette, of razor-blade fame, up into the canyons to inspect their mining claims. They had an engineer, Murl recalled, who said he had stood in the canyon during an earthquake and only one side moved. I wondered if the Bureau of Reclamation had designed Hoover Dam for that contingency. Murl took politicians and engineers up to inspect the site of the dam that was to be built in Black Canyon. Of what had become the river he said, "I wouldn't give you a tinker's damn for all those lakes."

Hoover was not the first dam on the lower river nor, in fact, was Laguna Dam, a low diversionary structure built north of Yuma in 1909. The Indians had been the first to divert the lower Colorado. Laguna Dam was designed to raise the level of the river just high enough to form a still pool so the water could be siphoned off into a canal for irrigation purposes. It was the first large "Indian weir" type of dam built by the newly established Reclamation Service. The dam was also designed to serve as a sediment trap, the thinking being that silt would settle in the quiet water behind the dam and not clog the canal. But with the full weight of the Colorado River and its tremendous silt load descending on the low structure, the settling basin was filled to the lip of the dam within weeks of its completion. Its effectiveness as a sediment trap was reduced more quickly than expected; as was to be the case with Imperial Dam, just upriver from Laguna, whose 85,000 acre-foot capacity would shrink to 1,000 acre-feet. And Imperial Dam was completed thirty years after Laguna and three years after Hoover, when the silt load on the river had been greatly diminished. Once permanent barriers were erected across the river, silt —its uneven accumulation and need for eventual deposition off the river —would become a major problem and, in turn, the cause of yet another major public-works program. The Indians handled the silt problem by erecting temporary brush and earth-filled weirs that were easily breached by heavy river flows, thus allowing the accumulation of silt to be washed downriver.

First Hoover, then its downriver progeny were constructed—Davis,

Parker, Headgate Rock, Palo Verde Diversion, and Imperial dams in the United States and Morelos Dam just below the border in Mexico. Palo Verde Diversion Dam was the last, completed in 1958. The progeny, besides causing siltation problems, had additional offspring; the All-American Canal departed from Imperial Dam for its journey to the valley of the same name in the California desert, and Lake Havasu, the reservoir behind Parker Dam, is where coastal southern California draws off its water through the Colorado River Aqueduct and the location of the intake for the Central Arizona Project.

The Mohave Valley stretches some forty-plus river miles below Davis Dam. Before the construction of Hoover Dam the Colorado River was described thus by the Bureau of Reclamation: "Unharnessed it tore through deserts, flooded fields, and ravaged villages. It drained the water from the mountains and plains, rushed it through sun-baked thirsty lands and dumped it into the Pacific Ocean—a treasure lost forever." Once the river was tamed, the dams spawned their own particular brand of devastation. The town of Needles flooded in the 1940s, forcing the Santa Fe Railroad to close its shops and move operations west to Barstow, an economic blow from which Needles never recovered.

What had happened to Needles was that, with construction of Hoover Dam and other structures, clear water was being released from the reservoirs. The reaches below the dams were scoured by the clear, fast-moving water, which gradually acquired a new sediment load that was deposited in the slower water at the upper reaches of the next reservoir downstream. As the silt was deposited and the riverbed rose during the early 1940s at the mouth of Topock Gorge at the head of Lake Havasu, normal flows of water were backed up into the town of Needles. The bureau's dredge *Colorado* was placed into operation at Needles in February , 1949. What began as a simple emergency operation evolved into a complete Lower Colorado River Management Program, whose ultimate purpose was to remake the lower river into an efficient ditch. What eventually resulted, in the bureau's own words, was a change in the channel "from a natural-looking stream to something resembling a canal in appearance."

The lower Colorado became the ultimate plumbing system. A river that had once gently meandered across the valley, shifting—usually gradually, sometimes violently—from one course to another, was now restricted between rigid banks and reservoirs for most of its length. Thirty-two miles of river through the Mohave Valley were channelized, and the program was slowed only when dredging started in scenically spectacular Topock Gorge in 1968. Conservationists and state fish-and-game agencies then raised a hue and cry. By one estimate only eighty-seven miles of river below Lake Mead remained in its natural state.

The dams were built first. Then the channel of the river was stabilized to counter the downstream effects of the dams. The next step was to make sure only the bare, minimal amount of water was lost between the place of storage and the point of application. Water was getting very short indeed. This meant a program of water salvage. It was hoped, probably overoptimistically, that 350,000 acre-feet could be saved yearly by straightening out the river and clearing its banks of vegetation. The river had to be rid of water-sucking plants and evaporation-prone backwaters. In the process herbicides were sprayed, fires were set, and plants were ripped out or buried by bulldozers. To replace tamarisk, less water-consumptive cottonwoods were planted. Cottonwoods had previously disappeared from the riverbanks, cut to fire the boilers of steamboats. There was a certain inconsistency in all this activity: the tamarisk had originally been imported to halt riverbank erosion, and the cottonwoods, in other parts of the basin, were being destroyed to save water.

In the face of criticism, the bureau would later admit to a single-purpose program—a bureaucratic sin when the governmental catchword had become "multiple-purpose." Besides the complaints of the conservationists and state wildlife agencies, the Indians did not like being cut off from the river by large levees and land stripped bare of riverine vegetation. The regional director of the bureau, Arleigh B. West, lashed back at the critics in 1968:

> There has been controversy concerning our river program. The uninitiated assert that the river must be left in its natural state. The basic truth is that the lower Colorado River is now a stream wholly controlled by man. It was in its natural state, or nearly so, prior to closure at Hoover Dam. Try as anyone will, it cannot be characterized other than a fully controlled river. . . . The river's flow can be manipulated in the same fashion as the garden hose on the tap outside your home, and is.

So true. Nevertheless, through the 1970s the bureau would decrease its river-management activities, and what was accomplished would be undertaken with a greater emphasis on enhancing or preserving the few remaining "natural" features of the river.

Topock Gorge leads into Lake Havasu, another long, thin body of still water where the cacophony of the internal-combustion engine is incessant. This is not surprising, since the founder of Lake Havasu City on the eastern bank of the river-lake was a manufacturer of chain saws who was looking for a place to test outboard engines. To promote the sale of subdivision lots, Robert P. McCulloch held motorboat races and flew in planeloads of potential buyers. I approached the lake in May, winding down the long alluvial grade on Interstate 40 into the outskirts of Needles, where I camped that night among a few drops of unseasonable rain. It was the year after the drought had ended, and it could not rain enough in the West.

The approach to the Colorado River Basin from the relatively lush coast is always a lesson relearned in sparseness: an open, lean land unfolds before the viewer. It is lean because of the lack of water. The sense

London Bridge, Lake Havasu City, Arizona.

of space can be overwhelming, what Joan Didion has referred to as "that vast emptiness at the center of the Western experience, a nihilism antithetical not only to literature but to most other forms of human endeavor, a dread so close to zero that human voices fade out, trail off, like skywriting." I was awakened by outboard motorboats, which had already begun their daily river dance. Dune-buggy tracks led through the dense stand of riparian vegetation to the water, where the riprap had been partially camouflaged by the greenery.

It was not too long a drive to the Sunday flea market at Lake Havasu City, where the discussion among resident retirees and the visitors to one of the seven wonders of the West centered on the Miss America contest seen the night before on television. Polyester flare pants, tank tops, cutoffs, and one-piece jumper suits predominated. Some women carried parasols to ward off the sun, while one youth used a Frisbee balanced atop his head for the same purpose. Sunglasses blotted out eyes. May is the season for strong winds and dust storms in the Southwest, but the Sunday-morning walkers who paraded by the stands set up in the parking lot of the shopping center gave off the aura of water-washed cleanliness, with skin polished dry like the desert stones about my sleeping bag the night before. The desert is essentially a neat, clean place with little to clutter up its spaces. And the people reflect this sparseness.

The other parade on Sundays, an entirely unorganized but seemingly compulsive affair, is under London Bridge in a motorboat or pedal-driven craft. There the wonder squats, stolid and gray in the blinding desert light. McCulloch, the chain-saw king, needed a focal point for Lake Havasu City and he found it in London for the price of $2,460,000, with the water from the Colorado River which flows under it thrown in for free. McCulloch chose to endow his instant town with a symbol of water-related antiquity.

The plan at Lake Havasu City was to install a wave machine for surfers, something that had already been established on the outskirts of Phoenix for some years. One wave, five feet in height, would be manufactured every ninety seconds. I also learned—from the tour guide on the boat trip I took underneath London Bridge and a short distance out into the lake —that water, pumped under 500 pounds of pressure, was sprayed in a fine mist from the tops of palm trees at the beach in front of the Nautical Inn, making it the world's only air-conditioned beach.

The river seems to attract superlatives. "By the way," added guide Kristy Neubaur, "*The Guiness Book of Records* lists the bridge as being the largest antique ever sold." Not too far away, she explained, was Parker Dam, the deepest in the world. The river, always the river.

. . .

Carl T. Hayden was Arizona. His life came as close as anyone's to spanning the Anglo history and various interests of that state, foremost among which was the acquisition of ever-increasing amounts of water. Hayden's productive years were of the twentieth century. His father's life, of which the senator was very cognizant, spanned Arizona's Anglo interests in the previous century. Charles Trumbull Hayden arrived in Tucson in 1858. Carl Trumbull Hayden died in Mesa, Arizona, in 1972. Rarely has such a decisive role in so much western history been compressed into two generations of one family. *Time* magazine, in its trenchant way, described Senator Hayden in the early 1960s as "a last link between the New Frontier and the real one." Charles Hayden emigrated from Connecticut and, after spending the first of his Arizona years in Tucson, founded Hayden's Ferry, now known as Tempe, on the south bank of the Salt River in 1871. Modern-day Phoenix now lies just across the dry riverbed, but in those early days there was virtually nothing in that desert.

From the start, the fortunes of the Hayden family were tied to water, or to be more exact, its diversion onto the land. The elder Hayden hired Jack Swilling to construct a ditch to serve the Tempe area. Swilling was no stranger to digging irrigation ditches. A few years earlier he had built a ditch on the north bank of the river with the help of twenty laborers, marking the start of modern irrigation in the Salt River Valley. The first crops were harvested in 1869. While building his ditch, Swilling noticed mysterious mounds arranged in parallel lines. They were the outlines of the ancient irrigation ditches constructed by the Hohokam Indians. The routes of Swilling's and Hayden's ditches either paralleled or exactly duplicated these earlier canals. There are just so many ways to make water flow downhill. Swilling disappeared from Arizona history a few years later, after he robbed a stagecoach and was sent to Yuma Territorial Prison, where he died. The Hayden family, choosing the path of respectability, survived and Charles Hayden was on hand at the 1871 picnic at the Pueblo Grande ruins where the drunken Englishman proclaimed, "As the mysterious phoenix rose reborn from its ashes, so shall this new city rise on the foundation of a dead civilization." Tempe was named by the elder Hayden for a sylvan valley in Greece through which ran the ancient river of Peneus at the foot of Mount Olympus. Perhaps the residents thought that classical names, like London Bridge, would lend an aura of antiquity to the new territory.

Carl Hayden was born in 1877 in a small adobe structure on the banks of the Salt River, now the popular steakhouse La Casa Vieja, located across busy Highways 60 and 89 from the Hayden Flour Mill, which is still operating. Hayden's boyhood home had the only running water in Tempe at the time. The Salt River, a tributary to the Gila, is now dry through this stretch, having been totally dammed and diverted above except for occasional, sometimes devastating, flood flows. But when Carl Hayden took over his father's businesses in 1899, the flour mill was run by water power. The younger Hayden ran for his first political office in 1902, winning election to the Tempe City Council. He was to remain an elected official for the next sixty-six years, fifty-six of them in Congress. He moved from the House of Representatives to the Senate in 1927. During Hayden's unprecedentedly long congressional career, he served under ten presidents. When Hayden first went to Washington, D.C., Lyndon B. Johnson, the last president Hayden was to serve under, was three years old. They both left the capital for good at the end of 1968. The year Hayden took his seat in the House as Arizona's only representative, the territory of Arizona had just become a state in 1912.

Hayden, then the county sheriff, had won election on the basis of the notoriety he had gained from the capture of two train robbers in 1910. It was the first time an automobile had been used to catch crooks in Arizona. The Southern Pacific train was robbed by the Woodson brothers, who escaped with $295 after pistol-whipping one passenger. The sheriff rounded up some Indian trackers and persuaded a local hotel owner to drive him in his new $3,000 automobile after the robbers, who were heading south across the desert to Mexico. The posse had already started out on horseback, but the car quickly overtook it and some deputies climbed on board. The chase resumed. The robbers were overtaken and Sheriff Hayden faced one of the brothers down with an unloaded pistol. It was later determined that the bandit's gun was loaded. After that incident, Hayden was easily elected to Congress.

In his first term, Hayden displayed his keen interest in and understanding of the importance of water to his state; he obtained congressional authorization for an investigation that led to the eventual construction of Coolidge Dam on the Gila River and the San Carlos irrigation project. Hayden played a key role in the passage of the 1919 act that created Grand Canyon National Park. The Ashurst-Hayden Diversion Dam is located on the Gila River near Florence, and the visitor center at Glen Canyon Dam is also named after the venerable Arizona senator, an honor occasionally bestowed by a grateful dam-building bureaucracy on congressional friends of Reclamation. After Hayden died there was a short-lived effort by fellow Arizona Senator Barry Goldwater to rename the Central Arizona Project the Carl Hayden Project, but that failed. The Goldwater and Hayden families were close, both being of pioneer stock and heavily involved in civic and mercantile interests. Goldwater courted his wife in Senator Hayden's house. In 1968 Goldwater took over Hayden's senate seat when the latter retired, but the junior Arizona senator never played a leading role in obtaining Arizona water, although he was among the first to run the river. Goldwater's interests were wider in scope, and because of his unsuccessful try for the presidency in 1964 and his absence from the Senate for the next four years, he missed the years of crucial congressional debate on the Central Arizona Project. Hayden had pursued a different role in the Senate. Throughout his long career, which included the chairmanship of the powerful Senate Appropriations Committee, Hayden was to concentrate on those needs peculiar to the West, like improved highways and water projects. As President John F. Kennedy said of Hayden, "Every federal program which has contributed to western irrigation, power and reclamation bears Carl Hayden's mark." He was a man who understood the need for water, said a grateful senatorial colleague who had benefited from Hayden's largess. And that perhaps is the greatest tribute possible for a western lawmaker.

The first bill to authorize the Central Arizona Project was introduced in the Senate in 1947. The Senate, under Hayden's quiet yet forceful prodding, passed CAP bills in 1950 and 1951 but the House, more prone to the pressure of the dissenting California delegation, refused to act. California, which wanted as much water as it could get and wanted the power from any future dams for its burgeoning population, was successful in 1951 in getting a motion through the House Interior and Insular Affairs Committee that the Arizona water project not be considered until the water was divided in the lower basin. An embittered Arizona filed suit in the Supreme Court in 1952. Twelve years, $5 million, 50 lawyers, 340 witnesses, 25,000 pages of testimony bound in 43 volumes, and more than 4,000 exhibits later, the court ruled. In the end, Arizona won the court case, but much of what it won was to be soon lost in the succeeding legislative battle.

Following the Supreme Court ruling in 1964, the action moved back to the legislative arena, where it soon became apparent that Senator Hayden was not going to be able to push through a central Arizona bill single-handed. He could dominate the Senate on this issue, that was never any problem, but his power and the political debts owed him did not extend far into the opposite house. Besides, as the debate intensified and the senator approached the age of ninety, he began to slip into senility. Hayden was ailing in his last few years in office, and his staff handled much of his work while trying to protect him. He became a sad relic of the seniority system. He was in and out of the hospital, and once a press conference was called to quell rumors of his death. Hayden's hearing and eyesight worsened, and his aides shuddered at the close calls he had with automobiles while crossing streets on Capitol Hill. A stooped figure by now, the senator used a cane to wave people on and off elevators, and on occasion he would brandish it at the young man his staff had assigned to follow him around the Capitol corridors to make sure he did not fall on the hard marble floors.

When Lyndon B. Johnson signed the Colorado River Basin Project Act on September 30, 1968, in the East Room of the White House, the President, who recalled having hauled some irrigation pipe in Texas, handed the pen to his former Senate colleague. Since Senate passage of the conference-committee report on September 12, and for some time before then, when the bill's passage seemed assured, Hayden had been deluged with glowing tributes—a fitting, indeed the greatest climax of a long, distinguished congressional career that was about to come to a close. He was called the "father" of the Central Arizona Project, but this description was not quite accurate, since a number of different people and interests had been hovering around at the conception and the child was raised to maturity mainly by others. Actually, Hayden was more the grandfather of a child of doubtful parentage who was raised during the critical years by its doting uncles, Wayne Aspinall and the Udall brothers. Needless to say, there were some differences within the family on which direction the child should take on reaching maturity. But there was little doubt that Hayden was the family's patriarch.

The 1968 act provided for much more than a Central Arizona Project; its significance to the Colorado River and the West lay beyond the boundaries of any one state. To begin with, it was basinwide in concept, and sought to be comprehensive—a status it never really achieved, since that word implies a more rational process than taking separate wish lists and molding them together by political compromise. Another way to describe its far-flung nature was to say that each state got a little bit of something and in the process a political coalition, a sort of mutual defensive alliance for Colorado River matters, was formed that would last through the 1970s. The coalition came together—where a disparateness had reigned before—not only to authorize and fund the physical structures that would ultimately divide the waters but also on the premise, deemed absolutely necessary, that more water from outside the basin would be obtained at some point in the future. Further, the obligation to furnish Mexico with water should not rest with the seven basin states alone but with the nation as a whole. All this was necessary, it was determined, because once the physical structures were in place and the surplus water in existing reservoirs was depleted, there would be a shortage. Assurances, again based on the plumbing-system concept, were made for the future.

To be sure, there was bickering and hard bargaining along the way, but in the end there was cohesiveness—and that was what Wayne Aspinall strove to achieve and was his greatest contribution to the West. The Colorado River Storage Project Act of 1956 was the forerunner. In 1968 the basin states would construct a much grander piece of legislation than that first, rather primitive model, reaching south across the international border and out of the watershed to the north. It was a classical case of distributive politics on the grand scale, where a coalition is formed among local and state interests seeking to benefit from federal funds. Conflict is removed by trading and bargaining, and a unified front is formed to present to a wider audience, such as the full House. The process was described by Congressman B. F. Sisk of California, a key player in the House in the 1970s: "You do it by living and letting live. You do it by cooperating. People say that's a bad way to legislate—that means you're going to vote for anything that Bill Smith wants, regardless, in order to get his vote for what you want. Well, I suppose that's putting it in its most brutal and coldest way." Twenty years earlier, the senator from urban Illinois, Paul Douglas, had described the trade-offs between eastern and midwestern harbor-improvement and flood-control projects and western water projects as "a process of mutual back-scratching."

House and Senate members outside the affected areas understood little of the proposed bills, which made it incumbent upon the congressional leadership in these matters to either generate trust or use naked power. Once an eastern congressman, after listening to a highly technical explanation of the possibility of water shortages in the San Juan River, commented at a subcommittee hearing: "At times I feel something like an eastern lamb who has strayed into a western fold." Then there was the midwestern congressman who said he voted for the 1968 bill simply because he trusted Aspinall. Both Hayden and Aspinall were trusted, they were members in good standing of their respective clubs; but it did not hurt, either, that they were powerful committee chairmen who knew how to use their leverage. Such proposals begin to flounder under sustained outside scrutiny and well-directed emotional attacks—the chief problem being that uninitiated congressmen become restive. A calm atmosphere is needed for the passage of such laws. Hence, they seem to germinate and come into fruition in an atmosphere of regional isolation, which certainly was the case with the 1968 legislation once the controversial Grand Canyon dams had been eliminated. As one authority, Dean E. Mann, wrote of the process, "The taxpayers, who must pay the bill, seldom have an effective voice in the political bargaining. The principal actors in this political process tend to be local and state interests and public agencies, federal bureaus and committees of Congress."

California had no single, powerful committee chairman; but what it did have was a large congressional delegation, some of whose House members were strategically placed and rising in seniority and influence on key water committees. It obtained from the 1968 legislation most of what it lost in the Supreme Court. The court had given Arizona a clear victory, apportioning Colorado River water in the lower basin thus: 4.4 million acre-feet per year for California; 2.8 million for Arizona, and 300,000 for Nevada. Arizona won its key contention that the 1 million acre-feet of Gila

River water was not to be counted as the state's share of the Colorado, thus gaining water from the main stem of the river at the expense of California. Surpluses and shortages were to be divided by the Secretary of the Interior. Throughout this time and into the early 1970s, California was consuming more than 5 million acre-feet a year and would probably continue to use more than its share of the river until it had to cut back when the Central Arizona Project went into operation in the mid-1980s. It had foreseen this eventuality and constructed the California Water Project in the 1960s to carry northern water south. Southern California, against northern resistance, sought to increase these flows in the late 1970s to further compensate for the pending cutback in Colorado River water. This meant construction of the Peripheral Canal, which would increase the efficiency of the delivery system. California's water project was the counterpart to the Central Arizona Project, since it moved large amounts of water from one end of the state to another. But it was accomplished by the state, not the federal government, and it tapped another river system. In the drought year of 1977 these two watersheds were linked in the greatest demonstration yet of the interchangeability of the plumbing system.

What California wanted from the 1968 bill was a firm guarantee that any shortages—the amount below 4.4 million acre-feet—would not be shared equally but would be taken out of Arizona's entitlement to the Colorado. Its price of support for the bill was met. In a time of declining river flows and increasing Indian demands for a share in the water, a guarantee against shortages was of more meaning than a paper commitment to a certain amount of water that may not exist when all claims are taken into account. Congressman Morris Udall of Arizona stated, "In effect, we gave away much of our 'paper victory' in the Court to get our aqueduct built."

What Arizona got was the realization of its long-held dream of a system of aqueducts and dams that would carry the state's remaining share of Colorado River water from its sparsely populated borders to the burgeoning central and southern regions. This was the water that farmers, urban interests, suburban land developers, and such industrial users as the copper industry had been counting on—what they had anticipated while depleting underground reserves and what was, at best, a temporary bailout until once again the ultimate savings account had to be dipped into.

There were also some benefits for the upper basin states in the legislative package. Colorado obtained five small water projects. Aspinall directed state officials to select them on the following basis: "The list of projects is just a list so far as I am concerned. You can add to it or take from it or leave it alone. That is your decision, but there must be equity in various parts of the state." The projects were the Animas–La Plata, Dallas Creek, Dolores, San Miguel, and West Divide, all in western Colorado. The Bureau of Budget raised questions on the economics of all five projects, but was overruled. The Carter Administration was later to question some, and even Aspinall would admit that not all the projects were deserving of authorization. But in 1968 they were thought to be a necessary part of the legislative package. New Mexico obtained the promise of Hooker Dam, or a suitable alternative; and Utah got the Uintah Project, a unit of the overall Central Utah Project, and the Dixie Project in the southern portion of the state. Nevada got what it wanted via a separate bill. Only Wyoming, which had no immediate wish, was left out.

For the Northwest there was a moratorium on any studies specifically aimed at importing water into the Colorado River basin from that area —a moratorium that undoubtedly would remain in place as long as Senator Jackson was the chairman of the Senate committee dealing with water

Morenci, an open pit copper mine in eastern Arizona.

matters. Undoubtedly, one day there would be an assault on the Northwest's water mounted from the south, but that would have to await a more favorable congressional climate. The beginnings of a development fund to finance such massive public works were included in the bill. But more than physical works, it was finding a way to shift the burden of the Mexican obligation that generated the greatest consensus and had the most importance for the future.

Everyone who had anything to do with the bill assumed that at an unspecified time not too far into the future there were going to be shortages. As Morris Udall pointed out to a California audience in the fall of 1967, "The fact is, my friends, we will all be in trouble—guarantee or no guarantee—win, lose or draw—unless and until we take steps to make augmentation a reality." On the other side of the ledger, such a knowledgeable congressional water leader as Harold T. Johnson of California would comment that "There is no reasonable chance that the Colorado River will supply enough water to meet the foreseeable demands of the area which relies on it." Aspinall had his own reservations about the availability of water should all the authorized projects be constructed without augmenting the flow. These congressional estimates were backed up by figures supplied by the state and federal water bureaucracies. The figures and statements, of course, had a political purpose—and that was to get the concept and beginning steps toward virtually free augmentation written into the bill. But they also contained a large degree of accuracy. The augmentation schemes, those ideas that were discussed to bring more water to an arid land, took on some fantastic and bizarre shapes.

The channelization and water-salvage programs along the lower river and the attempts at cloud-seeding in the upper basin were tangible efforts directed at augmenting the river's flows by the Bureau of Reclamation in the last two decades. They were all the bureau could do, given the constraints of the moratorium written into the 1968 bill. But there were wilder schemes floating around the West. North of the Grand Canyon, in an area called the Arizona Strip, the Bureau of Land Management experimented with waterproofing the earth in order to increase runoff and cut down on seepage. The material used was paraffin, the same that seals the top of homemade jars of jam or jelly. The paraffin was sprayed onto the ground during a hot day from a tank mounted on a two-wheel trailer, much like roofers' tar pots. Said a bureau publication: "Although the BLM is still experimenting, enthusiasm is running high for the paraffin method of catching rainwater." Before the waxlike substance was used, concrete, sheet metal, plastic, and artificial-rubber membranes had been laid down to increase the runoff into small reservoirs and catchment basins used by livestock. To cut down on evaporation under the hot sun, synthetic rubber rafts, polystyrene strips, and paraffin have been dumped into round metal stock tanks. Again, cows were to be the main beneficiaries of the increased water supplies.

On a much grander scale, western water planners in the 1960s cast their eyes as far north as the Yukon and Mackenzie river systems in Alaska and northern Canada. The North American Water and Power Alliance plan advocated transporting water from these sources as far south as the three northernmost Mexican states. A watershed of 1.3 million square miles would have been tapped and the water transported to a 400-mile-long regulating reservoir stretching along the Rocky Mountain Trench into northern Montana. Besides Mexico, the gigantic system would have fed the Canadian prairies, the Midwest, and the Southwest. The cost in those years was estimated at something like $200 billion.

Needless to say, strong objections to the plan were raised in Canada. A modified version of it, dreamed up by a retired Bureau of Reclamation engineer, would have tapped the Liard River in western Canada and then dumped the water into a huge reservoir located in the Centennial Valley of southwestern Montana, from where it would have been fed into the Colorado, Rio Grande, and Missouri river systems. Closer to home was the plan to place a fiberglass pipe under the Pacific Ocean to transport water from northern California rivers to southern California. The bureau even spent some money to study this scheme. A related proposal was to divert Columbia River water south by means of an undersea hose. Coming closer to being barely feasible was a proposal by Los Angeles water interests to divert water from the Snake River in Idaho through an aqueduct to Lake Mead. There were a number of engineering variations on this theme to bring Columbia River water south into the Colorado River Basin. In all, the Western States Water Council counted twenty-four water-transfer proposals in 1969, both interregional and international in scope. The council, set up by western governors, was careful not to endorse any of the proposals.

Shortly before the ten-year moratorium on importation studies imposed by the 1968 bill was to expire, the Los Angeles County Board of Supervisors in that drought year called upon the federal government to study how Columbia River water could be diverted south. The supervisors, uninformed and inept in the ways of western water politics, had stumbled into the proverbial hornets' nest. From Interior Secretary Andrus on down to the California Department of Water Resources and the Colorado River Board of California, the state's two top water agencies, the resolution was repudiated. This was not to mention the unified opposition to such a proposal voiced in the Northwest, where an environmental writer, Daniel Jack Chason, commented, "While the prospect of dumping Columbia River water into distant deserts has always seemed absurd to Northwestern politicians, the prospect of dumping it onto local deserts has always seemed attractive." Some cogent reasons were given by California water interests for not diverting the Columbia at this time, including the huge monetary and environmental costs along with the expenditure of large amounts of energy to move the water. These interests also knew the time was not yet politically ripe to move in that direction.

If not by land, well then, by sea. At the 1972 meeting of the American Farm Bureau Federation in Los Angeles, Reclamation Commissioner Ellis L. Armstrong endorsed the idea of hauling giant icebergs from the Antarctic to such cities as Los Angeles to increase their supplies of fresh water. It was the most significant endorsement the idea had yet received since being broached some twenty years earlier by a University of California professor. Subsequently, two federal government researchers calculated it was feasible; and researchers at the Rand Corporation, a southern California think tank, suggested the icebergs had best be wrapped in plastic to prevent excessive melting and hitched together in a fifty-mile-long train to be propelled north by means of a floating nuclear power plant. Then Prince Mohammed Al-Faisal of Saudi Arabia gained publicity for his iceberg-towing scheme at an Iowa State University conference sponsored in part by the National Science Foundation. One of the prince's aides pointed out that between 1890 and 1900 small icebergs had been towed 2,340 miles along the coasts of Chile and Peru. From time to time such an iceberg story appears in the *Los Angeles Times,* sometimes on the front page. One Los Angeles huckster figured that from such publicity spinoffs as films and souvenirs an iceberg landed in that city could yield between $50 million and $100 million in commercial benefits alone, not to mention its value as fresh water. "No other destination for the pilot program," he concluded, "would yield comparable revenues."

The Colorado River Basin Project Act was a very fragile house of cards, with one structural member interlocking with the next and the whole edifice threatening to collapse should any card be substantially altered or eliminated. As the time neared for debate and a vote on the House floor, it was evident that the most vulnerable portion of the act pertained to who would pay for furnishing Mexico with water—the seven basin states or the whole nation. Around this issue had been woven the most delicate consensus. If the little-noticed provision making the future supply of water to Mexico a "national obligation" did not make it through the legislative process, there was an excellent chance that the whole bill would be lost. This was because Arizona could accept California's 4.4 million acre-foot guarantee as that state's price of support for the Central Arizona Project as long as providing Mexico with water was made a "national obligation," an obscure phrase to outsiders but crucial to those who knew its real meaning.

It was the making of western water law at its most arcane and byzantine. What it meant was that when shortages began to appear, the Colorado River basin states would not have to shoulder the whole cost of augmenting the river to satisfy the commitment of furnishing Mexico with 1.5 million acre-feet. The commitment was sealed in a 1944 treaty with Mexico. The Colorado River Compact of 1922 made it the responsibility of the basin states to furnish Mexico with water. The bill placed the cost and responsibility on all fifty states. Simply put, it meant an additional, almost free 1.5 million acre-feet of water, the beneficiaries of which would be the seven basin states whose insatiable thirst was to cause the shortage. There was little sympathy for the obscure maneuver outside the basin, and also little knowledge of it. An amendment to delete the provision narrowly lost in the House committee by a 12-to-10 vote—the closest any amendment unfavorable to western water interests came to acceptance—and the bill at last was sent to the floor of the House.

In the words of Chairman Aspinall, it was time to shift the burden of furnishing water to Mexico "from the backs of the seven children to the entire family." He argued that the treaty was national in origin and scope and did not represent the interests of the basin states. It was the same argument the seven states were to use successfully a few years later when they escaped having to pay for the deteriorating quality of the river's waters, which were by now stifling Mexican crops. Representative John Saylor of Pennsylvania, who had gained a reputation as a conservationist on western issues, represented the loyal opposition. Saylor, the ranking minority member of the House Interior committee, respected Chairman Aspinall. He always played by the rules—a type of opposition the chairman could count on and deal with. Saylor maintained, "This shallow reasoning covertly attempts to saddle the other states of the Nation with the costs of studying, planning and augmenting the water supply of the Colorado River as a federal responsibility and not a responsibility of the seven basin states." Saylor cited one estimate that such a provision could mean a $2.5-billion gift—the equivalent of two "cash register" dams in the Grand Canyon. Along with revenues from the sale of power from other dams and the Navajo coal-burning power plant, this federal contribution would provide a substantial subsidy for any future augmentation works. Saylor lost. The bill, neatly packaged and guided with skill through the House by Aspinall, was approved by a voice vote. The strategy called for avoiding a roll-call vote, the fear being there would be opposition by some members who would feel obligated to vote against a $1.3 billion water project at the height of the Vietnam War, should their votes be publicly recorded.

Would those who benefited from such water projects do so at an unreasonable expense to the nation's taxpayers? The matter of subsidies for federal reclamation projects in the West is a touchy subject for both critics and defenders of the system. Defenders tend to minimize the extent of the free ride and often point a recriminatory finger east at federal expenditures for flood control, harbor improvements, transportation, welfare, housing, and education. The claim is that many of these costs—subsidies if you will—are not repaid to the federal government, as are some of the costs of reclamation projects. Critics poke holes in the defenders' arguments, mostly on the basis of there being substantial hidden costs which are not repaid. During the debate on the floor of the House, Aspinall stated that 90 percent of the cost of all the projects in the 1968 act would be repaid, the total amount then being estimated at $1.3 billion. When the project came under scrutiny by the Carter Administration in 1977, the price had risen to $2 billion. The Bureau of Reclamation said there was a 25 percent subsidy, while reviewers of the bureau's figures within the Administration put the "true" subsidy at something closer to 60 percent. Critics of federal involvement in western water projects forget there was once a very valid national policy of westward expansion which, although no longer cogent, has never been repudiated. Perhaps it would make more sense to offer a subsidy to use less water, rather than to divert more.

There are some basic facts to consider about Arizona. About one-half of the state receives less than ten inches of rain in a year, some areas receiving as little as three or four inches. About 60 percent of the state's water supplies come from underground sources, and the pumping of such water greatly exceeds its replenishment—by as much as 2.2 million acre-feet a year. Arizona's basic allocation of Colorado River water is 2.8 million acre-feet a year. About 1.2 million acre-feet are expected to be used in western Arizona along the river, and the remainder, an amount estimated at 1.6 million acre feet, with some loss due to evaporation, will be delivered to the central and southern portions of the state in the beginning years of the Central Arizona Project's operation. State water planners say this amount will drop to 1.1 million acre-feet by 2020, while critics of the project maintain it will be closer to 600,000 to 800,000 after the year 2000. Federal officials estimate an average 1.2 million acre-foot delivery for the project's lifetime. Increasing upper-basin development, runoff averaging less than 1922 compact allocations, and growing Indian claims to water will eat into Arizona's share. Against the amount of water that might be available, municipal and industrial water users have asked for 600,000 acre-feet of project supplies in 1985 and 1.9 million acre-feet in 2020. Agricultural interests, which use about 90 percent of the state's total water supplies, want 4 million acre-feet. In a vast understatement, the State Water Plan commented on the imbalance, "Obviously, all requests for Central Arizona Project water cannot be satisfied."

In Maricopa County, within which lies the Phoenix metropolitan area, about 2 million acre-feet of underground water are pumped each year, nearly 1 million of which are lost forever. This is where residential water use is 50 percent above the national average and a fountain drawing attention to a desert subdivision shoots water nearly 600 feet into the air. The Carter Administration review team of the Central Arizona Project stated in its 1977 report, "The declining water table has been one cause of wide fears that Arizona at some point not too far off might simply run out of water or that its growth might have to be severely curtailed." But the phenomenal growth rate continued into the 1980s. Right before the last decade ended, the Phoenix area, which refers to itself as the Valley of the Sun, underwent a five-day smog alert—the longest yet for the region that claims its climate is perfect for curing respiratory ailments.

About halfway between Phoenix and Tucson on Interstate 10, not far from where the Hohokam civilization once flourished, is the small, dusty agricultural town of Eloy, whose main distinction is that it is sinking, as is much of the land surrounding it. The U.S. Geological Survey mapped a 4,500-square-mile area in south-central Arizona, where 60 percent of the state's population and a little over half its agricultural lands are located. Here, in the sixty years since 1915, more than 109 million acre-feet of groundwater were extracted. And the land both sank and opened up into jagged fissures. In the 120 square miles surrounding Eloy, the land has dropped more than 7 feet since 1952, reaching a maximum of 12.5 feet south of town. Fissures in the earth opened up, eroded rapidly during infrequent rains, and became trenches more than 1,000 feet long and as much as 10 feet wide and 10 feet deep. One fissure 9 miles long intersects the interstate southeast of Eloy, requiring constant repairs to the highway's surface. Railroads, utilities, irrigation canals, sewage systems, farmland, and private homes have all been damaged—mostly in rural areas. Tucson shifted some of its underground pumping activities to rural areas to avoid subsidence in more populated sections. But the expanding populations of both metropolitan areas were encroaching on the sinking land. The survey's report had little effect in Arizona.

I stood atop Theodore Roosevelt Dam seventy miles to the east of Phoenix on the Salt River, a tributary to the Gila and the river once used to power the Hayden flour mill in Tempe. Roosevelt is the bureau's first multiple-purpose dam (power generation, storage for irrigation, flood control, recreation); the handful of other projects completed earlier, like Laguna Dam, were for such single purposes as the simple diversion of water. The term "multiple-purpose" was to be used as a selling point for subsequent water projects, the idea being there would be a little something for everybody. But Roosevelt was first. It is also the world's highest rock masonry dam, a type of construction no longer used. Thus, it is an engineering landmark, and a publication of the American Society of Civil Engineers comments upon it: "The size and beauty of the masonry arch design, coupled with the natural setting of the Tonto Basin, make Roosevelt Dam one of the most pleasing early engineering efforts in the United States." Dedicated by the former President himself in 1911, the dam, like Hoover and Glen Canyon, is one of the stations of the cross for a movement.

In 1977 giant earth-moving machines and graders, operating in well-coordinated fleets, tore apart and then precisely smoothed a serpentine

The partially completed Central Arizona Project, north of Phoenix.

path through the desert that would be overlaid with gleaming concrete. It stretched east from Lake Havasu in gradually expanding increments. Theodore Roosevelt Dam, a graceful arch sprung against the pressure of the body of water behind it, was the first large structure built to serve the Salt River valley surrounding Phoenix. The Central Arizona Project would be the last. In the early 1900s dams were conceived of in terms of sculpture. Stonecutters brought from Europe to Arizona cut and smoothed the granite blocks for the face of the dam and set them in mortar. As Roosevelt Dam grew, so did the costs, winding up far in excess of the original estimate. There were disputes over who would get the water, and far more requests than could be fulfilled by one water project. Six years after it was completed, Representative Carl T. Hayden was present at the meeting with the Secretary of the Interior at which it was agreed to hand over the operation of the dam from the federal government to the Salt River Valley Water Users' Association, now the Salt River Project Agricultural Improvement and Power District, which draws its power from as far away as a coal-burning power plant near Craig.

Like Hoover Dam, the construction of Roosevelt allowed its progeny to flourish; next on the Salt was Mormon Flat Dam, followed by dams at Horse Mesa and Stewart Mountain, and Bartlett and Horseshoe dams on the Verde River, a tributary to the Salt. Just below their confluence is the Granite Reef Diversion Dam. Orme Dam, a controversial part of the Central Arizona Project and named after one of the early promoters of Roosevelt dam, would be built just upstream from Granite Reef. The resulting reservoir would back water up almost to Stewart Mountain Dam and the Salt River, which now runs dry, would become a series of descending pools, more intensely compressed than those along the lower Colorado.

From the top of Roosevelt, one can hear the sound of dripping water and watch it spread in a dark swath down the steps of granite. It reminded me of the seepage at Glen Canyon Dam and, given the length of time the Colorado River system has flowed toward the ocean, I was struck by the essential impermanence of these imposing structures. Along the top of the dam, as the pink light in the west darkened, I noticed that the bulbs in the old-fashioned light standards had been shot out. Had they functioned, there would have been a definite European cast to the scene, enhanced by the three small turreted structures placed along the curving roadway. As it was, stripped by random gunblasts of any pretension, it was just another western scene.

2

Nowhere on the lower Colorado River does the play get more intense than along the Parker Strip, extending 14.4 miles downstream from Parker Dam to Headgate Rock Dam. This quiet-flowing but raucous portion of the river is also known as Lake Moovalya. Headgate Rock Dam, completed in 1941, diverts the water onto some 100,000 acres of the

Dark swaths mark seepage on Roosevelt Dam.

Colorado River Indian Reservation and creates the stable pool behind it where all types of urban madness occur on weekends in the blazing heat.

This is Johnny Cash country, where a million cans of beer can be consumed in one long holiday weekend. It is the river as theater, where each of the 125,000 weekend visitors compressed into the fourteen miles, and huddled as close as possible to the cool water, can act. To be a protagonist, rather than part of the supporting cast, one needs a sleek water-ski boat in the shape of a tapered arrowhead and powered by a jumble of chrome. Such glittering jet-thrust boats can reach speeds of 95 mph and in the ear-splitting process create tingling ecstasy or instant death. Both seem intermingled on this stretch of the river. Glistening bodies are energized. They throb to the disco beat of waterside taverns, whatever stimulants one chooses to consume, and the large amounts of vibrating, sensual horsepower at one's personal command.

The Colorado is a charnel river at this point—one of the nation's most dangerous inland waterways, says the Coast Guard, which patrols it. Thirty-seven persons died along the strip in that drought year, half of them from boating accidents. Two men died and a third was seriously injured in one hit-and-run boat accident. The lone survivor managed to scrawl the registration number of the fleeing boat in blood on a piece of wreckage. People die, drink, sleep, and make love in their boats along this stretch of the river. All this happens in or upon the water that began to flow pure at its sources, then was laid upon fields or compressed into penstocks along the way, its eventual destiny to be churned by speeding hulls or stained with blood and beer.

On a number of separate occasions I have passed by this section of the river, but I have never lingered for long, so it was with some shock that I found myself caught in one of those turbulent mass outdoor recreation scenes that are seasonally scattered about California. I had seen other such scenes, along the coast and in the mountains, but the lower river had such a deceptive tranquillity about it and the late spring heat seemed too formidable a barrier. Not here, I thought. But I was wrong.

It was further on down the river from the Parker Strip and on a May day that a friend and I launched a two-person kayak into the Colorado at Walter's Camp, eighteen miles south of Palo Verde. My friend was from New York City, freshly arrived the day before in Phoenix, and we were looking forward to a quiet float and paddle trip some forty unchannelized miles through a national wildlife refuge and a state recreation area. Scenically, the stretch from Walter's Camp to the Imperial Dam is the most spectacular on the lower river, with the possible exception of Topock Gorge. It is where the river wanders unhindered, but screened by a thin veil of vegetation, across the desert and through mountains and marshes. Lieutenant Joseph C. Ives had traversed this section of the river

120 years previously in his journey of exploration upriver in a steamboat and remarked, "The mineral wealth of the country somewhat atones for its animal and vegetable poverty, and in a geological point of view possesses a high degree of interest. . . . The mountains passed today—Chimney Peak, the Spires, and the Chocolate Range—have exhibited a rare diversity of outline, colors and tints; and the brilliancy of the atmosphere heightens the effect of every shade and line." A gold mine and stamp mill were eventually built at Picacho. The mill's ruins can still be seen a short distance off the river and not far from the small store in the state park where we stopped to buy beer. I wondered how Ives and his men ever made it up the river without any beer. The two seem to meld together in one continuous flow. It was mid-January when Ives reached this point, but still the days were warm.

We got an early start. The mosquitoes had soon made sleep impossible under the tamarisk trees where we had dropped our sleeping bags late the previous night. Before departing I made a brief stop at Walter's to purchase some last-minute needs and overheard a conversation about a regatta among the early-morning patrons who were already guzzling beer. Exactly what they said did not register; I was too preoccupied with the chores of departure.

The first few miles on the river were utterly peaceful. The current kept us moving, with just a few strokes of the paddles, and there was that evenness in the temperature of the air and water that made both inviting. Then we heard a faint buzzing sound, which quickly grew in intensity. It was the forerunners of the Blythe Boat Club's annual regatta—billed as the largest freshwater regatta in the United States. The lead boats in each succeeding wave carried a red flag with the sign "pilot" emblazoned on their sides. Eventually some three hundred motorboats of every description were to pass us, sometimes in narrow stretches of the river where sandbars impinged on both sides.

The people waved from their fast-moving boats and wondered why we did not wave back. To have done so meant losing headway, a precious commodity when rowing or paddling, since we would have had to drop our paddles and in the process lose our meshed rhythm. Besides, we were the oddity. Then there were the boats' wakes to contend with, bad enough on the initial pass through but raising a confused chop when they rebounded from the river's banks to meet fresh wakes. Lastly the cacophony, including every possible pitch the internal-combustion engine could make, intermingled in absolute dissonance, making neither of us overly happy.

As the final insult, a water skier approached us silently from the rear after his noisy towboat had passed and on one ski cut an arc not two feet from the kayak, sending a sheet of water upon us. It was like dumping cold water on a teakettle that was about to boil. I was barely able to contain my anger and was sorry my double-bladed paddle had not been fully extended so I could have cut the skier off at the knees. I later read a boating safety pamphlet which stated that Colorado River accidents "are usually the result of improper or illegal skiing practices and carelessness." Had my fantasy been realized, I am sure the jury would have acquitted me.

I think the trip was ruined at that early point for both of us, and for

Heavy weekend recreation use along the Parker Strip, which borders Arizona and southern California.

the rest of our time on the river we would try to find what little tranquillity we could in the midst of madness. For the remainder of the trip my friend seemed dazed—out of his element, but not certain what new element he was having to deal with. For myself, I felt a keen loss at not being able to experience what I had looked forward to—a quiet trip down the last few navigable miles in this country of a river I had followed from its sources. But I realized what I was witnessing was a more realistic scene for this portion of the river than what I had hoped for. It was like a rock music beat; it finally got to you no matter how much you wanted to avoid it.

The next day was a repetition of the first, except the motorboats were returning upriver and their wakes were easier to contend with since they quartered on the bow of the kayak. We stopped at Fisher's Landing to watch the swirl of boats depart. Others had arranged their folding lawn chairs on the raised banks. It was like having a box seat at a freeway interchange. My friend seemed more remote, more withdrawn from what was happening around him. I was beginning to feel the throbbing rhythm.

The sun was more incessant on the second day, making it necessary to cover ourselves in a hot climate as we climbed back into the kayak. We differed briefly about what route to take and I won, since I controlled the rudder. The last few miles of paddling were against a headwind, which we tried to avoid by sticking to the backwater channels of the sediment-filled reservoir behind Imperial Dam. Imperial Oasis, a rude collection of parked recreational toys, emerged suddenly into view from around a bend in the channel on the Arizona side of the river. We were not displeased the river trip had come to an end. A scratchy, dry wind blew off the Chocolate Mountains to the west.

About 85,000 people live in Imperial County, while the Metropolitan Water District, which wholesales Colorado River water to southern California cities, serves an area containing 11 million people. Yet 80 percent of California's entitlement to the Colorado goes to four agricultural districts, of which the Imperial Irrigation District is by far the largest. The district is also the single largest consumer of Colorado River water, annually diverting an amount of surface water far in excess of what any entire state outside of California uses from the mainstream.

In the not untypical year of 1976, California diverted 5.2 million acre-feet of water, an amount just about equal to the state's average diversion for the five previous years. Of that amount, 2.9 million acre-feet went to the Imperial Irrigation District and 800,000 to the Metropolitan Water District of Southern California. The irrigation districts serving the Coachella, Palo Verde, and Yuma project areas accounted for the remainder of the diversions. About 535,000 acre-feet of water drained back into the river mostly from the Palo Verde Irrigation District, to be used again downstream. Thus, California's net or consumptive use of water was about 4.7 million acre-feet in 1976. The Imperial Valley is the Imperial Irrigation District, spread over some 490,000 exceedingly fertile acres.

It was George Wheeler who guided me through the district. To Wheeler, it is The District. He has been employed by it almost his entire adult life, having started in 1946. His father worked for The District before him. "The District means something to me," he explained. "It's not perfect. But it's my life, my sweat." Wheeler, who worked for an irrigation district with nearly 1,000 employees near the end of the river, reminded me of Harv Stone, the single part-time employee of the Canyon Ditch Company, which diverted the first water from the Green at the headwaters of the entire Colorado River system. He had the same solidness and sense of purpose, revolving around the beneficial use of water. Stone saw the land being carved up into small lots around him, while Wheeler sensed a loss of unity in the Imperial Valley. The people in the towns, rather than the rural farmers, now controlled The District, and he felt Los Angeles would eventually get the water.

I would hear echoes of this theme throughout my travels in the West —urban-industrial interests about to eat into the traditional use of large amounts of water for agricultural purposes. Wheeler belonged to the Rotary Club, where he had a thirteen-year record of perfect attendance. He neither smoked nor drank. One of his fondest memories was of how everyone had pitched in to repair the damage to The District's facilities caused by a tropical storm a few years ago. There had not been as much community cooperation since the 1940 earthquake. The valley once suffered another natural disaster that was enhanced by one of the greatest engineering mistakes of the twentieth century. It had spent the next seventy-five years trying to insure itself against a repetition.

Around 1850 the idea that Colorado River water could be made to flow into the Imperial Valley occurred separately to two men of different temperaments. Dr. Oliver M. Wozencraft, an Ohio physician who had come to California during the gold rush, traversed the valley in 1849. He was a dreamer, but his vision had some practical aspects to it. Wozencraft saw that the valley could be irrigated by diverting the water through an overflow channel of the river, called the Alamo River, which ran north across the border from Mexico. And he wanted to own all 1,600 square miles of the valley, then called the Colorado Desert or Salton Sink. The California Legislature went along with the doctor, who also served as an Indian agent, but Congress balked at the proposal to hand over the lands in the public domain, and Wozencraft's vision was shattered, to be unsuccessfully pursued until he died. He had made the first appeal to the federal government to aid in the development of the valley.

A geologist, William P. Blake, passed through the valley in 1853 while making a railroad survey, and noticed that the land sloped gradually downhill as he proceeded north. He took barometric readings, which confirmed that the desert was below sea level, and noted what looked like the indentation of an ancient lake's shoreline along the foot of the mountains. Blake correctly deduced that the Colorado had been diverted periodically from its path to the Gulf of California and filled the ancient lake bed, now known as the Salton Sea. He voiced some doubts about any irrigation scheme that would divert the Colorado; they proved to be prophetic: "It is indeed a serious question, whether a canal would not cause the overflow once more of a vast surface, and refill, to a certain extent, the dry valley of the ancient lake."

Wozencraft's dream, minus the part about having the land taken out of the public domain and ceded directly to the promoter, was given new life when Charles R. Rockwood formed the California Development Company in 1896. Rockwood's choice for a canal route was the same as the unfortunate doctor's, but he had the promotional know-how and the management skills of a man who had worked for the Southern Pacific Railroad. Rockwood enlisted the services of George Chaffey, the well-known promoter of two successful southern California communities and an ill-fated irrigation scheme in the Australian desert. It was Chaffey's idea to give the desert the imposing name of the Imperial Valley. In such a way has many a western land-development scheme been launched. Advertisements depicting the potential lushness of the bare scrub-and-sand desert flooded the country. It was the time when the Reclamation movement was about to crest with the passage of the 1902 act; its chief booster, William E. Smythe, wrote of the valley's promise in a 1900 issue of *Sunset,* a magazine that has historically promoted the idea of vegetative fecundity in the West:

This vast plain of opulent soil—the mighty delta of a mighty river— is rich in the potentialities of production beyond any land in our country which has ever known the plow. Yet here it has slept for ages, dormant, useless, silent. It has stood barred and padlocked against the approach of mankind. What is the key that will unlock the door to modern enterprise and human genius? It is the Rio Colorado. Whoever shall control the right to divert these turbid waters will be the master of this empire. Without the right and the ability to use this water nothing is possible.

The first settlers arrived in January, 1901. They were three families from the Salt River Valley who, hearing of the pending development, loaded their goods into three wagons and set out for the short trip west driving five cows and one bull. They also had two Fresno scrapers, ten crates of chickens, and seventeen children between them. They built a crude raft to ford the muddy Colorado and upon arrival in the valley of sand—actually Colorado River silt, 20,000 feet deep in places—went to work for the California Development Company building the first canal which would provide the means to put a wafer-thin layer of vegetation upon those miles of accumulated river sediments. The first water reached the Imperial Valley on June 21, 1901. Within months the land boom was on, as it was to be elsewhere in the West where water was either promised or actually delivered to dry land. The succeeding years were marked by quick growth and fraudulent land practices. Water was money, and land was the only tangible medium which expressed it. As Helen Hosmer wrote in *The Grand Colorado* of Chaffey's departure from the company in 1901, "But Chaffey had given them more than a crude canal and wooden headgate. He had perfected a magic formula for debasing the intent of the Homestead and Desert land acts. He had designed a mechanism for turning water appropriation into land appropriation."

But even Chaffey's headgate and canal were not to survive for long against the uncontrolled thrusts of the Colorado River. Not only were the crude headgate and beginnings of the canal subject to the vagaries of the river but, in order to avoid the sand hills descending to the border, the canal had to be routed through Mexico. There were problems. The canal was blocked with silt on occasion, thus halting deliveries of water to the valley, and the Mexicans demanded concessions in the form of water from the canal. In the spring of 1905 there was a series of floods on the river, and attempts to hold the intake to the canal proved vain. Finally in August the swollen river breached the intake, widening it to a half mile, and the entire flow of the engorged Colorado washed across the Imperial Valley to settle in its old resting place, the Salton Sea. The silt canal banks gave way much like a sugar cube does under the deluge of warm, running water in a teacup. The paper value of the California Development Company dissolved almost as quickly.

The Southern Pacific Railroad took over the battle to seal off the river from the valley, and finally gained control in February, 1907, after the expenditure of tremendous amounts of manpower, machinery, and money. The river then resumed its previous path to the gulf. But Blake's prognosis had proved accurate. It was one of the greatest engineering mistakes of the century, to be equalled in southern California only when the Saint Francis Dam north of Los Angeles ruptured in 1928, sending a wall of water through Ventura County that killed 385 people in its path to the ocean. Those, along with the Teton Dam disaster in Idaho in more recent years, showed that water could take away just as well as it could give life.

From this disaster in the Imperial Valley grew the residents' conviction that a canal with a permanent headgate that could not be broached or clogged with sediment and a route entirely within the United States—an All-American Canal—was needed, was, indeed, an absolute necessity if the valley wanted permanent security. In 1911 the Imperial Irrigation District was formed with these eventual goals in mind. The pattern of federal aid coming to the valley's rescue was established when the railroad refused to halt a 1910 flood and Congress appropriated $1 million.

The idea for a canal entirely north of the border dated back to 1876, when the U.S. Army Corps of Engineers took a quick look and judged it impractical. The newly formed Reclamation Service later examined such a project, but found the cost too high. The cost was a factor in the Imperial Irrigation District first backing off from such a plan in 1912; but the district later reversed itself, when faced by a rival valley group, and asked the Secretary of Interior in 1917 for a survey of a canal beginning at Laguna Dam. Two years later the district's voters endorsed such a proposal, along with looking at a much larger project that would include major storage facilities on the river. The Department of the Interior and Congressman Carl Hayden favored the second approach. Hayden cited the flood-control benefits for irrigation projects on the Arizona side of the river, and the department and the Reclamation Service saw a more comprehensive way to deal with the river's problems and establish it as their fief. Imperial Valley interests then swung behind this approach, which would result in the authorization of Hoover Dam and the All-American Canal in 1928. As Norris Hundley, Jr., wrote in *Water and the West*, "In effect, the Reclamation Service's desire for a program of comprehensive development and the valley's demand for an All-American Canal had produced a powerful alliance. Though informal and uneasy, that alliance would alarm and frighten leaders throughout the Colorado River Basin."

To get a large storage dam on the lower river authorized by Congress, a near-consensus among the seven basin states was needed. There would be no such consensus until each state was assured a share of the river's waters—not only assured for the moment, but for all time against its neighbors' greed. Thus the idea for an interstate compact. Hundley wrote, "Essentially the situation called for bartering. The lower basin wanted a dam. The upper basin wanted protection, and each concluded that they could probably best reconcile their interests in a compact."

Los Angeles swung its support behind the Imperial Valley's desire for a high dam in either Boulder or Black Canyon. Thus, the most potent political forces within California that had an active interest in the Colorado River were unified. Los Angeles wanted Hoover Dam for power initially, and only later would it want the water. Although the compact was signed in 1922 by the states' representatives to the negotiations, its ratification by the separate state legislatures dragged on. (One of those who played a major role in drafting the compact was Winfield S. Norviel, Arizona state water engineer and cousin to Wayne Norviel Aspinall.) California was the specter. It badly wanted the dam and canal and, because it seemed likely that Arizona would not ratify the compact, it sought to assure Arizona by limiting itself to a 4.4-million-acre-foot share of the lower basin's allocation. Such a limitation was written into the Boulder Canyon Project Act, and it was this limitation that the Supreme Court upheld in 1964.

The Boulder Canyon Act was passed in 1928, and its provision for six-state ratification of the compact to make it binding, thus bypassing intransigent Arizona, was not satisfied until Utah became the sixth state to approve it the following March. The act was officially proclaimed effective on June 25, 1929, by the newly elected President Herbert Hoover, who had participated in the compact negotiations as Secretary of Commerce. Hoover's Interior Secretary, Ray Lyman Wilbur, was to play a decisive and controversial role in shaping the valley's development in the dying days of that Administration. It was Wilbur who changed the name from Boulder to Hoover Dam, a surprise move that did not stick until Congress made it permanent years later. Actually the dam was in Black, not Boulder, Canyon.

Regardless of the name, it was the dam that would smooth out the flows of the lower Colorado, thus preventing unwanted floods and storing the water for downstream use by the Imperial Valley and others. It was a fabulous free gift for the valley. Advertisements throughout southern California at the time proclaimed: "Buy land in Imperial Valley now; speculate on Boulder Dam." Again, the land rush was on with the promise of a permanent, reliable water supply. As a senator commented in the debate on the act, "Then it is apparent that the residents of Imperial Valley will have the benefits of flood control, storage water, the certainty of getting an equated flow, and will be required to pay for nothing except the cost of the All-American Canal." The interest-free canal would be paid for out of power revenues collected by the Imperial Irrigation District from generating facilities located along drops in the canal; power sales in coastal southern California would pay for the dam.

But the bill seemingly contained one obligation. It was an obligation spelled out in the Reclamation Act of 1902 and in a clarifying bill passed in 1926, with both being incorporated into the Boulder Canyon Project Act—and that was to limit to 160 acres (320 acres in the case of a husband

The power plant at Hoover Dam.

and wife) landholdings benefiting from federal reclamation projects. The valley's efforts to free itself from this obligation would extend over fifty years and, in the process, assume the proportions of a classic western struggle.

Shortly after the act became effective in June, 1929, the valley began its long campaign to exempt itself from the acreage limitation and residency requirements of the 1902 law. The 160-acre provision and the requirement that farmers live on or near their land that would receive water from federal water projects were the closest this country was to come to agrarian land reform, to halting the monopolization of land and water resources in the West. Senator Paul H. Douglas of Illinois, an enthusiastic advocate of the 160-acre concept, testified in 1958 before a hostile subcommittee consisting of five western senators, "Probably in the press of affairs here, we rarely can pause to realize that in this 160-acre restriction in reclamation law, we have the only vestige with which I am familiar on our Federal statute books of land reform that has become such a burning issue throughout the rest of the world and that has swept with such passion through many nations that bloodshed and revolution have been the unavoidable result."

The provisions of the 1902 law were only sporadically enforced by the federal government; the trouble was that, while the intent was commendable, the realities of the West did not fit the almost mystical belief in the sturdy yeoman farmer that was nurtured in the humid East and later institutionalized as a rigid number in the arid West. As Walter Prescott Webb stated in *The Great Plains,* "The land laws were persistently broken in the West because they were not made for the West and were wholly unsuited to any arid region." The demonstrable need for a larger unit of land, or a flexible limitation, has forever clouded the issue of land reform in the arid regions. There was no question but that something workable was needed to deal with the problem of the increasing size of large family and corporate landholdings in the irrigated West.

The idea of limiting the amount of land that would benefit from federal water projects had its roots in the European cultural tradition that romanticized the concept of farmers working the soil. The life of a peasant was beautiful, wrote the poets, when reality lay somewhere else. From this celebration of the pastoral and bucolic, which was essentially passive, the small farmer became a symbol of more aggressive republicanism, still wedded essentially to the virtues of the land but now determined and sturdy. The American Revolution brought this concept to the East Coast, where Benjamin Franklin celebrated the virtues of the hardy frontier farmer. But it remained for Thomas Jefferson, writing from his well-watered and -manicured Virginia estate, to become the chief spokesman for the agrarian ideal that was to be transported west. "The small landholders are the most precious part of a state," wrote Jefferson. As President, Jefferson embarked on a policy of westward expansion that was to give his ideas a practical outlet. He dispatched Lewis and Clark to the Pacific Ocean and consummated the Louisiana Purchase. Historian Henry Nash Smith, who referred to Jefferson as the "intellectual father" of the westward movement, wrote of the agrarian ideal in *The Virgin Land:* "The master symbol of the garden embraced a cluster of metaphors expressing fecundity, growth, increase, and blissful labor in the earth, all centering about the heroic figure of the idealized frontier farmer armed with that supreme agrarian weapon, the sacred plow." The region would suffer periodic legal and political convulsions because of the inapplicability of Jefferson's ideas to a dry landscape and the hostility of the natives to outsiders' attempts to impose alien doctrines incompatible with local customs.

The western embodiment of the Jeffersonian ideal lay in the limitation of each landowner to 160 acres of public-domain land, written into the Homestead Act of 1862. The act provided the land, but not the water to make it fruitful where rainfall was slight. Such a limitation, it was thought, would ensure a region of family-sized farms, all imbued with some measure of prosperity.

The figure 160 or its integral fractions and multiples, which could range from 40 to 640 acres, were embedded in a number of western land laws prior to passage of the Reclamation Act of 1902. It was a derivative of the rectangular survey—640 acres being a section of land, a square mile—which was much better suited to the flat, evenly watered lands to the east than to the wildly fluctuating extremes of the West, where the only unifying factor was the irregularly shaped watershed. But the 160-acre figure wound up in the 1902 law without much debate as to its relevance. President Theodore Roosevelt wanted it included, as did the bill's chief proponent, Congressman Francis G. Newlands of Nevada, who stated: ". . . and the very purpose of this bill is to guard against land monopoly and to hold this land in small tracts for the people of the entire country. . . ." Newlands and other westerners perceived that the bill would not get the President's support, nor that of many eastern congressmen, unless such a provision was incorporated into it. The act, when passed and signed by President Roosevelt, read, "No right to the use of water for land in private ownership shall be sold for a tract exceeding 160 acres to any one landowner, and no such sale shall be made to any landowner unless he be an actual bona fide resident on such land, or occupant thereof residing in the neighborhood of said land. . . ."

Eastern opponents of the bill had some doubts about the adequacy of its antimonopoly provisions. As embodied in the later Boulder Canyon act, the land restriction was so hedged about with qualifying language that it seemed to have been deliberately obscured—as witness subsequent court decisions. This obfuscation fit the interests of both those for and those against the acreage limitation in the Imperial Valley; each side could read into the legislation what it wanted. The ultimately successful struggle of the remote, hot valley to rid itself of the limitation and its undue influence on policy set in Washington, D.C., are a fascinating example of the failure of land-reform concepts bred elsewhere and attached to western water law.

Shortly after the Boulder Canyon act became effective, negotiations began between the Imperial Irrigation District and the Department of the Interior on a contract for the All-American Canal. Under the terms of the act, the federal government would build and finance the canal—interest-free—and the district would eventually pay for it. There would be no charge for water. It was the custom for the 160-acre limit to be incorporated into such contracts, as was done in a later contract with the Coachella Valley water district, which would be served by a branch of the All-American Canal. Early in the contract negotiations with the Imperial Irrigation District, there was a possibility the acreage limitation would be included in the contract. But that was not to be. The first draft of the contract contained such a provision. Northcutt Ely, who was then executive assistant to Interior Secretary Wilbur, answered a legal query thus: "I see nothing to do but enforce it unless the Imperial Irrigation District can get new legislation. In any event, enforcement of this requirement would undoubtedly have a salutary effect on suspected speculation activities in that locality." Not only Ely, but the assistant commissioner and chief counsel of the Bureau of Reclamation, Porter W. Dent, thought the limitation should be in the contract. But the irrigation district and its attorneys went to work on federal officials, who badly wanted to build the project, and subsequent drafts of the contract contained no mention of the acreage limitation. It had been reluctantly dropped by federal officials after relentless pressure from the district, which argued it really did not apply to the valley after all and its inclusion would make it difficult to obtain ratification of the contract by valley voters. From the start, political expediency was the dominant factor.

Landowners in the valley overwhelmingly approved the contract in February, 1932; but there was one other matter to deal with before construction could begin on the canal. The contract had to be confirmed first by a court decree—not just any court, but the highest appellate court, if the lower court decision was appealed. At the time there happened to be a troublesome case pending in Imperial County Superior Court. Charles Malan, one of the few dissident landowners in the valley, claimed the contract was invalid because it did not contain the acreage-limitation provision. Malan's legal brief, prepared by attorney Marvin W. Conkling, maintained: "Not in one, but in some three places in Reclamation Law it is distinctly provided that no water shall be delivered from such canals to the owner of more than 160 acres of land. Counsel for the District admit that if such provision applies here the contract is invalid." The Coachella Valley water district had also challenged the contract in court, but for different reasons. The Imperial Irrigation District was worried about the Malan suit, so it approached Interior Secretary Wilbur through an intermediary who sent one of those periodic communications dealing with sensitive Colorado River matters that are marked "confidential." The district wanted a favorable ruling from the department on the omission of the limitation provision, to impress the local court. Such a formal ruling was desired, wrote the bureau's district counsel in Los Angeles, Richard J. Coffey, only if it was favorable to the district's interests. Coffey's letter to Dent, his superior in Washington, added, "He [the district's attorney] doesn't want any formal ruling, of course, if the Solicitor were to hold that the limitation applies so far as Imperial Irrigation District is concerned."

The Department complied with the district's request for a favorable ruling, which did not exactly surprise the district: it had already been assured in informal conversation that the solicitor, albeit reluctantly, would supply such a ruling and Ely would recommend that Wilbur sign it. These hurried actions took place in the last days of the Hoover Administration. Ely directed Assistant Reclamation Commissioner Dent to draw up the letter and on February 24, 1933, Interior Secretary Wilbur signed it.

Although the district had requested a formal opinion, what it got was an informal letter which had less weight. This was a conscious decision by the department, which had been rushed by the request and about which there were some lingering doubts. It was this letter the Imperial Irrigation District and valley residents would ceaselessly and vociferously claim exempted them from limiting the size of their farms—completely disregarding its self-serving aspects. The valley would persist in its claim, sometimes stridently, and the federal government would not enforce the law despite one later formal administrative opinion to the contrary. The 1977 opinion of a federal appeals court stated: "The Department of Interior itself began to have doubts about the legal soundness of the Wilbur letter's conclusions shortly after water deliveries through the All-American Canal began in the early 1940s but continued to adhere to the practice of nonenforcement of acreage limitations in the Imperial Valley because of its previous practice of nonenforcement." The letter, said the San Francisco court, was "itself an informal opinion that is legally incorrect," and as for the issue of nonenforcement, "Inaction based on previous inaction cannot be elevated into an administrative determination to which the courts should defer." But what should be done? Should the economy of a whole valley be disrupted?

By itself, the Wilbur letter probably had little to do with deciding the Malan case. Most likely, the local judge was in any case favorably disposed to the district's arguments. It would have been political and social suicide to be otherwise in that tight little community totally dependent on the existence of a reliable canal system built from the Colorado River. The letter did not even address a principal legal issue, a key section of the Boulder Canyon act, but rather placed all its reliance upon an interpretation of the 1902 law. It was the Boulder Canyon act on which were based both the 1977 appeals court decision and the Supreme Court's 1980 reversal of that decision.

In later years the Department of the Interior itself would abandon the Wilbur letter as a means of justifying its failure to seek compliance with the acreage-limitation law. Shortly after the letter was issued, the attorney for the district expressed dissatisfaction with it. The local court not unexpectedly ruled in 1933 that the acreage limitation did not apply to the valley, hence the contract was valid. But a different ruling might be expected from a less friendly appeals court located outside the Imperial Valley, and the strategy was to halt such an appeal—a similar strategy under like conditions being used to halt another such court appeal nearly forty years later.

The Imperial Valley called upon its friends in Washington. The Bureau of Reclamation warned the Coachella water district that it would get no water unless the appeal were dropped. Coachella complied, leaving Malan as the only obstruction, along with his steadfast lawyer who wanted to press the appeal. Conkling, who came from a prominent valley family, knew water law and the valley, having helped organize the Imperial Irrigation District. He had also served as its attorney in its early years, and had been a superior court judge in Imperial County. Altogether, Conkling was a dedicated, formidable opponent in any case involving water law. The districts' strategy—by now Coachella was allied with Imperial—was to concentrate on Malan, the owner of 210 acres. Conkling complained to the new Interior Secretary, Harold Ickes, about the pressure being put on his client to drop his appeal. He also pointed out that a state-court decision was not binding on the federal government, a technicality that was to be overlooked. Both water districts felt confident they could pressure Malan to change attorneys. Coffey wrote Reclamation Commissioner Elwood Mead in February, 1934, that ". . . the two boards of directors feel confident that they will be able to force dismissal of that appeal. His attorney won't do so, and arrangements are now being made to substitute attorneys, the new attorney to take appropriate action as soon as he has been substituted for Judge Conkling, his present attorney."

By the end of the month, both appeals pending before the California Supreme Court had been dropped, and the lower-court decision thus made final. By the end of March the $6 million that had been appropriated to begin construction of the canal, and impounded by the Comptroller General until the outcome of the court appeal, was released. It was the worry about losing the appropriation that had contributed to the districts' haste in the matter. The valley would soon have its assured water supply.

The first trickle of Colorado River water reached the valley through the new canal in 1940, but it was not until 1942 that the canal would carry a full flow. Fourteen years later the Bureau of Reclamation reported, "the increase of farm crops in the Imperial Valley has been phenomenal." Comparing 1954 with 1940 the bureau reported a 1,122 percent increase in the value of field crops, 274 percent in livestock, and 167 percent in vegetable crops—with only a 9 percent increase in the amount of area irrigated. The per-acre yield would be the deciding factor in the huge increases, now that there was a dependable, year-round source of water. Yet in the same period there had only been a minute increase in population, indicating that a number of new farms had not materialized in the valley along with the increased yields. Indeed, it seemed the process of consolidation of farmholdings had speeded up with the opening of the new canal. Land speculation was once again rife in the valley.

Two years after the valley began to receive all its water via the new canal, Northcutt Ely, by now in private law practice in Washington, testified before a Senate subcommittee that acreage limitation was simply being ignored in the Imperial Valley. That would be the inevitable result, he said, as long as the law remained on the books. The next year, 1945, Ely went to work for the Imperial Irrigation District on a retainer basis, dealing with problems arising out of the Mexican water treaty. In the next few years Ely, who retained excellent contacts within the Department of the Interior, would aid the district in unsuccessfully seeking a legislative exemption from the 160-acre-limitation law, indicating that valley interests did not feel all that secure with the Wilbur letter. The Washington lawyer successfully represented the district in 1980 before the Supreme Court. In the intervening years Ely represented California water interests in a variety of forums pertaining to the Colorado River.

Also testifying at those same Senate subcommittee hearings in 1944 was William E. Warne, assistant commissioner of reclamation, who, after some initial confusion, referred to the Wilbur letter as being the reason why the acreage limitation provision did not apply to the Imperial Valley.

It was the last time there would be a clear reference by the department to the letter as a defense for noncompliance; the argument later turned, instead, to whether it would be fair to yank the carpet out from under valley residents who, after all these years, had put their reliance on the letter—regardless of whether its arguments or the way it was solicited were valid or not. Doubts within the department as to the letter's validity arose in late 1944 when negotiations were started for a supplemental contract with the Coachella Valley County Water District, the original 1934 contract having contained no reference to the limitation. The doubts led to a formal opinion the next year by Solicitor Fowler Harper that the 160-acre limit did, indeed, apply to Coachella. Of the Wilbur letter, Harper wrote, "Purposely, the letter of Secretary Wilbur never took the form of a formal decision. It was written solely for the purpose of giving partisan help to the Imperial Water District, as the delay of the final confirmation of the contract held up the construction of the All-American Canal." This was the first formal ruling, but it was peripheral to the issue of the Imperial Valley as it was directed at the Coachella Valley, although supplied by a branch of the same canal.

By 1962 Interior Secretary Udall could report that the Coachella Valley was in compliance with the acreage limitation but in the neighboring Imperial Valley, where no exact records had been kept because of the Wilbur letter, ". . . we would assume from general knowledge that there are considerable large holdings and that they have been increasing." In 1974, according to the Department of Agriculture, the 715 farming operations in the Imperial Valley averaged 716 acres per farm, with 335 being under 180 acres and 380 over that figure—73 farms were between 1,000 and 2,000 acres, and almost an equal number exceeded 2,000 acres.

Consistency in the application of the law was lacking. The Coachella Valley was made to comply, the Imperial Valley was not. In Utah there would be a preponderance of small farms receiving water from federal reclamation projects, while in Arizona large corporate ventures benefitted from such largess, and there was no assurance that those getting water from the Central Arizona Project, when it was completed, would be made to lessen their holdings. In 1957 Solicitor General J. Lee Rankin argued in the California-Arizona Supreme Court case that "no conclusion seems permissible other than that" the limitation should apply to private lands in the Imperial Valley. On the other hand, said Interior Department Solicitor Elmer F. Bennett the next year in a letter to Rankin, "the time has long since passed when it is realistic and practicable" to reexamine the issue. But in December, 1964, another departmental solicitor, Frank J. Barry, issued the first formal opinion on the Imperial Valley, and he came down solidly on the side of applicability: "The fact the department has failed for over 30 years to enforce acreage limitations in

Imperial Valley cannot legitimize a violation of public policy contrary to the spirit and the letter of the law." Again, there were differences within the department, with Undersecretary John A. Carver complaining to Udall about the opinion's legal insufficiency. But the secretary stuck with Barry, his former congressional campaign manager and close friend.

All hell broke loose in the valley. The water was going to be turned off, it was feared. Land reform Russian and Cuban style, said some. Economic ruin, said others. And within a week after the opinion was issued, two old men traded blows at a public meeting of the irrigation district's board. The two were protagonists—Harry Horton, the district's longtime counsel and defender of the status quo, and Dr. Ben Yellen, a more recent arrival and crusading land reformer.

Feelings were high, and tempers flared. Glasses askew, Yellen shouted, "He just committed an assault," after having accused Horton of being "a stooge for the big farmers." That was more than the lawyer could take, and he replied, his face getting redder, "You've been getting away with this for years," whereupon he lunged at Yellen declaring, "I'm seventy-two years old but I'll still clean your wick." The combatants were pulled apart, but Yellen would remain an irritant to the valley—in fact, its chief nemesis.

There should have been no fears that the flow of Colorado River water through the canal would be halted. The weight of history was on the side of the valley. Water would keep running despite the valley's refusal to agree to dispose of excess lands by January 1, 1966. That deadline passed and the federal government decided to go to court, filing suit against the district in January, 1967. Thirteen years later, on appeal to the Supreme Court, the suit was finally decided, thus eclipsing the twelve-year length of the California-Arizona suit. Yellen would join in the suit and along with his attorney, Arthur Brunwasser, and a retired University of California economics professor, Paul S. Taylor, form a triumvirate that would manage, through various obstacles thrown up by the water interests, to get the case appealed. Taylor would provide the philosophical framework, Brunwasser the legal expertise, and Yellen the sheer drive and determination, along with the ability to garner publicity for what seemed like a lost cause. All three were outsiders—Taylor and Brunwasser from the San Francisco Bay area and Yellen more a product of his native Brooklyn, New York, than his adopted Imperial Valley. Besides conservationist organizations, the triumvirate would be the only outside group to effectively challenge western water interests. For this reason, and because it demonstrated the cohesiveness, lasting power, and political muscle of basin water users, their story deserves telling.

Ben Yellen was the principal force, the bundle of raw, raucous energy that propelled the challenge forward. It was a legal challenge aimed at the very heart of the valley's economic structure. Because of his persistence and grating manner, Yellen would remain a permanent outsider—a loner and a dedicated revolutionary among the landed gentry. Yellen got attention by his outrageous behavior, which masked a native canniness. Considering that he was a doctor to the poor in a small town with what seemed like a lost cause, Yellen was a tremendously successful manipulator of the media. Yelling Yellen. His voice beats upon one, then lowers to a conspiratorial tone, then comes on again rasping, accusing with the inflammatory phrase, punctuated by the excited gesture, the poking finger in the chest. When the San Francisco appeals court still had not ruled three years after oral arguments were heard, Yellen traveled to that city from his hometown of Brawley to picket the court, declaring simply, "The court is constipated." A newspaper published the story, with a picture of Dr. Yellen declaiming in that drought year; the court ruled, perhaps by coincidence, shortly thereafter.

Yellen's newsletter, mimeographed in the doctor's office in the best pamphleteer tradition, went into greater detail about the court's lack of bowel movements. Other issues of his newsletter, printed on single yellow sheets (Yellen's brand of yellow journalism), have taken on the large landowners, that one being titled *The Chiselers;* absentee landlords in *Tax Shelter U.S.A.;* directors of the Imperial Irrigation District in *Kick Out the Crooks;* and the federal district court judge in San Diego who ruled against acreage limitation requirements for the valley in *Judge Howard Turrentine Is a Polluter of Justice.* Yellen's personal assessment of the impact of the Turrentine newsletter was, "I gave him the hot foot." Yellen was sued once because "I called a guy an ignorant jackass." But the yelling paid off. He was interviewed by Mike Wallace of CBS and Calvin Trillin of *The New Yorker,* plus a multitude of other national and regional publications.

Yellen, who came to the valley the same year the All-American Canal delivered its first full load of water, was not getting around as much as he used to when I visited him in his slightly decrepit but serviceable office crowded with welfare patients. In his seventies, the doctor was suffering from heart and asthma problems. On the wall was a reproduction of the Norman Rockwell painting of a small boy pulling up his pants after just getting a shot and examining the doctor's diploma on the wall. A sign in Yellen's office stated, "Welfare patients must pay $2 at each visit for the shots as the government will not pay for the shots." Dr. Yellen shouted from an inner office, there being no receptionist, "Number thirteen, please," and a young Mexican woman led four small children into the examining room. Yellen's office offers a perspective on the downtrodden of Imperial Valley, and it is from this vantage point that the doctor has launched his crusade. It is a point he frequently makes, knowing the underdog is more popular: "It was one man upsetting the applecart, you know. It was a challenge. For a few dollars I could have some fun. What else am I going to do? Go out and hit a golf ball like other doctors? This is how I get my kicks. I've got the truth. All I've got to do is tell the truth." Yellen could be anyone's kindly old grandfather, except that he has never married and is possessed by a magnificent obsession.

Paul S. Taylor was another old man, perhaps best described as a scholar, rather than an activist like Yellen. Through the years Taylor had become the conscience of the Reclamation movement, which he followed more as a social critic rather than an economist, but with the detachment of an academic. Yellen found Taylor at a time when the doctor was feeling particularly isolated in the Imperial Valley, his practice having been endangered because of his interest in protecting the medical welfare of Mexican farm laborers. Yellen wanted a way to get back at the large landowners, who he felt were the powers behind a boycott of his practice. He wanted to "fight the boycott by ruining the boycotters," said one of his newsletters. Yellen visited the Department of the Interior in Washington in 1961 and heard of the professor, who was an authority on reclamation law, from Gilbert G. Stamm, then an assistant but later to become commissioner of reclamation from 1973 to 1977. Taylor recalled, "Stamm told Yellen if he came to me he could find out what the reclamation law was all about. What I told him was that the law was being eroded." It must have been a referral that Stamm later deeply regretted. New vistas opened up for Yellen. Taylor had a wealth of knowledge and was willing to share it with Yellen, the street-wise activist. Additionally, Taylor had the prestige that went along with publishing regularly in reputable journals and having served once as a consultant to the Department of the Interior. Taylor's articles would later be cited in court opinions dealing with the acreage-limitation law. In his office on the Berkeley campus of the University of California, there is that same clutter of miscellaneous papers and documents that can be found in Yellen's office. The professor has never visited the doctor in Imperial County, but he once wrote, "The winds of politics have bent the law in Imperial Valley for a long time."

Yellen next recruited Arthur Brunwasser to the cause, pointing out that Professor Taylor had already done much of the research and was willing to help further with a legal challenge to Imperial County interests. Brunwasser, young and energetic and eager to make his legal mark, had been involved in Cesar Chavez's farm labor movement in the early 1960s. The two men crossed paths then, and Yellen later wrote the lawyer asking legal questions about the reclamation law. Then Yellen, who knew his man, made his move. "He said to me, 'Mr. Brunwasser, I won't make you rich but I'll make you famous,' " the lawyer recalled.

With such limited, but dedicated, resources the triumvirate was ready to take on the combined might of a large, well-financed and politically well-connected entity: the Imperial Irrigation District and the large landowners in the valley, who would, in turn, hire southern California's most prestigious law firm, O'Melveny and Myers. For a while, Warren M. Christopher would work on the case for the valley interests, before going on to become deputy attorney general in the Johnson Administration and later deputy secretary of state in the Carter Administration. Lining up on the side of the district and large landowners would be such political heavyweights as Governor Ronald Reagan and Senators George Murphy and John V. Tunney. As a congressman, Richard M. Nixon had campaigned for the Senate in the Imperial Valley, making a strong statement in favor of its exemption from the acreage limitation. Governor Jerry Brown and Senators Alan Cranston and S. I. Hayakawa would similarly support the valley's efforts to get a legislative exemption, along with whatever congressman and state legislators represented the area. Behind them all was the constant editorial support of the *Los Angeles Times,* the most influential publication in the West. Publishers of the *Times* have had a long involvement with large-scale farming in the Imperial Valley, elsewhere in California, and in Baja California, Mexico. It was a gathering of the traditional forces interested in water development, western style. It was this cohesive grouping that the doctor, the professor, and the lawyer faced. It was laughable, or almost so. Following their 1977 appeals-court victory, the *Imperial Valley Press* editorialized: "But, admittedly, he [Yellen] has been underestimated and should be taken seriously." In 1980, after the Supreme Court overturned the appeals-court decision, Yellen vowed to continue the fight. But there seemed little further that he could do.

There were two suits pending in the San Diego federal district court that pertained to the reclamation law and the Imperial Valley. One was filed by the federal government on the acreage limitation and the other by Yellen on the residency requirement. Yellen also sought to intervene in the first suit, so as to be in a position to appeal it should the government decide not to do so. Brunwasser felt that Judge Turrentine, recently appointed to the court by President Nixon on the recommendation of Senator Murphy, was not sympathetic to the acreage limitation; lawyers for the irrigation district, significantly, expressed pleasure when they heard that Turrentine had been assigned the case. At a pretrial conference, the jurist indicated concern about the acreage limitation's potential effect on property values in the Imperial Valley. As expected, he ruled against the federal government and Yellen's petition for intervention.

In the first few months of 1971, after the judge's decision, all the attention focused on whether the federal government would appeal the case. Ostensibly if it decided not to appeal, the case—as in the Malan situation—would rest on the decision of the lower court and the valley would then possess a more solidly based exemption from the provisions of the reclamation law than the Wilbur letter. As early as May, 1969, the district and its large landowners had mapped out their strategy. It was to get a favorable district-court ruling, then convince the federal government not to appeal. The president of the irrigation district's board, Carl Bevins, wrote Norman B. Livermore, Jr., head of the California Resources Agency and a member of Governor Reagan's cabinet, "We believe that we will be successful in the Federal District Court. Assuming we obtain the judgment of the District Court that the limitation does not apply to this area, we would then ask assistance in convincing the Attorney General of the United States that no appeal should be taken and that judgment of the District Court be permitted to become a final adjudication of the matter." Bevins was greatly disturbed because Livermore unwittingly had asked the Department of the Interior to drop the pending suit, thinking this would be in the best interests of the district. Not so, said Bevins, because without a favorable court ruling "there is no reason why some future administration would not again endeavor to impose acreage limitation on this area." What was needed was a favorable lower court decision that would not be appealed.

When the judge issued his ruling on January 5, 1971, the valley was ready to lobby the Nixon Administration not to appeal. Indications were that the government would indeed not appeal, but no chances would be taken. Robert F. Carter, the district's general manager, taped a telephone conversation with Arleigh B. West, the Bureau of Reclamation's director for the Lower Colorado region. Said West, "Oh, I'd be surprised if we do [appeal]. However, I certainly can't predict." To West, who would soon retire, the whole matter of implementing Solicitor Barry's opinion had been "a sad performance from the start." His boss, Reclamation Commissioner Floyd E. Dominy, had voiced similar reservations about Secretary Udall's decision to attempt enforcement, stating at the 1965 dedication of an Imperial Irrigation District building, "I know of no other agency in government required by law to attempt to reverse the trend toward large farms." As in 1977, the more entrenched bureau employees would side with their traditional clients, rather than top Interior officials who came and went with the change in presidential administrations.

Now began a veritable caravan of traveling suppliants from the valley to Washington. It would peak in March, shortly before the deadline for the government to make its decision. The suppliants were preceded by a thirty-seven-page memorandum put together by O'Melveny and Myers that ended, "Surely, in these circumstances, the time has now come when the books should be closed on this controversy. The Imperial Irrigation District and its landowners should be permitted to go in peace." Carter and a valley landowner met with a special assistant to the Secretary of

Agriculture. There were similar meetings arranged by the valley's congressman, Victor V. Veysey, with Interior Secretary Rogers C. B. Morton and Deputy Attorney General Richard Kleindienst. A delegation also visited Attorney General John N. Mitchell. One member of the delegation that saw top Department of Justice officials was Steven H. Elmore of Brawley, whose family owned thousands of acres in the Imperial Valley and whose father, John, was a neighbor of President Nixon's at San Clemente and a good friend of Morton's. Indeed, it had been Steven Elmore who had arranged the sale of 2.9 acres of the family's San Clemente holdings to the President, who would later sell the acreage at a great profit. Governor Reagan and his top aides wrote letters urging that the government not appeal the case. A background memorandum on the 160-acre issue prepared by the Southern Pacific Company, the largest private landowner in the state and holder of big blocks of agricultural lands in the San Joaquin and Imperial valleys, had been furnished the governor's staff and the Department of the Interior. A lonely voice requesting Attorney General Mitchell to appeal the case was the California AFL-CIO.

The decision was, as expected, not to appeal. Solicitor General Erwin N. Griswold stated, "I considered the matter carefully and thoroughly, and over a considerable period of time. As a result of my consideration, I became convinced that (a) we would not win the case in the court of appeals and (b) we should not win it."

There was jubilation in the valley, which thought it was now home free. Almost six months later a visiting judge from Montana sitting in the same San Diego federal district court ruled in favor of Yellen's attempt to impose the residency requirement of reclamation law on the valley. Thus, the same court had issued two conflicting opinions based on the same law. Faced with the contradictory opinions, and a legal end-run by Brunwasser around the earlier ruling by Judge Turrentine that had seemingly blocked Yellen from appealing, the San Francisco appellate court joined the two cases for a common hearing. In one of those non sequiturs of jurisprudence so confusing to the layman, the appeals court in 1977 would overturn the lower-court ruling on acreage limitation—which Solicitor General Griswold had said could not be won—and reverse the district-court opinion on the residency requirement, declaring that Yellen and the group of valley residents who had joined him did not have standing to sue.

The large landowners also pursued their cause in the legislative and public relations arenas. Congress first considered a moratorium on enforcement, then an outright exemption for the Imperial Valley. The valley's interests were well represented in Washington. Robert H. Meyer, a Brawley cotton grower who became an Assistant Secretary of Agriculture in the early days of the Carter Administration, would resign under pressure and return home a local hero. Meyer had used his influence within the Administration to plead the case of large valley landowners. Tom Rees, a former liberal congressman from a southern California coastal district, was hired for $5,000 a month to lobby the liberal congressional bloc, which might be expected to support the concept of land reform. A political consulting firm was hired, and soon advertisements promoting the concept of "fairness" for valley farmers were flooding the national media. For all its remoteness, the valley was quick to seize upon the sophisticated and expensive techniques that are used to influence political decisions. And with that proper homespun touch, rallies were held and tractors paraded at the county fairgrounds and in front of the hotel where President Carter was staying in Los Angeles. So that passengers aboard airliners to and from San Diego might get the message, one farmer used his tractor to scratch out in the stubble of his field, "No 160 acres."

In June, 1980, the U.S. Supreme Court overturned the appeals court decision and ruled in favor of the valley's large landowners. They had at last rid themselves of that onerous obligation. The unanimous ruling, written by Justice Byron R. White, cited the contract and Wilbur letter as validating the valley's contention that the 160-acre limitation did not apply to it. The opinion also recalled the successful prosecution of the Malan case. But the court's reasoning was based primarily on a narrow interpretation of the 1929 act; one of its confused provisions having freed the valley from another provision which seemingly applied the acreage limitation. Pointing out that it was not until 1964 that the Secretary of the Interior "officially" repudiated the department's position, the opinion stated: "It is also a matter of unquestioned fact that in the ensuing years the Secretary has delivered water to the District pursuant to its contract and that the 160-acre provision of the reclamation laws has to this date never been an operative limitation with respect to lands under irrigation in 1929." The large landowners had successfully persevered. It was too bad, said Yellen, but at least he had gotten some "monkeyshines" out of the unequal contest.

It comes down in the end to The District's man in the field—the ditch tender or rider, or, as he is known in the Imperial Valley, the *zanjero.* It is he who provides for the delivery of water to 450,000 irrigated acres through a complex system of 6,000 headgates and 3,000 miles of canals and drains, which eventually flush the water into the Salton Sea. The ditch tender is the unsung hero of the West, unfairly eclipsed by the more glamorous cowboy, and neglected by popular chroniclers. But it has been this man—first on horseback, then in a horse-pulled wagon, next in a Model T, and now in a four-wheel-drive pickup or on a trail bike—who has been most closely attuned to the distribution of the West's most valuable asset. Here, near the end of the river in this country, I thought it would be interesting to spend a day with a *zanjero,* to follow the water I had first encountered along the Continental Divide to its final destination.

The day began with a flashlight in Leonard Schaffer's hand, which he used to inspect a headgate as the palm trees gradually separated themselves from the morning sky. A cock crowed at a nearby farmhouse. The water, the exact amount having been ordered the previous day and an estimate made as much as a week in advance, began to arrive, I would like to think, via the following route: Knapsack Col, Peak Lake, the Green River, the Colorado River, the All-American Canal, the Central Main

Spreading the water on the Imperial Valley, the place where it finds its greatest single use.

Canal, and the Elder Canal on whose banks we stood in the chilly morning air.

There would be forty cubic feet per second more water in the canal today than there had been the previous day, an amount determined by how much the farmers had ordered. It would be enough of an increase to keep Schaffer running through the morning hours. We departed from the meeting place in a tan pickup with the district's decal on its sides and a water cooler in back. This is a region where less than three inches of rain falls in an average year, and where summertime temperatures can climb toward 120 degrees. There has been only one recorded snowfall of consequence in the valley: in 1932 a little more than two inches fell. It was winter now, and as the orange band grew in the east, Schaffer turned off the heat in the cab. It would not be long before we would be in shirtsleeves.

It was near the end of the lettuce season, and along the run that day we would pass fields of asparagus, canteloupes, tomatoes, wheat, barley, and, of course, alfalfa. (A few days earlier, while taking photographs, I had been unceremoniously kicked out of a lettuce field by two grower's representatives who thought I was a United Farm Workers organizer, and would accept no explanations to the contrary. This is Cesar Chavez country.) Cotton is also a major crop in the valley. Some of these crops would find their way into vegetable-starved homes on the East Coast, others would be eaten within 24 hours at a Michelin-starred restaurant in Paris or by dairy cows near Los Angeles.

Schaffer stopped frequently to adjust gates, making sure the water flowed evenly and each farmer got his desired share. "I've been ten years with the district and all that time running water. After a time you get a feel for how the water works," he explained.

A farmer hailed Schaffer, asking, "You got me set for nine?"

"Nine point eight now."

"That's all right. I run a little strong."

Another farmer complained about the late delivery of water to his field. "Pretty persnickety," murmured Schaffer as we drove away.

There was a quick understanding built around the common knowledge of how water works between the gringo *zanjero* and a Mexican irrigator, although neither spoke the other's native language. Schaffer will cover about seventy-five miles in an average day. They would be tediously driven miles, short spurts along rutted dirt paths atop canal banks.

Schaffer was worried now. The head of water was catching up, threatening to overflow the sides of the canal—a primal sin. The *zanjero* hustled, opening, closing, and adjusting the various headgates, following and anticipating the path of water downhill, turning it aside, trying to keep ahead of it.

The threat abated, and it was time for a cup of coffee at Bessie's Cafe in Seeley. Then back onto the dirt roads that paralleled the delivery system which in turn matched the rectangular plots of land. It was impossible for a stranger to know who owned what and how much was excess, but the benefits from federal water projects were very evident that day as the water spread down the plowed rows in uneven, gleaming tendrils.

It was Schaffer's hope that not too much water would flow into the New River and thence to the Salton Sea after the last irrigator had taken his share. It was a delicate balancing act, matching what irrigators had ordered the previous day to what they actually took. By noon Schaffer was back in the division office, placing the orders for the next day, and, after a quick lunch, I would be on my way again. It had been a long journey from Harv Stone and the Canyon Ditch, but there was still a little further to go.

Lettuce harvest, Imperial Valley.

3

The Colorado River below Imperial Dam ceases to be a river, becoming, instead, a drainage ditch for most of the twenty-seven miles to Morelos Dam in Mexico, where the remaining water, dumped back into the river just north of the international boundary, is diverted by the Alamo Canal to the Mexicali Valley. Both the Gila River, entering the main river fifteen miles downstream from Imperial Dam, and the lower Colorado contain little else but the well-used drainage flows from local fields that have found their way back to the weed-and-sediment-choked beds of the two rivers. Most of the water destined for Mexico is diverted at Imperial Dam, run through the All-American Canal for a short distance, and then, separated from the vast bulk of the river which winds up in the Imperial Valley and Salton Sea, is directed through a power plant and returned to the lower Colorado just above the border before being shunted aside for the last time at Morelos Dam.

On the Arizona side of the river an agricultural drain carries waste water from east of Yuma all the way to the Gulf of California. In fact, there is a plethora of drains, canals, flumes, and unlined ditches in the Yuma area, so many and so intertwined that they resemble a complex freeway interchange, a sort of climax of major interstates just before the border. More of the same is planned, as part of the political solution backed up by the physical works that will supply Mexico with better-quality water, while surrendering none of the same north of the border. The freeway interchange can be viewed as a liquid sleight-of-hand trick—here you see the water, there you don't. In this hand, good quality water; in the other, poor quality water. Guess which hand. Once the current proposal to upgrade the plumbing system is completed, it will be the greatest shuffle of water in the basin, and perhaps the world. It will represent the apogee of modern technology imposed upon a river.

Below Morelos Dam, a scant one mile south of the border, the Colorado River is a dry riverbed during most seasons of the year. Follow it a few miles further down as it parallels the border, and it begins to get wet with the return flows from Mexican fields; until it is once again the drainage ditch that it was just north of the border. The river, in this stretch, "is no longer important as a channel for carrying irrigation water, and inadequacies have developed in its capacity to carry floodflows," stated the Bureau of Reclamation in a survey of river-control work along the lower Colorado. When I visited the area it was a difficult task to locate the dry riverbed—that channel navigated by steamboats a hundred years ago. What had been a river was constricted between two levees, seem-

ingly useless but perhaps of some help when intense local storms threatened to flood the area. Cattails choked the riverbed and swallows had
made their mud nests in the concrete overhangs of Morelos Dam. The
water had a musky smell—much like wet, rotting grass. There was trash
lying about, and a general feeling of decay pervaded the dry riverbed
below the dam.

There is a curious fact about the mapping of the Colorado River that
speaks volumes about how the seven basin states and their nominal
patron, the Bureau of Reclamation, regard the last users of the river's
waters in Baja California and Sonora, Mexico. Almost all the maps drawn
by the federal water agency and copied by the western states in their
various publications show nothing, or at most, the lone detail of the river
flowing the last few miles through a featureless foreign land. It is as if
Mexico did not exist as a viable country, except for the single blue swath
inaccurately extended all the way to the Gulf of California. Depending on
the length of the page or the whim of the cartographer, which seems to
be mainly shaped by jurisdictional limits, the river might halt at the
intersection of California, Arizona, and Mexico or be extended for a few
more miles in order to depict the small segment of river shared in common by Mexico and Arizona. Then the inevitable blue line ends too
abruptly.

There is one type of rendering of the Colorado River in Mexico containing no omissions or wishful thinking; and that is photographs taken
from satellites, particularly infrared photographs, which can pick up the
nuances of plant growth. The infrared photograph of the inverted triangular area between the head of the gulf, Mexicali and Yuma tells all in
a rather dispassionate and cruelly exact manner. There is no river seen
flowing through to the gulf. But even more astonishing are the thick
clusters of vibrant red rectangles in the Imperial Valley and the absence
of such an intensity of color connoting rich plant life in the Mexicali and
San Luis valleys. There are a few scattered red rectangles south of the
border, but in greatly diminished numbers and much paler of hue than
in fields where the same crops are grown to the north.

The desert meeting between two nations of such divergent cultures
and standards of living has never been graphically portrayed in so striking
a manner. It is the same soil, the same river delta. But it is not the same
river water, nor are there the same means to deal with its differing quality.
The difference speaks eloquently of the riches a river can bestow or take
away, as well as of a country's ability, or lack of it, to employ expensive
technical solutions to deal with the problem of salinity that has plagued
so many past desert civilizations.

Up until the mid-1930s the Mexicali Valley was little more than an American agricultural enclave—an extension of the same process of turning a barren desert into an irrigated paradise that was taking place in the Imperial Valley. Hoover Dam benefited the Mexicali Valley as it aided the Imperial Valley, and both valleys used the same canal up until 1942. Through the latter half of the nineteenth century, following a number of disillusionments with the American policy of Manifest Destiny, Mexico

Mexican boy helping self to water on U.S. side of international boundary, which is the fence. The boy crawled back through the fence with two full buckets.

left its border area undeveloped. It served as a desert buffer strip that protected the populated interior of the country. Gradually the aggressive Americans pushed across the border to fill the vacuum; in the case of the Mexicali Valley, they followed the flow of water. In the last century, there was little visible representation of Mexican sovereignty to respect. To cross back and forth across the border was an extremely casual affair. With or without Mexican permission, steamboat traffic cruised up and down the river, and Americans established a shipyard and transfer facility at the head of the gulf. Early travelers never mentioned any official border and later, to float down the river to the gulf meant, at most, a brief and perfunctory stop at the Mexican customs house.

American dominance of the Mexicali Valley increased greatly when water was first diverted to the Imperial Valley. The Mexican government forced concessions from the California Development Company in 1904, demanding and receiving half of the water diverted from the Colorado River in Mexico and transported across Mexican soil to the Alamo River and thence to the Imperial Valley. But the major beneficiaries of the water in Mexico were American landowners. As water began to be diverted, a Los Angeles business syndicate, most of whose members had strong ties to the owners of the *Los Angeles Times,* eventually purchased 840,000 acres —virtually the whole Mexicali Valley. The Colorado River Land Company brought in large numbers of oriental laborers to work its huge ranch. With the 1910 Mexican Revolution came turbulent times for foreign landowners in Mexico, and the syndicate, headed by Harry Chandler, attempted to use its influence in Washington to get federal troops dispatched to protect its Mexican holdings. No United States soldiers were sent, but Mexican units did arrive to serve the same purpose. Chandler, who was to become publisher of the *Times,* and some of his cohorts were indicted by a United States grand jury for attempting to foment a revolution against the new government in Baja California that was unfriendly to the syndicate's interests. It was feared the new government would break up the ranch and form small farming units. A map of a proposed new country, with one of the states named "Otis," ostensibly after *Times* publisher Harrison Gray Otis, father-in-law of Harry Chandler, was presented as evidence to the grand jury. But, amid indications of jury-tampering, a subsequent Los Angeles trial resulted in the acquittal of Chandler and his associates.

After the upheaval of the revolution had settled down, the ranch prospered. It became, according to *The New York Times,* "the largest cotton-growing enterprise under a single management in the world." Eight thousand Mexican workers were employed on the ranch, and there was a large number of Asian tenant farmers, primarily Chinese. Succeeding Mexican governments looked with favor on the American holdings in the valley, and the syndicate did all it could to curry goodwill in Mexico City. So it was not until the mid-1930s that the Mexican government, acting under the provisions of the 1917 constitution, moved to expropriate the foreign landholdings. The process of compensation dragged on to the late 1940s, but the Chandler heirs eventually profited greatly. The authors of *Thinking Big,* a book about the family's newspaper dynasty and Southern California, wrote, "The Chandler involvement in Mexico stands as a classic case of foreign profiteering in an undeveloped country." It showed, as other examples would also demonstrate, that the smart money in Los Angeles knew how water would increase the value of land. In the 1960s, as the issue of salinity came to preoccupy Mexican–United States relations, Mexican resentment at having been dominated for so long by Americans would be turned against its neighbor.

Mexicali eventually became a city of the West, almost more American than Mexican in its economic outlook. In the early years the city's growth paralleled the agricultural growth in the valley. Water brought the people. A dusty hamlet of 462 inhabitants in 1910, by 1940 Mexicali could count 18,775 residents. By 1950 there were three times that number. Then its growth took off, matching the succeeding importance of cotton, tourism, and industrialization, and finally as a departure point for illegal entry into the United States. By 1955 the Mexicali Valley was the leading cotton-producing region in the country. Cotton was King, and Colorado River water was making its growth possible. A few years earlier Morelos Dam had been completed, thus giving the valley a better point of diversion than the old Alamo Canal heading. Mexico had obtained a treaty with the United States in 1944, assuring it of 1.5 million acre-feet of water a year, although there was no explicit guarantee of quality. Nevertheless, the Mexican government went ahead and almost doubled the land under irrigation in the Mexicali Valley in the years after World War II, an amount equaling 7 percent of the irrigated land in the whole country. Referring to the principal crop of cotton, one authority on the Sonoran Desert commented, "The profits from this single crop during a few short years gave more impetus to the development of agriculture in the Mexican Northwest than any other economic factor aside from irrigation itself."

The monocrop economy began to lose its vigor in the early 1960s, with the growing salinity of the river in Mexico, a sharp decline in prices, and an infestation of the pink bollworm. In recent years valley farmers have sought to diversify, and more wheat and barley is grown. The Mexicans never put as much effort into raising alfalfa as their meat-eating neighbors to the north did, preferring to let livestock fend for themselves in desert areas. The inevitable result in that land of little rainfall and great sparseness was overgrazing. Still, agriculture was the primary cause for the imbalance in water use—just as it was across the border. The 1.5 million acre-feet of water from the Colorado River went almost entirely towards irrigating 490,000 acres in 1975, almost the same amount of land irrigated by the Imperial Irrigation District with 2.8 million acre-feet of water. The city of Mexicali, whose population had climbed to 400,000 in 1976, used 32,428 acre-feet of water. Some additional agricultural water, about 1 million acre-feet, was pumped from underground sources; and some domestic water was trucked to Mexicali from wells in the Yuha Desert in Imperial County. A 76-mile aqueduct was being built from Mexicali to Tijuana on the ocean side of the Baja California peninsula in order to transport Colorado River water to that notorious border city.

The whole border area was booming. American industrial firms, seeking cheap labor, were locating below the border. With a burgeoning population, high unemployment, and low wages, workers from interior

Mexicans looking across the border from Mexicali to Calexico, in the U.S. The highest fence is the border.

Mexico were flocking to the border cities, preparatory to making the greater leap into the prosperous United States. Tijuana and Mexicali were two of the fastest-growing cities in the world in the last two decades. What was happening in those two border cities was happening throughout the interior West. Granted Mexico did have its own unique variations on this theme; but it, too, was coming up against the ultimate limiting factor of no more water. (One such variation in Mexico is the distribution of water by irrigation districts for social purposes. They are instruments of a land-reform program, dating from the mid-1930s in the Mexicali Valley, where slightly more than half of the irrigated land is in *ejido* holdings, federally owned but individually farmed, and the remainder is privately owned. There are about 11,900 farmers in the valley on plots averaging 50 acres in size.) The Mexicali Valley could boast of having almost everything going for it that its neighbor possessed—everything including a growing salinity problem.

Salinity results when minerals are dissolved in water—mainly calcium, chloride, bicarbonate, magnesium, sodium, and sulfate, plus small amounts of other constituents. Their unit of measurement is either parts per million (ppm) or the nearly equivalent milligrams per liter (mg/l). Both refer to the count of total dissolved solids (TDS). The numerical system of measuring pollution, western style, is an excellent indicator of the river's progression downstream. At the headwaters of the river system, the salinity count is less than 50 mg/l. For the year 1974, there was the following salinity count: Green River, Wyoming, a town on the Green River, 324; Glenwood Springs, Colorado, on the Colorado River, 300; Bluff, Utah, on the San Juan River, 490; on the Colorado River in the Grand Canyon, 668; just below Hoover Dam, 751; at Imperial Dam, 861; at the north end of the international boundary, 1,000; near the end of the river, after the last Mexican diversion, between 4,000 and 5,000; and in the Salton Sea, about 39,000. The salinity of sea water averages 35,000.

The sources of salinity within the Colorado River basin are about equally divided between natural and man-made—but it is the latter that has caused the rise in the river's salt load. As more development occurs on and off the river, its salinity increases. The natural background remains about the same. Natural sources are divided between point sources, such as hot springs and uncapped wells, and diffuse sources, meaning water running over or through saline deposits extending over a large area. Irrigation is by far the largest man-caused source of salinity, accounting for 37 percent at Hoover Dam. The salts are leached out when water is applied to a field and the return flow causes the river's salinity level to rise. Next are reservoir evaporation, 12 percent, and exports from the basin, 3 percent. Evaporation results in water being lost and salts retained. The transfer of water out of the basin from the headwaters results in the loss of good-quality water, meaning more salts will be concentrated downstream. In contrast to rivers in humid regions, municipal and industrial wastes—the traditional form of river pollution—contribute only 1 percent to the Colorado River system.

High levels of salinity do bad things. On the farm, crop yields decline and plants and trees can die when the saline waters reach their roots. It is the higher-value crops that are most susceptible. As a countermeasure, lower-value, more salt-tolerant crops can be planted. Or the farmer can add more water to leach out the salts and additional amounts of fertilizers. An expensive system of tile drains can be installed to draw off excess water. The drains must be cleaned periodically. Land can be leveled more precisely, irrigation ditches lined, a sophisticated sprinkler system installed, or water ordered and applied more exactly. Such advanced technologies as computers and laser beams are involved, meaning money and scientific know-how—luxuries that are not generally available in such countries as Mexico. In urban areas hard water can mean using more soap, corrosion on distribution pipes and plumbing systems, scaling on hot-water heaters, expensive water-treatment facilities, and a boom in the bottled-water business, which is very evident in the Los Angeles area. Colorado River water used for domestic purposes in southern California varies between 700 and 850 mg/l. The recommended public-health standard is 500. There are some health consequences to drinking highly saline water, such as its laxative effect. Then there are damages expressed in monetary terms—rather speculative estimates seemingly aimed more at influencing policy than pinpointing the actual cost. The Bureau of Reclamation estimated losses from the river system's salinity in 1973 at $53 million per year, and predicted a rise to between $122 and $165 million by the year 2000 if nothing were done about it. But the bureau and the basin states had a clear idea what to do about it.

It can be said that the Colorado River became an international stream of importance when George Chaffey first surreptitiously tapped the river just north of the border and diverted part of its flow fifty miles through Mexico in order to bring the water to the Imperial Valley. Five months later the Mexican ambassador in Washington lodged a strong protest, noting that "a change in the course or complete exhaustion of the Colorado River" might result from this unauthorized diversion. This, of course, happened in 1905, much to the detriment of any downstream Mexican users. But the Mexican government did not press its protest, since it did not want to awaken the aggressive spirit of the neighbor that from time to time had coveted Baja California. In its defense the United States relied on the Harmon Doctrine, a rather self-serving opinion rendered a few years earlier by Attorney General Judson Harmon on a matter involving the Rio Grande. The doctrine asserted the principle of absolute control over water originating upstream from a foreign country. The basin states relied on this doctrine in 1941, when they laid out their position for negotiations on a treaty with Mexico to the Department of State. According to the states, Mexico had no legal right to water from the Colorado River. The Harmon Doctrine was the height of international arrogance involving the sharing of rivers and was later abandoned, although its residue would remain, in different forms, to exacerbate relations between the two countries to this day.

The basin states skirted the Mexican issue in the 1922 compact and the Boulder Canyon Project Act of 1928. The Colorado River Compact made a vague reference to the possibility of a treaty with Mexico at some future date, and that country was simply assigned any surplus waters that were above and beyond the claims of the seven states. In case of shortage, the upper and lower basins were equally to supply an unspecified amount to Mexico. Because of differences in accounting with the lower basin, the four states in the upper basin refused to acknowledge any obligation to furnish Mexico with water. The surpluses did not appear and Mexico simply got whatever water flowed down the river or its entitlement from the Alamo Canal. To Herbert Hoover, who helped negotiate the compact, the matter was simple. Said Hoover, "We do not believe they ever had any rights." The 1928 act echoed the Harmon Doctrine: "nothing in this act shall be construed as a denial or recognition of any rights, if any, in Mexico to the use of the water of the Colorado River system."

In 1929 brief negotiations between the two countries stalled when the United States offered 750,000 acre-feet and Mexico asked for 3.6 million acre-feet a year. The negotiations were resumed again in 1939, when Undersecretary of State Sumner Welles wrote the Mexican ambassador proposing a swap of Rio Grande water for Colorado River water. In later years the thought of having lost water from the Colorado River to Mexico in exchange for Texas getting additional water from the Rio Grande rankled the basin states, and they took a tougher stance with the Department of State over negotiations on the quality of water Mexico was to receive. The states were never to trust the federal government on this matter. As early as the first round of negotiations in the late 1920s Mexico had insisted on tying the two rivers together, thus gaining a tactical advantage. By the time the second round of negotiations began it was apparent such a package had a better chance of getting Senate ratification since Texas would greatly benefit, and it happened that the chairman of the Senate Foreign Relations Committee was Senator Tom Connolly of Texas. At the peak of the debate on the Central Arizona Project in March, 1968, Morris K. Udall circulated a memorandum to fellow members of the House Interior and Insular Affairs Committee pointing out why augmentation, which would first go toward furnishing Mexico with its share of water, had to be a "national obligation." It stated: "Although never officially recognized, it seems to be no secret that Senator Connolly of Texas was the chief proponent of a new treaty with Mexico relating to these rivers. . . . So the 'trade' was made by which the Mexican government gave up a big part of its claim on the Rio Grande—in exchange for doubling Mexico's supply on the Colorado." The gist of Udall's argument was that the Colorado basin states, therefore, were not responsible for the treaty, nor should they be held responsible for its implementation. Udall neglected to point out that the majority of basin states, including his home state of Arizona, supported ratification of the treaty, at the time seeing it as a way to set a limit on Mexico's claims and then get on with distributing the water above the border.

If the basin states felt themselves being unwillingly led into a treaty with Mexico because of Texas's desire for more water and the Department of State's wish to cement its Good Neighbor policy, there were also pressures operating on the Mexicans. In 1942 the Mexicans could look north and see Hoover Dam completed and the first deliveries of water being made through the All-American Canal, thus making the Alamo Canal and water deliveries via that route obsolete to Imperial Valley interests. As one of the Mexican treaty negotiators noted,

> The situation in 1942 showed us how well founded were our fears because that year, during several of the hottest weeks, there came down from the great American dams constructed on the Colorado River only a small volume which did not permit of filling the requirements of irrigation in Mexico. And with this came the clamor of the public landholders, the small owners, and colonists of our Colorado irrigation district, who saw their crops lost for lack of water. But there is even more, for at the end of the summer, there came from Boulder Dam a great flow of water which overflowed in Mexico, inundating cultivated lands and ruining crops of other thousands of hectares.

In the spring of 1943 there was a drought, and farmers in the Mexicali Valley had to purchase water from the Imperial Irrigation District at what were termed exorbitant prices. The United States had the decidedly advantageous position of being upstream and controlling the dams and canals. It also possessed greater technology and wealth. "There was quite definitely an inequality in the bargaining positions of the parties, above and beyond the situation of the extreme pressure of the drought which ultimately compelled Mexico to come to agreement on a basis not chosen by it," wrote one analyst of the treaty. But Mexico did have some things going for it, besides controlling the drainage into the Rio Grande. The matter could be taken to the World Court, or submitted to international arbitration, where Mexico would probably win, and the United States desired good relations with its neighbor, especially during World War II.

The treaty that resulted, and was ratified by the Senate in 1945, set the quantity of water owed Mexico at 1.5 million acre-feet a year, but the quality was left ambiguous. Negotiators from both sides reported different interpretations of the same treaty to their respective senates. As Northcutt Ely told his California clients, "This is not to say that the Mexican negotiators were right and ours were wrong, in reporting what the treaty accomplished, but . . . they could not both be right." American negotiators reported that their counterparts in Mexico fully understood they were to accept water of any quality flowing across the border, even if it was not usable for irrigation. On the other hand, Mexican negotiators told their Senate the treaty provided for water of good quality, or at least the same quality as that used by the Imperial Irrigation District. Undoubtedly both parties knew they were talking about return flows, since no unused water existed so far downstream, but the treaty did not deal in specifics, only such generalities as water "from any and all sources . . . whatever their origin . . . for any purpose whatsoever." Probably what happened was that the issue was left deliberately fuzzy, since ratification, which both sets of negotiators desired, would be made that much easier. Each side could then read into it what it wanted—a familiar tactic.

With the Imperial Irrigation District leading the way, California opposed ratification of the treaty, fearing the loss of water. The irrigation district also wanted compensation for improvements to the Alamo Canal. Nevada straddled the fence, but the other basin states favored ratification, including those in the upper basin, which, after it was determined how much water Mexico was going to get, would shortly put together their own interstate compact. Stanford University law professors Charles E. Meyer and Richard L. Noble wrote, "Only California had all that she was ever going to get from the Colorado, and only California, among the affected states, opposed the treaty." The Senate ratified the treaty by a 76-to-10 vote on April 18, 1945, six days after the death of President Roosevelt and partly as a tribute to his Good Neighbor policy.

Throughout the negotiations the Colorado basin states, operating as advisors to the Department of State through a group called the Committee of Fourteen, insisted that Mexico accept return flows as her share of the eventual allotment. The two Stanford professors said the insistence of the basin states on this and other issues was "more troublesome than the Mexicans" to the negotiators. It was a complaint that was to be repeated by Department of State negotiators who worked on a 1973 salinity agreement with the Mexicans, the basin states playing a much more decisive role in these later negotiations. That Mexico knowingly agreed in the 1944 treaty to accept water unusable for irrigation was "inconceivable," according to the two professors and "highly foolish," in the words of Norris Hundley, Jr., another authority on the treaty. When the salinity levels of the river rose to a high of 2,700 mg/l in 1961 because highly saline agricultural waste waters were being dumped into the Colorado just upstream from the border, the Department of State told Mexico that the treaty contained no provision for the quality of water. It was the administrative response the basin states had hoped for, and was the last official echo of the Harmon Doctrine.

The drainage waters came from the Wellton-Mohawk Irrigation and Drainage District a few miles to the east of Yuma. It was difficult to imagine an area with a worse record of irrigated agriculture. That crops are still raised on some 65,000 irrigated acres along the Gila River is a tribute to the persistence of the Reclamation ethos through one disaster after another. The lesson that should have been learned was that the area was not suitable for such agriculture and should have been abandoned to the surrounding desert many years ago. But this was not the western way. It would have meant the repudiation of an institution that was judged to be basic to the West—an institution around which a number of myths had grown, including the myth of its own inviolateness.

Pima Indians first lived along the Gila River bottomlands of the Wellton-Mohawk valley, then came the first white homesteaders in 1857. By the 1880s two canals were diverting water from the Gila, but upstream irrigators used so much water that soon there was not enough for surface diversions in the Wellton-Mohawk area, and farmers there began to turn to wells. By 1931 about 11,000 acres were irrigated by underground water supplies. As the pumping increased, the wells began to go dry or to bring highly saline water to the surface, measuring 6,000 mg/l. A magazine writer described the failing agricultural situation in 1952: "The water was getting scarcer every year. And the scarcer it got, the saltier it became. Livestock and people wouldn't drink it—neither would most plants. Farms went out of production. . . . Dead mesquite trees, their roots deserted by the falling water table, add to this oft repeated drought story of western agriculture. Where lush fields of alfalfa once thrived, only the hardiest of desert shrubs and weeds remain. White salt deposits glistening beneath a blistering desert sun, explain all too vividly what has happened." What had happened was that the farmers in the valley had poisoned their own crops; the ground water, having been recycled too often, had become too saline. It was the classic case, although much speeded up, of an oasis civilization dying because of overuse of its most precious resource.

But a rescue operation was mounted. In 1947 Congress reauthorized the Gila Project, providing for the construction of canals that would bring the valley a fresh supply of water from the Colorado River. The first river water arrived via the twenty-four-mile canal system in 1952. The desert was about to bloom again. As the fresh supplies of water were poured on the land, a basaltic plug, a virtual underground dam at the head of the valley, prevented the ground water from flowing back to the Colorado River. The underground water table rose, pushing up with it the accumulation of salts that had been buried in the aquifer after many years of irrigation. As the saline waters came in contact with the root zone, crops once again began to fail. Thousands of acres went out of production, and water even rose to the surface to inundate a section of the state highway. The solution concocted by the Bureau of Reclamation was to build a drainage channel to the nearby Colorado River and pump the underground water into the channel. It was like bailing out a sinking ship. Eventually, 109 wells were placed into operation for this purpose. The pumps went to work and in 1961 the waste water flowed into the drainage channel, thence to the river and eventually Mexico. Unfortunately, very little fresh water was flowing down the Colorado that year, because Lake Powell was being filled behind the newly completed Glen Canyon Dam. The average yearly salinity of water delivered to Mexico quickly shot up from 800 to 1,500 mg/l, with maximum salinity levels hitting 2,700. In the Imperial Valley successful agriculture is dependent on an extensive and costly drainage system and salinity levels of 850 mg/l. The Mexicali Valley did not have such a drainage system, nor water of such quality.

What had happened was that the Wellton-Mohawk valley had simply transferred its problem downstream to Mexico. This did not unduly upset the district, its attorney remarking that the outcry coming from Mexico was only an attempt by that country to get a larger quantity of water than that allotted by the treaty. "Quality is only a convenient peg to hang their hat on," said W. M. Copple. The Department of State held to its rigid interpretation of the treaty. But the Mexicali Valley reacted with mass indignation. There was a march on the United States consulate in Mexicali. American diplomats in Mexico warned that the Communists were capitalizing on the issue and gaining popularity in Baja California. There was a great deal of emotional impact, said one reporting foreign officer, "where the facts lend themselves so easily to criticism against the U.S." The Mexican foreign minister hinted that where Mexico had been cooperative in the past in furnishing sites for American space tracking stations, this might not be so in the future unless there was a satisfactory response to the salinity problem. Clearly President Kennedy's Alliance for Progress was in trouble just below the border, and Mexican President Lopez Mateos said the salinity issue was the greatest diplomatic problem confronting the two countries. Senator J. W. Fulbright, chairman of the

Senate Committee on Foreign Relations, declared the position of the United States "leaves something to be desired." And there was a threat by the Mexican government to take the case to the World Court. If that should happen, it was feared, not only worldwide public opinion but legal opinion would side with Mexico.

The Department of State reversed itself, and negotiations were begun for the "permanent and effective solution" demanded by both presidents. An agreement was reached in 1965, but in time it proved to be neither permanent nor effective. Known as Minute No. 218, the document provided for a mechanical solution that, it was hoped, would give Mexico better-quality water while still allowing the same level of irrigation in the Wellton-Mohawk district. The drainage channel to the Colorado River would be extended to allow the highly saline waters to be placed in the river either above or below Mexico's last diversion. Mexico could then choose when it wanted to mix the Wellton-Mohawk water with better-quality river water flowing downstream from Imperial Dam. It would all count towards Mexico's 1.5 million acre-foot allocation.

But despite the expenditure of $11 million, the problem persisted. Mexico was getting water that averaged 1,240 mg/l, while a short distance away at Imperial Dam it was 850. The farmers of the Mexicali Valley were still unhappy—"irrational" and "discourteous," in the words of the American ambassador who visited the valley in the spring of 1970. The Committee for the Defense of the Mexicali Valley, a group representing almost all the major interests in the valley, was pressing the Mexican government to take the issue to the World Court. The valley's salinity problem was again becoming a major political issue within Mexico. In the United States, the Committee of Fourteen, made up of two representatives of each of the basin states, had been reconstituted to advise the Department of State, but the latter and the Department of the Interior saw the problem differently. State was thinking primarily of the international damage to the United States if the issue should be taken to the World Court and was pressing Interior to improve the quality of water. Interior, according to a State memorandum, was resisting "measures that would require the sacrifice of considerable quantities of stored water or otherwise incur ill will on the part of U.S. water users and thus considerable political liability."

Domestic political considerations in both countries were to dictate the second "permanent" solution to the salinity problem in the early 1970s. And the solution would be one further abuse to the Colorado, Big Red, the raging bull, the tranquil summer stream. What little was left would be a turgid, algae-choked remnant of a river. This was the accurate price, a price not widely known or fully evaluated when the important decisions were made, to be paid in exchange for a flourishing desert civilization in both countries. From start to finish I had followed the river, and it never ceased to amaze me that the quantity of water seemed the same at both ends. A river should start small and end big. The Colorado, instead, started small and ended small. A natural progression was lacking. I felt a certain dislocation standing at Morelos Dam, or later at the extension of the drain to the Santa Clara Slough near the gulf, watching the foamy discharge of Wellton-Mohawk water.

Mexico was in an agitated state of transition, quite possibly not equaled since the Revolution, when President Luis Echeverría Alvarez took office in December of 1970 and not too much later dramatically escalated Mexico's demands for an improvement in the quality of water furnished the Mexicali Valley. The quality had been improving slightly over the preceding few years, and the issue had subsided between the two countries. But suddenly it was raised to a fevered pitch, because of internal upheavals within Mexico, just as the response by the United States would be dictated by the domestic concerns of the Colorado River basin states. It was President Echeverría's task, as one writer commented, to "restore the revolutionary mask," which had slipped badly in recent years. The 1910 Revolution is venerated by successive governments in Mexico from the safe distance of many years. The trick is to avoid inciting a new revolution by using the political rhetoric that recalls the past one. What Echeverría had to do was bring the ruling Institutional Revolutionary Party (known by its Spanish acronym PRI) closer to the center from where it had strayed to the right. The PRI is the dominant force of government in Mexico, with no other political party able to challenge it effectively. In the 1968 Baja California elections the minority National Action Party (PAN) attained the unthinkable by winning in Tijuana and Mexicali, partly on the salinity issue. The election results were annulled by Echeverría, then Minister of the Interior, and another election was conducted that assured the continuation of the one-party system.

As the top law-enforcement official in the administration of President Díaz Ordaz, his immediate predecessor, it had been Echeverría's responsibility to deal with the student unrest of 1968. There was violence, and violent repression. In a country that prided itself on its record of land reform, there were takeovers of land by landless, unemployed farm workers. A guerrilla movement bombed cities, robbed banks, and kidnapped the United States consul general in Guadalajara for four days. Most unsettling for politicians in a one-party country was the massive absenteeism in the 1970 presidential election in Mexico City. Not only within Mexico, but also within the Nixon Administration, there was a desire to restore order in what had been the most stable and prosperous of Latin American nations. President Echeverría, billed as the "new Echeverría" to distinguish him from his Minister of the Interior days, moved aggressively to establish order by shifting the government slightly to the left. He also sought the leadership role of Third World countries, making a one-month world tour in 1972. At home he projected an image of social concern and liberalism, being particularly concerned about agrarian problems. The salinity issue was a natural for President Echeverría. It would play well with Third World countries, it had national appeal and a clear villain, and it was amenable to a quick, popular, political victory that involved no cost to Mexico.

All that was needed was a willingness in the United States to find a solution, and this Mexico found in the Nixon Administration. Henry A. Kissinger, President Nixon's security-affairs advisor, liked Mexico, had honeymooned there, and was a close friend of Emilio O. Rabasa, President Echeverría's foreign minister. After being presented with an eight-page memorandum from the Department of State on the salinity problem, Kissinger told Interior Secretary Rogers C. B. Morton and State Department Secretary William P. Rogers in April, 1971, "The President has directed that vigorous efforts be made to negotiate as expeditiously as possible a practical settlement of the Colorado River salinity problem with Mexico." The secretaries were instructed to consult with the basin states and their congressional representatives. Much more so than previously, the negotiations conducted by the Department of State and the special ambassador later appointed by President Nixon would be on a dual-track basis—one track being the Mexican government and the other, parallel track being the basin states, represented by the Committee of Fourteen. The Department of the Interior was allied with the interests of the states, although nominally in accord with Administration policy. There were two ultimate forums the negotiators had to keep in mind and seek to avoid or placate. They were the World Court, a possible and constantly threatened last resort of the Mexicans, and Congress, the extension in water matters of the will of the western states. Any physical works that resulted from an agreement had to be authorized and funded by Congress. So the Department of State, that seeming citadel of the eastern Ivy League establishment, would eventually be at the mercy of some western farmers. It was a highly unusual situation, and was relished by those westerners who participated. They knew where all the political levers were that made the water run or stop. Mexico also knew its neighbor well and liked to watch it jump a little bit whenever it mentioned the World Court. The button was pushed enough times so that the Department of State undertook an analysis of the makeup of the World Court and determined that its justices would probably rule unfavorably on such a case, since so many were from downstream nations.

When President Echeverría took office in December, 1970, the two countries were operating under a one-year extension of Minute No. 218, the previous government not wanting to bind the incoming Echeverría regime to any long-term commitment. Negotiations began with the new government in early 1971, and the United States, with the consent of the Committee of Fourteen, offered Mexico an agreement based on salt balance, meaning the tonnage of salt in drainage water flowing from all lands below Imperial Dam would not exceed the tonnage of salt in the water applied to those lands. This would result in Mexico getting somewhat improved water, but not the same quality as downstream American users, since the concentration of salts would be greater, there having been some water loss during irrigation. During the negotiations a basic shift took place in Mexican policy. No longer was it satisfied with merely having the quality of water improved; it wanted water of the same quality as that used in the Imperial Valley.

The change in policy was basically a political decision. Negotiations on a new minute, which were very close to being successfully concluded, were abruptly broken off by the Mexicans. The American Ambassador in Mexico City, Robert H. McBride, was told the proposed minute was "insufficient politically" for Echeverría to endorse as the only achievement on this problem during his administration. (The term of the agreement would have overlapped into the six-year term of the next Mexican president.) The proposal being considered, Mexican officials agreed, was a definite improvement over the existing situation but it was "insufficient for Mexican public opinion and particularly for the inhabitants of the Mexicali Valley." The solution was to extend the old minute with a minimum amount of publicity. What had been proposed by the United States was dismissed by Rabasa as "peanuts," and the foreign minister pressed American officials hard for a settlement prior to Echeverría's visit to Washington in June, 1972. Rabasa, recalling past territorial ambitions of the United States to acquire Baja California, pointed out that had this happened, most probably farmers on both sides of what was now the international boundary would be receiving the same quality water. Why, it was asked, after one more threat had been made to take the issue to the World Court, did the United States spend billions of dollars in Vietnam and relatively little to solve a serious problem of a friendly neighboring country? Ambassador McBride reported to Secretary Rogers: "Our assessment is that the Mexicans believe that they can drive a hard bargain with us and they have to some extent seized on the Echeverría visit to Washington as a lever of time and circumstance to use upon us."

On the eve of President Echeverría's visit, Kissinger was apprised of the situation. While subject to "real domestic pressures," the Mexican President "had consciously agitated the problem and lent official support to exaggerated claims apparently in an effort to arouse the country and make it a national issue—and presumably to bring added pressure on the

U.S.," according to a confidential Department of State memorandum sent to Kissinger at the White House. Presumably Kissinger passed this evaluation on to President Nixon before his meeting with the Mexican President. While in the United States, Echeverría managed briefly to raise the issue to the level of national concern in this country in an address to a joint session of Congress and an appearance before the National Press Club. To Congress, he declared, "The artificial salinity of the Colorado River is the most delicate problem between our two countries. . . . It is impossible to understand why the United States does not use the same boldness and imagination that it applies to solving complex problems with its enemies to the solution of simple problems with its friends." After their meeting, Nixon pledged to improve the quality of water and appoint a special representative to seek "a permanent, definitive and just solution" to the problem. Echeverría returned to his country via the Mexicali Valley, his fourth visit there in three years, where he told residents, "We will no longer endure deterioration of our lands. We will no longer be subject to arbitrary decisions from abroad." He had turned on all the taps and, as Rabasa was to admit one month later, expectations in the valley had gotten somewhat out of hand. From this point on, the Mexican government would be more a victim of its own campaign than a perpetrator of it. Rabasa would constantly plead with American officials to hurry up the process of settlement. The foreign minister wanted the special representative to be named quickly and once his recommendations were made to President Nixon, he wanted the report as soon as possible. So far the government had been able to control public reaction but, warned Rabasa, "It was a time bomb which could go off any time." The goal of the Mexican government was to obtain a settlement in time for President Echeverría to announce it midway through his term.

It was no easy task to find the ideal special representative. As one westerner involved in Colorado River matters commented, "This is an extremely difficult assignment, particularly for someone who may be unfamiliar with the complexities of the problems." The qualifications set forth by the Department of State for the assignment, later to be elevated to ambassadorial rank, were truly remarkable—if not almost self-eliminatory—in their all-inclusiveness. Such a person should not be from the basin states, because of their competition for water. To satisfy the Mexicans, he or she should possess prestige, intellectual probity, and independence. On the domestic front, the special representative needed to be able to deal effectively with western congressmen. On top of all this, the special representative should possess technical, diplomatic, political, and legal skills, with the emphasis on the latter. Yet the representative needed to perceive the problem and its solution not in a narrow legal sense, but in its broader perspective. Four names were suggested. One of them,

Herbert Brownell, was appointed by President Nixon. They had worked together in the Eisenhower Administration, Nixon as vice-president and Brownell as attorney general.

Brownell turned out to be the right man for the complex assignment. A partner in the prestigious New York law firm of Lord, Day and Lord, Brownell was a native of Nebraska who had a lifelong interest in the development of the West. He knew somewhat, and could quickly learn more completely, what it meant to live in an arid land. (Samuel D. Eaton, Brownell's top aide and a career foreign-service officer, after a six-day tour of the Colorado River basin made the following report to Brownell: "The trip was highly useful. The task force was able to see for itself the aridity of the area; to sense at first hand the consequent intensity of feelings over water; and to observe the detailed management of a river that often seems only a poor, muddy stream harassed and cajoled by man into accomplishing more for him than ever anticipated.") Brownell specialized in international law, but had dealt with Colorado River matters as attorney general. As campaign manager for Thomas E. Dewey's two abortive tries for the presidency and Eisenhower's first successful attempt, Brownell had dealt with the West on political matters. He had the necessary prestige and breadth of concept. Such an appointment from the outside was unusual and could easily have been resented by the State Department bureaucracy. But Brownell gained its respect. "His personal prestige, his negotiating skill, and the authority of the White House behind him enabled him to knock bureaucratic heads together among the U.S. agencies involved, to persuade the Basin States and their congressional delegation of the fairness of his proposals, and to convince the Mexicans of the limits beyond which he would not go. Without his intervention on the scene, we would almost certainly be still quibbling over the rights and wrongs of both parties to the dispute," wrote H. Freeman Matthews, Jr., in an analysis of the negotiations for a State Department senior seminar on foreign policy. Rabasa would give most of the credit to his friend Kissinger, but it was really Brownell who pulled it all together.

Using a staff of four and an interagency task force, Brownell completed a comprehensive report containing his recommendations within the time limit and sent it to President Nixon in late December, 1972. Brownell was unable to get a personal appointment with Nixon, who was becoming increasingly preoccupied with Watergate, and said later that he did not think the President ever read the report. In any case, within the White House Kissinger was shepherding the matter, and Nixon approved Brownell's recommendations in time for Secretary of State Rogers to present them to President Echeverría on May 13, 1973. All through the process the Mexicans had been apprised of "the rather formidable domestic problem" (Brownell's words used during a background session with Mexican media leaders) with the western states, and Secretary Rogers repeated the warning about this potential stumbling block to Echeverría. His problem, the Mexican President emphasized in turn, was how to issue a press release indicating there had been some favorable movement toward solution of this highly visible political problem, yet keep the details contained in the report secret so as not to jeopardize negotiations that were to begin the following month. An innocuous statement was issued, and *The New York Times* and *Los Angeles Times* dutifully repeated it without probing any deeper. But that was not enough for the Mexican press, which pressured the Echeverría Administration for more. Two days later Rabasa, with the permission of the United States, issued a longer yet equally innocuous statement, which was used extensively in Mexico.

Starting in June, the negotiations proceeded quickly. The Mexicans repeatedly stressed that Echeverría wanted to make an announcement of their successful completion in his September State of the Union address. When Rabasa proposed what seemed like unreasonable demands, Brownell pointed out the difficulty the western states and Congress would have in accepting them. The Foreign Minister then pulled back, asserting in private to Brownell that they did not represent Mexico's final position. The Mexicans dropped their insistence on the quality of water being equal to what was delivered to the Imperial Valley. Agreement was reached on Mexico receiving water no greater than 115 ppm more saline than what was diverted at Imperial Dam. This represented a compromise between the original Mexican position of parity and the salt-balance concept of the western states, which would have meant water with a difference of about 280 ppm. The 115 figure, plus or minus 30 ppm to allow for the different methods of measurement employed by the two countries, represented what the increased downstream quality of the river would be without the highly saline flows from the Wellton-Mohawk district, it was thought. How the quality was to be improved to reach the specified differential was an internal problem of the United States, Mexico maintained, although they understood the plan was to build a huge desalinization plant near Yuma. By not endorsing a specific technical solution, the Mexicans could still insist on better-quality water, whether or not the solution chosen by the United States worked. As it turned out, this was a wise decision. Under the minute existing at the time of the negotiations, Mexico was receiving water with an average salinity of 980 ppm, which was about 130 ppm above the level at Imperial Dam.

The negotiations were completed in three short sessions during the summer of 1973. On August 30, after meeting with President Nixon at San Clemente, Brownell held a press conference. He emphasized the frequency of contacts with members of Congress and the Committee of

Fourteen before and during negotiations. Then the Ambassador got his principal point across: "This is a project that is based on dollars and not on water. I told the western states at the beginning of the negotiations that nothing would be done and nothing has been done as a result of this agreement which would adversely affect the orderly development of the western states. There are no limitations in the agreement which would adversely affect any of the planned programs for the development of the natural resources of the basin states." The basin states would disagree with Brownell's assessment, and then go to Congress and get everything they wanted.

Brownell later recalled, "I had been in this thing five minutes when I realized getting an agreement with the western states would be the most difficult part of the job." When he took on the assignment, there was already existing in the White House an assessment of the domestic political considerations involved in any settlement of the salinity problem. Furnished to Brownell and marked "Secret," it read, in part, "Wasted water is money and the economic and political forces of the Basin will quickly be marshalled to protect their pocketbook, not their aesthetic interests, if they perceive a threat to supplies or inequitable water quality distribution." The memorandum pointed out the entire basin would oppose any solution "if only the Mexicans were to benefit." Water interests in the basin desired, along with any solution dealing with Mexico, a Bureau of Reclamation program to control upstream sources of salinity, primarily in the upper basin, "particularly if there is no cost-sharing and the obligation is totally Federal." In other words, what was wanted was the same as the "national obligation" provision of the 1968 act, which laid on the national taxpayer, not the western user of Colorado River water, the responsibility to pay for augmenting the river with enough water to furnish Mexico with the amount promised by the 1944 treaty. Salinity-control projects in the upper basin would allow development to continue there under the provisions of federal water-quality standards then under consideration. Another package deal was beginning to emerge. It would cost basin water users very little. The memorandum, remarkable for the accuracy of its perceptions, continued with an analysis of the western institutions and personalities in water politics at the time:

> In assessing the political situation in the Basin, it is necessary to distinguish "political leadership" from the "water leadership." The water leaders tend to be very protective and defensive of the water interests of their State and to counsel the political leadership at both the State and National level on a technical and narrow basis with regard for their preeminent concern. . . . The Congressional leadership can be described as powerful, volatile, parochial and very protective of the

Basin's water which is both precious and in short supply. . . . The Governors of the Basin States are in most instances strong individuals who will fiercely protect their perception of their State's water interests.

The memorandum then dealt with key congressional water leaders, among them: John Rhodes of Arizona, House Appropriations Committee, "Aggressive and recognized leader of Lower Basin, defends Lower Colorado Basin against excessive Mexican demands"; Harold T. Johnson of California, chairman of House Interior public-works subcommittee, "Both powerful and effective, although district not in basin"; Clinton Anderson of New Mexico, Senate Interior committee, "His staff wields his considerable power"; Frank E. Moss of Utah, Senate Interior committee, "Will get into the fray if politically expedient." The governors of the seven states were then evaluated, among them: Ronald Reagan of California, "Has excellent staff support and tremendous interest in Colorado water for southern California"; and Calvin Rampton of Utah, "Strong and aggressive in behalf of state economic development."

Already John Rhodes, who was working up to a position in the House Republican leadership and was to serve as the western states' main conduit to the President, had contacted Nixon to stress that the United States "has no legal duty" to improve the quality of water flowing to Mexico but, "I feel it is to our best long-range interests to do so if we can without irreparable injury to our people." Brownell had his job cut out for him. In a sense, it was to be a wide-open football game—Rhodes could end-run Brownell to the President, and the western water interests, represented by the Committee of Fourteen, could pass over his head to congressional water leaders who were the ultimate keepers of the score. Moreover, Brownell and the Nixon Administration were to deal with the unusual phenomenon of basinwide unity that had existed only since 1968, and before then only briefly in 1922. As Wesley E. Steiner of Arizona, chairman of the Committee of Fourteen, commented after the bill was signed by President Nixon in June, 1974, "We were successful in achieving enactment of our legislative package because we were united. We must remain so if we are to realize implementation. Our work is not finished." Because of this unity and its resounding success, the techniques used to gain passage of the Colorado River Basin Salinity Control Act of 1974 would serve as the exemplar of western water politics and an indication of the cohesiveness of the forces President Carter would face three years later when he challenged the system.

Brownell was being squeezed. First, there were the Mexicans who wanted better water. Then there were the states pressing hard for a solution whose cost would be solely in terms of money, not water. There was a third countervailing force, this one within the Nixon Administration. The President's budget watchers (the Office of Management and the Budget), the Council on Environmental Quality, and the science adviser to the President voiced "deep concern" over the high cost and the environmental effects of building a desalinization plant, which was emerging in the fall of 1972 as the key to the solution in Brownell's mind. The three entities within the Administration wrote a memorandum to Brownell objecting to the expense of the plant and to the federal government's paying for it, since "the water quality problems occurring now are largely the result of irrigators in the Colorado River basin." Brownell shot back, "The memorandum fails to cover the attitude of the Basin States. This is a critical point, because without their support one does not have a solution to the problem with Mexico. Their Congressional delegation could hold up appropriations . . . and their public reaction could undermine the credibility of our negotiations with Mexico." Brownell's view prevailed within the Administration. It was not the first time that economic and environmental objections had been overridden by political realities in the basin.

Actually, Brownell had a second solution which he considered briefly but discarded because it would be futile to try and sell it to the western states. The idea was simply to close down the Wellton-Mohawk district and buy it out. "I was very much intrigued with this idea. After all, they were the ones who had caused the trouble. It would have represented a handsome profit for them. But the Arizona people were violently opposed to it," Brownell later recalled. It would have been strictly a monetary solution and no water would have been surrendered, its use just being transferred elsewhere within Arizona. What little input there was from conservationists on the Mexican salinity problem centered on this solution. Some academics favored it from a straight dollar-for-dollar standpoint; so did California Congressman Craig Hosmer, who declared, "We could spend $65 million and buy it up, shut it down, break it off and float it down the Colorado and sink the whole misbegotten scheme in the Gulf of California . . . to the same water quality end that would be a lot cheaper in the aggregate than spending $50 million to build a reverse osmosis [desalinization] plant and $10 million ad infinitum to run it. But I do not expect America's bureaucrats to go for such a simple solution as that." The colorful Hosmer, a veteran of the congressional water wars, also knew his colleagues would not go for such a solution. Actually, $65 million was only the federal investment in the district. The Bureau of Reclamation came up with a total cost of $266 million, which included acquisition of all the private property in the area. This price did not look too bad in 1979 when the cost of the desalinization plant, still not completed, had risen to $333 million (based on 1977 prices) with an annual operation and maintenance cost of $14 million. When Brownell gave his recommendation for the plant to the President in December, 1972, its estimated cost was $42 million with an annual cost of $6 million to run it. William E. Martin, a professor of agricultural economics at the University of Arizona, wrote in 1974 when Congress was working on the authorizing legislation for the plant, "Thus, it is somewhat frustrating to watch the results of poor past planning and present political necessity bring about enormous public expenditures on structural remedies at this time."

In addition to the economic argument, there undoubtedly would have been a fair amount of social disruption if the 75,000-acre district, whose main crop was hay, had been closed down. As a sop to the advocates of this alternative, the 1974 legislation directed that the district be reduced by 10,000 acres. All reduction costs would be absorbed by the federal government. As the bill was implemented, almost half the acreage that was retired was federal lands that had not been in production. Clearly, no one was to be hurt by the denial of water. The solution to close down the entire district was never seriously considered because of the political repercussions. "Strong opposition to such a measure would probably occur throughout the Basin," the Bureau of Reclamation predicted. It was just not the western way.

Throughout Brownell's search for a solution in the fall of 1972 and negotiations with the Mexicans in the summer of 1973, a period of time during which the ambassador met frequently with the Committee of Fourteen, the western states continued to be unhappy that he had not acquiesced to all their demands. Again and again the states asked that the federal government be responsible for replacing the reject brine from the desalinization plant, that the upstream salinity control projects be authorized simultaneously with the desalinization plant, and that "implementation of the proposal be without cost, impairment, or injury to landowners under the Wellton-Mohawk project." What is more, the committee constantly badgered Brownell to clear any changes in his negotiating position with it. Brownell resisted these demands, insisting they were more domestic in nature and should be dealt with separately at another time. But the political reality of the situation was that unless these domestic concerns were made part of the overall package, they stood very little chance of being authorized on their own.

There were veiled threats, and counterthreats. Committee members told the ambassador that he should not underestimate their power. Congress had supported finding a solution to the international aspects of the problem, so far, because the western states, in the words of one committee member, "had paved the way" with western congressmen. If the committee's attitude changed, "Congress' attitude would change," this key member asserted. In turn, Brownell threatened to break off negotiations, the alternative being an unfavorable World Court decision and the surrender not only of money, but most probably of water. There were hard words, followed by apologies. But ultimately Congress would resolve the issues in favor of the western states. There was little doubt of the outcome after fifty-two western congressmen wrote the President in July, 1973, repeating the demands of the Committee of Fourteen.

Following the signing of Minute No. 242, the basin states concentrated on drafting legislation that would flesh out the international aspects of the solution with the domestic concerns. Later they would briefly consider, then drop, the idea of hiring a New York law firm to challenge the international agreement. It was dropped when their bill was assured of passage by favorable committee hearings. During the fall of 1973, the states plotted the strategy that would be necessary to get both the Administration's and the basin states' bills referred to western-dominated Interior committees in the House and Senate, rather than the more inhospitable foreign-affairs committees. It was in the Interior committees that western water interests could control the destiny of the two bills. In February of 1974 the Administration bill was referred to the House Subcommittee on Water and Power Resources, whose chairman was Harold T. Johnson, and Steiner gloated about the Administration's predicament: "Sooner or later we're gonna getcha."

And, indeed, the federal government was outmaneuvered by the basin states. The hearings were decidedly tilted in favor of the bill drafted by the states. Four salinity-control projects were included in the bill, even though the necessary feasibility studies on them had not been completed. Johnson, referred to as one of "The Three Magi" for his ability to obtain authorization for water projects, told an Administration witness at the subcommittee hearings, "It is not our intention to try to hold up anything or hold anyone at gunpoint. We would, however, like to have consideration and recognition given to our problems. We will try and perfect you a good title I to take care of the international problem, and we would like to have a title II in the bill that would help give us a little boost on the problems of the American side of the border that are of concern to the basin states." It was probably one of the most visible holdups ever staged at a congressional hearing.

Introduced at the start of the year, hearings conducted in March, passed by the House in a 403-to-8 vote on June 11 and in the Senate by a voice vote the following day, the Colorado River Basin Salinity Control Act of 1974—as envisioned and drafted by the basin states—was one of the speediest legislative resolutions of a western water problem. It was the unity of the basin states, despite Administration resistance, that had made it possible. "The achievement of unity within the entire basin has made the salinity control coalition a powerful one indeed. It has been successful thus far in achieving an approach to the solution of the salinity problem that promises extensive benefits to traditional beneficiaries of such an approach with very little cost," wrote Dean E. Mann, a professor of political science at the University of California at Santa Barbara. Although he was not completely happy with it, President Nixon signed the bill on June 24. It was the only way he could fulfill the commitment made two years earlier to President Echeverría before a July 1 deadline, which had been informally agreed to during the negotiations. The salinity-control projects, Nixon said, were premature and the large federal contribution that would go toward their construction was contrary to the Administration's policy of "placing most of the financial responsibility for pollution abatement on those who are causing the pollution problem." But the bill did contain the "permanent and definitive solution" to providing Mexico with better quality water. Or did it?

As the years passed and the costs increased—115 percent for the desalinization plant and 123 percent for the four salinity-control projects within five years—congressional critics of the plumbing-system solution became more vocal. The original cost estimates for the four projects, it was determined, were imprecise and hastily prepared, according to a General Accounting Office report, and their potential effectiveness "is questionable at best." As for the desalinization plant, to be the largest by far in the world and originally sold on the concept of advancing the state of the art, there were some who wondered if it would work as it was supposed to. It was certainly going to prove to be a large energy drain on the Southwest. It all made one wonder what temporary expedients the Hohokam and Anasazi Indians had devised to deal with the river's increasing salinity during their time.

The political coalition formed to deal with the Mexican salinity issue, which had originally come together to push the Central Arizona Project bill through Congress in 1968 and had lasted through President Carter's challenges to the West, would undoubtedly dissolve in time. There were too many diverse interests who wanted a piece of the river, and when the eventual shortages arrived, as they must, these interests would trample each other to retain their shares. They could only unite on the commonly perceived need for augmentation, which provoked the opposition of the Northwest, or on shorting Mexico either on quantity or quality of water. This, of course, provoked opposition from across the border. So the basin states were being squeezed from north and south; or, another way to look at it, they were becoming bloated with growth and impinging upon the rights of their neighbors. The way out of this seeming impasse depended on either one or the other or both neighbors weakening politically. Then their water could be captured.

Should the two neighbors hold firm, then the West would have to face the ultimate limiting factor on further growth—no more water. New technologies, such as desalinization, if they worked, could offer a short-term fix until their limits were also surpassed in time; but such solutions depended on a governing system still geared to providing the West with more water. There was no guarantee—indeed, there were indications such a policy would not survive much longer. Agriculture was surfeited with water, but it could only surrender so much to the cities and energy developments before the food supply became insufficient. Besides, the essential fabric of the West could not tolerate the massive dislocations that would result from too great a transfer. The ultimate savings account was the underground water supply. Maybe it was time to draw extensively on it. But such a decision implied an admission of a limited future. Going one step further, would it not be better to set meaningful limits on the surface supply of water while the basin states remained unified and politically potent, rather than continue toward the ultimate degradation and exhaustion of the Colorado River? In this way a desert civilization would have a better chance to survive intact, and perhaps once again the Colorado could bear some resemblance to a river.

These were some of the thoughts I was beginning to deal with the first time I traveled the length of the river. It was the summer that Ambassador Brownell was shuttling back and forth, negotiating with the Mexican government and basin states. On the map it all seemed so simple. I could clearly see the Colorado River and its tributaries were one river system; a whole, living entity like, say, a tree with roots and leaves to take in and disgorge nourishment.

But listening to and seeing how people used the river, the Colorado became a series of segments, each labeled "mine." My own happened to be rather movable, depending on where I camped each night. Most others were much more stable: a ranch, a power plant, an Indian tribe, an irrigation district, running whitewater, or slow greenwater. One neighbor's vision rarely penetrated into another's backyard with any great degree of compassion. The upper basin did not recognize its obligation to furnish half of the water due Mexico. The lower basin did. Both hoped to gain more water by their separate positions. A farmer in western Colorado's Grand Valley, where one of the salinity-control projects is located, did not recognize the fact that his overirrigating caused the salinity level to rise, thus affecting a farmer downstream in the Imperial Valley who had to install an expensive system of drains on his eight hundred acres and apply more water to leach out the increasing amounts of salt washed downstream. Someday there would not be this additional amount of water, above what is needed to irrigate a crop. The Imperial Valley farmer wondered why the impoverished Mexican farmer did not install a similar system of drains. When I visited him, the farmer in the Mexicali Valley was butchering a goat on a Sunday, with all his children and relatives gathered around him, in a country whose population has doubled in the last twenty years. Agrippino Castillo bemoaned the decreased productivity of his small plot of irrigated land and the fact that he had to get another loan from the bank.

The upper basin and the lower basin, to use the arbitrary designations settled upon for political and administrative purposes, were two different worlds as far as their perception of the salinity problem was concerned. The lower basin wanted to protect what it already had, and the upper basin wanted to exercise its right to more water. To listen to the voices, one would never suspect they were connected to the same river, or that the Colorado led to Mexico and the expectations of its multitudes, or, after being last used in that country, that it no longer emptied into the gulf.

CHAPTER SEVEN Ends:
Death in the Desert

1

I am not sure what awakened me shortly after 6:00 A.M. Perhaps it was an airboat, such as is used in the Everglades of Florida. We had seen some moored at the last fishing camp on the river and had heard what sounded like an airplane engine rev up, run at a high pitch for a few moments, and then die out. This had happened numerous times the previous afternoon, and now that I was awake, it could be heard again. Its rhythm was like repeated attempts to extricate a car stuck in snow, and it left the same sense of frustration hovering over the river. What woke me could also have been the mice that scampered through our food bag that night, chewing their way into an envelope of instant cocoa but leaving the bulk of our food supply intact. That was fortunate because we did not know if we had a few more hours or a few more days left on the river. A Mexican fisherman with a disabled outboard engine had groped his way up the river with a flashlight during the night, and another with a functioning engine had set off downriver in the early-morning mist. Coyotes were all around us, yapping and howling to an extent neither my companion nor I had ever heard before. Avian noises of every description added to the cacophony.

The Colorado River delta, where we were camped in Mexico while searching for the end of the river, is a land full of illusions. How appropriate, I thought, for the river to be enveloped in fog this morning. It hid the harsh realities of the most inhospitable terrain on the North American continent and made the realization of our quest in the two-person kayak as ephemeral as it seemed when perusing all the available maps, charts, satellite photographs, and literature prior to embarking on this journey. Like the elusive quality of the land, the printed material was either obscure, inaccurate, or obsolete. The latest chart I was able to obtain for the northern portion of the Gulf of California, dated 1975, relied on soundings made between 1873 and 1875 by the U.S.S. *Narragansett*, commanded by a naval officer named George Dewey who was later to gain fame at Manila. I had made one previous attempt to find where the Colorado River disappears seventy-five miles below the border and a few miles before reaching its historical outlet to the gulf, a 650-mile long arm of the Pacific Ocean. But locating the end of the river had proved to be a difficult goal.

There are two ways to approach the end; from the south via the diminishing gulf or down the sluggishly flowing river from the north. On the previous trip I had engaged a Mexican fisherman at the village of El Golfo on the Sonoran side of the delta to take me and a friend to the head of the gulf. We left early in the morning of a calm day, but soon a strong wind blowing across the shallow water of the gulf kicked up whitecaps. The outboard motor coughed, then quit. Fortunately we were close to shore and the rowboat had oars. We walked back along the beach to El Golfo, minus the money I had advanced for a day's worth of gas. On another voyage to an island in the gulf, this time further south, the outboard on another Mexican fisherman's boat threw a rod. There was a spare motor aboard, and this sputtered along on one of its two cylinders. It was just barely enough to get us back to shore from those fearfully isolated waters. I took no chances on this latest trip to the end of the gulf and brought along my more reliably propelled kayak, although this would mean no local knowledge would be immediately available.

Such knowledge would have been helpful, since the literature, maps, and charts give no firm answers and, in fact, all inaccurately portray the river as serenely flowing to the gulf. It has not done so, except infrequently, since 1961. A 1975 report by the Bureau of Reclamation, which surveyed the delta as part of the salinity control project, stated the Colorado no longer flowed into the gulf "but is restricted by an earthen plug . . . about 14 miles upstream from the Gulf of California." Maps accompanying the same report showed a solid blue line passing all the way through the delta to the gulf. Other bureau personnel, who periodically fly over the delta, state the sand-and-silt barrier is between five and six miles wide. Obviously its width varies, with the height of tides and local storm runoff.

Only the satellite photographs clearly demonstrated the reality of a river no longer attaining its historical outlet to the sea, a river so greatly diminished along its 1,700-mile journey from the furthest point on the Continental Divide that before reaching its proper ending the limp water simply evaporates from shallow ponds into the azure skies of the hottest region on the continent (average maximum summer temperatures for the area are between 104 and 108 degrees, with highs frequently near 120 degrees). Since the accidental diversion of the river into the Imperial Valley in 1905, the closing of the gates on Hoover Dam in 1935, and the increasing demand for irrigation water in the intervening years, the natural regimen of the delta has been irrevocably altered.

From 1950 when the Mexicans built Morelos Dam at the border to divert their share of what remained, virtually no water has flowed in the river channel south from the dam. The water that does return to the channel further south, to swell its flow so it can once again be recognized as a river, is the thickened return flows from Mexican fields. The water no longer seems fresh; in fact, it has not seemed so for some distance upriver. To follow the river from Morelos Dam to the gulf is a tricky business. First it is there, then it isn't, then it is, then it isn't. Its presence depends on when the toilet is being flushed.

A great river passing through arid lands has been wrung dry by man, but along its way the Colorado has nourished the heartland of the West. Given the emphasis on growth during the first half of this century—at its most fervid, a national policy of "manifest destiny" for the West; at the very least, the absence of challenge to a region's dominant economic thrust—the increasing use of Colorado River water was understandable. The questions that were first posed by conservationists about dams in the 1950s—the vague desire to stop, to limit, couched in aesthetic terms—and the economic justifications sought by the Carter Administration for water projects have barely detained this process.

Actually, very little remains to be built on the Colorado River, the primary projects already being in place, the remainder authorized, and only the cleanup chores left to be completed. But there is always the chance the final, big push will be made to import more water from elsewhere and, if this is done, it means another titanic struggle with all the traditional forces lined up as they have been in the past. Hopefully a wider public would be more aware of what was being proposed and less apathetic, as in the past, to the disposal of the West's most valuable natural resource.

Whether the policies that have governed the West as a separate province will ever be consciously reevaluated and another course embarked upon is doubtful, at least in the immediate future. It would mean a change in basic institutions, in basic ways of thinking that are endemic to the West and, above all, it would mean limits in an area that has little tolerance for them, be they the 160-acre law or restrictions on grazing on public lands. There first has to be a realization that the West is at its ultimate limit—meaning there is no more water available—before anything different will happen. The fact that it is fast approaching this limit has meant very little, witness how quickly the lessons of the recent drought have been forgotten. Nowhere else in my travels did I find an area more symbolic of what the West could become than the illusionary wasteland of the delta.

In a sense, there are now four ends to the river, just as there are four beginnings. There are the two interior basins called the Salton Sea and Laguna Salada, the point just short of the gulf where the water disap-

Shrimp boats anchored in the Gulf of California. Santa Clara Canal in foreground.

pears, and the Santa Clara Slough on the east side of the delta. A fifty-three-mile, concrete-lined canal transports the highly saline waste waters of the Wellton-Mohawk district across the border and dumps it into the slough, thus bypassing Mexican water users. Construction of the canal was part of the 1973 salinity agreement between the two border nations. The agreement unwittingly added one more end to the river.

Ironically, it may be through this combination of canal and slough that the Colorado River will once again find an outlet to the gulf. (When I last visited the delta in mid-1980, this was the case, following heavy runoff in the lower basin.) The slough is a slight 103,000-acre depression in the salt flats lying twenty-seven miles north of the gulf and abutting the delta's eastern escarpment. Before the waste water from Arizona was dumped into the slough's northern end, a small marsh of 75 acres, inhabited by the rare Desert Pupfish, was fed by an agricultural drain from local fields. There was another small marsh of 375 acres, fed by springs of no visible source. Then came as much as 180,000 acre-feet a year of Wellton-Mohawk waste water with the good possibility that it could not all be absorbed or evaporated in the slough and would eventually find a way to the gulf. We camped one night beside the Santa Clara Canal, a tidal creek where Mexican shrimp boats unload their catch and the most likely outlet to the gulf for the slough. Climbing the steep bluff that forms the eastern escarpment, I could see the two bodies of water lying widely separated. But it seemed likely one day they would connect. There was only a long, low rise of salt flat separating them.

Even at its termination, in this most inhospitable and thinly populated of all regions traversed by the river, the Colorado resembles more a vast plumbing system than a river. At the beginnings, high along the Continental Divide, those canals diverting water out of the Colorado River basin are part of the same system, the same ceaseless attempt in an arid land to transport water where it is needed, or in the case of the Santa Clara Slough, away from where it is no longer wanted, once it has been used.

After eating a breakfast of instant oatmeal and what cocoa remained, we packed the kayak. Our camping spot was on low-lying sand amidst tamarisk dripping with moisture from the early morning fog. We were in a small world, not the usual limitless, horizontal spaces of the delta. Everything about us was muted, lacking the usual shimmering vibrancy. We shoved the kayak into the water at 7:20 A.M. and headed for the end of the river.

Sitting in front was Gary D. Weatherford, a lawyer of precision and

care. We had spent some time working together in the frenetic atmosphere of the California Resources Agency, dealing with such problems
as how to reorganize the board representing the state in Colorado River
matters. Gary at one time had been a special assistant to the solicitor for
the Department of Interior and a participant in the Lake Powell Project,
a National Science Foundation study of the central portion of the
Colorado River. He lent a certain scientific respectability to our small
expedition. At the oddest moments Gary would confront me with the air
or water temperature, or while taking a water sample, he would state with
a slight smile, "There goes a blow for soft science." At the moment he
was plagued with a sore left shoulder which limited his paddle strokes to
fifty, then a rest period—all precisely measured.

The problem was that we did not know where we were on the river, the
maps not being accurate and the satellite photos not being detailed
enough. There was no high ground to reconnoiter from, nor any tree
strong enough to support our climbing it for a view. The fishermen we
passed at the last outpost on the river indicated we had ten or twenty
miles to go. Ten miles was supposedly to the canal drawing off most of
the water to Laguna Salada. Twenty miles was to the deserted site of Joe
Bourbon's fishing camp. Somewhere in between these two points, according to local knowledge, the river ended.

While paddling through the dense morning fog, I suddenly felt disori-
ented and experienced a slight case of internal panic. There was no
perceptible current, both riverbanks were out of sight, and the sky had
melded into the water. If it had not been for the dull globe of the sun,
there would have been no indication of what was up or down except our
own disjointed sense of equilibrium. But constant uncertainty in a strange
land was what we experienced the whole ten days spent either paddling
down the river or up the northern extremity of the gulf, camping at Isla
Montague at its head, or trudging the six miles across the hot, desolate
flats to a shipyard abandoned a hundred years ago.

About the only thing we did not experience was the tidal bore, a wall
of water which rushes upstream at the mouths of only a few rivers in the
world. The bore at the head of the gulf is much diminished from earlier
times, and we probably missed it because I had purposely picked a period
of minimum tidal range to explore the delta. With the range of tides
running to a maximum of close to thirty feet, much of the low-lying
sand-and-silt flats are exposed during spring tidal periods, making it
difficult to find a camping spot or to wade ashore at low tide through the
hip-deep silt that has the looks and consistency of gray grease. Bores are
more pronounced during periods of high tidal ranges. They form when
the funnel effect of an estuary and the shallow depth of water at its mouth
bear the correct relationship to each other. The bores at the head of the

gulf are much diminished now because no longer is there a river channel to rush up, nor a descending flow to oppose it.

The first Spaniard to explore the Gulf of California, then named the Sea of Cortés, encountered the bore at the mouth of the Colorado River in 1539. Francisco de Ulloa had been dispatched north in ships, like Coronado who was taking a land approach, to find the fabled Seven Cities of Cibola. Instead Ulloa discovered the mouth of the Colorado River. It is hard to imagine an area that would appear more worthless to a Spanish conquistador looking for mineral riches. Ulloa did not spend much time there. He landed on one of the two islands at the mouth of the river and then left quickly after being badly frightened by the tidal bore. He guessed that "some great river" might be the cause of the bore. Other early explorers also encountered the bore, their accounts of its ferocity depending on the range of the tide and their imaginations. In 1826 that early wanderer throughout the West, the trapper James Ohio Pattie, went to sleep on what appeared to be safe ground at the mouth of the river, only to be awakened by a "rushing noise." His campsite was quickly inundated by three feet of water. Lieutenant Joseph C. Ives, who in 1857 headed the first scientific expedition up the Colorado River, experienced a large bore at the river's mouth while aboard a schooner mired in the mud by the outgoing tide. His account of the phenomenon is vivid:

> About nine o'clock, while the tide was still running out rapidly, we heard, from the direction of the Gulf, a deep, booming sound, like the noise of a distant waterfall. Every moment it became louder and nearer, and in half an hour a great wave, several feet in height, could be distinctly seen flashing and sparkling in the moonlight, extending from one bank to the other, and advancing swiftly upon us. While it was only a few hundred yards distant, the ebb tide continued to flow by at the rate of three miles an hour. A point of land and an exposed bar close under our lee broke the wave into several long swells, and as these met the ebb the broad sheet around us boiled up and foamed like the surface of a caldron, and then, with scarcely a moment of slack water, the whole went whirling by in the opposite direction. In a few moments the low rollers had passed the island and united again in a single bank of water, which swept up the narrowing channel with the thunder of a cataract. At a turn not far distant it disappeared from view, but for a long time, in the stillness of the night, the roaring of the huge mass could be heard reverberating among the windings of the river, till at last it became faint and lost in the distance.

Gary and I camped one quiet night near the northern tip of Montague Island. A few miles to the west of us, across what was once the river's estuary but is now the most northerly extension of the gulf, lay what was known in Ives's time as Robinson's Landing. It was off this point, actually a shack on stilts that served as a crude hotel and navigation aid for ships approaching from the gulf, that Ives experienced the bore. Ten miles to the northwest and sixty-five years later at La Bomba, a collection of rude shacks serving as a port in 1922, the small gulf steamer *Topolobampo* was anchored in midstream at dusk on a November evening. Aboard the overcrowded steamer were 125 Mexicans on their way to work in the fields of the Mexicali Valley. The crew and passengers went to sleep. Author Frank Waters, who embarked from La Bomba on a similar voyage a few years later, described the scene: "Near midnight everyone on board was awakened by a terrifying roar. It sounded like a gigantic waterfall booming downriver. Terrified, the passengers rushed to the deck. Clearly in the moonlight they saw traveling upriver with the speed of an express train an immense wall of water nearly fifteen feet high. There was hardly time for a single frightened cry. The wave caught the *Topolobampo* squarely abeam, snapped the hawsers like threads, and rolled the ship over like a log." Of the 86 persons drowned, only the bodies of 21 were ever recovered.

That night on Montague Island I lay awake on the squishy ground for a long time, using a life preserver for a pillow. We had lifted the kayak high up on the bank of the island and anchored it securely on the semidry ground. Although it was the wrong time of the month and I knew the bores had supposedly lost their punch, I was taking no chances in that illusory place.

Two days earlier Gary had talked to two bottle hunters who had been searching the delta area for their particular brand of treasure during vacations the last ten years. The bottle hunters, from a Los Angeles suburb, had gotten to a great many places once they had adapted trail bikes to the special needs of the delta. Oversized tires with metal extensions to scrape off the mud were two of their improvisations. The bottle hunters said they had once seen a one-foot-high bore. Then they disclosed something more interesting. They said they had been to what remained of the shipyard near Puerta Isabel that had been abandoned a hundred years ago when the brief era of steamboats on the lower river began to come to a close. I had heard of the shipyard, but had no idea how to get there.

Early the next morning we headed northwest from our campsite near the road to El Golfo with plenty of water, a little food, and a compass. It was like heading into nowhere, except there were the tracks of the trail bikes to guide us part of the way. It was November, but the temperature was in the low nineties by midday. The land—actually I never had the feeling that it was quite solid enough under my feet to refer to it in such a definite way—was absolutely flat and completely devoid of vegetation. Small pieces of driftwood and wooden crates resembled trees or structures shimmering in the distance above the surface of a glass lake. Rowboats improbably sat in the middle of this ultimate desert, ripped from their moorings and deposited on the silt and sand by a combination of high tides and those brief, intense storms that swoop unexpectedly into the gulf region. The debris was scattered about with the randomness of an old graveyard. The scene was sepulchral.

As we walked, Gary asked me what I thought would happen if there was an earthquake. Given the evanescent nature of the place and my imperfect knowledge of geology, I replied that I thought the earth would liquefy and we would disappear from sight, slurped up into the thousands of feet of jellolike silt beneath us. The chance of an earthquake was not academic. Few areas are more earthquake-prone, since the delta lies within both the San Jacinto and San Andreas fault systems, consisting of a series of faults running northwest-southeast that have been the major factor in forming the Salton Trough. Hundreds of seismic events are recorded daily, but only a few rate being called earthquakes. Their frequency is such that an early observer noted, "There are so many as to receive scant attention from the local inhabitants." The Yuman and Cocopah Indians barely flinched when they felt the earth shake, but newly arrived soldiers fled to the parade ground at Fort Yuma in 1852 when a particularly strong quake shook the delta and sent a cloud of steam rising a thousand feet above the volcanically active area surrounding Cerro Prieto. The schooner *Capacity* of San Francisco was anchored in twelve feet of water at the mouth of the river. When that quake struck it suddenly found itself aground. The instability of the earth fascinated early explorers around Cerro Prieto, a volcanic cone on the west side of the delta where geysers, mud pots, steam vents, and hot springs bubbled and boiled. A shallow lake that collected the river's waters was called Volcano Lake. The Mexican government, recognizing a handy energy source, recently built a 75-megawatt geothermal power plant at Cerro Prieto. A 1940 earthquake produced forty miles of ruptured surface, some of which was horizontally displaced fourteen feet. I felt vindicated, although somewhat uneasy, when I later read that such an earthquake "may cause liquefaction and subsidence because of the deep, wet alluvial deposits."

The final mile to the abandoned shipyard was the most difficult. A thin sheen of salt water lay over the silt, whether from the last high tides or the waste water being dumped into the Santa Clara Slough I could not tell in that flatness. Our feet had to be forcibly torn from the gook step-by-step. The boots I wore were too heavy. High-topped sneakers that would have stuck to my feet when removed from the clinging silt would have been preferable. Huge gobs of mud, feeling like weights, formed on the bottom of the cleated soles and had to be knocked off every few minutes at a rest stop. It was getting hotter. A sickly sun was shining through a slight haze that seemed to magnify its intensity. Three-quarters of a century earlier, Arthur Walbridge North trekked north across this same desert in August and described the ordeal thus:

> Nature, herself, seemed bent on our undoing. The fiery shafts of a relentless sun beat down upon our heads. The hot, saturated earth again and again gave way beneath our feet. The air was stifling, murky. Mirages concealed the true horizon. Thickets and strange, weird objects arose at either hand only to disappear in the twinkling of an eye. The glassy surface of a broad lake glittered in the sunlight before us, its unstable shoreline ever receding, while a shimmering sea crept stealthily in our wake. Deceived by our eyes, hemmed in by the unreal, we came to doubt the stability in our minds.

North and his companion made it across the delta plain. A group of Chinese immigrants traveling the same route on foot a few years earlier did not have similar luck. The Chinese had been landed by ship at San Felipe at the southwestern end of the delta, and were bound for the Mexicali Valley, where they hoped to get farm work in 1902. The immigrants, so the story goes, paid $100 in gold to a Mexican to guide them across the desert now traversed by a single paved road. It was another August. Forty-three started the trip; only six, including the Mexican guide, completed it. A dark volcanic butte along the route of this death march has been named Cerro El Chinero. Those who survived conceivably went to work in the Mexicali Valley. Chinese and Japanese laborers built the levees and irrigation ditches that made it possible for an American land syndicate to prosper in a foreign country. Yet other Chinese, living in caves, helped construct the Grand Ditch at the start of what is now called the Colorado River. Referred to as "the oxen of the Pacific" and prized because of their cheapness and docility, the Chinese built many of the early waterworks in the West.

We finally struggled onto the slightly raised ground that signified the remains of Puerta Isabel. Most great rivers have large cities near their mouths: New Orleans on the Mississippi, New York on the Hudson, Alexandria on the Nile, and Karachi on the Indus. These cities grew in large part because of their strategic commercial location between ocean and river. The Colorado has no such port; but once it had just barely something. The schooner *Isabel* ventured into a small slough just to the east of the estuary in 1865 and a few miles up the serpentine tidal creek found an anchorage protected enough from the tidal bore so that freight could be safely transferred to barges for the trip upriver. Thus was born Puerta Isabel, at its height a community of between 300 and 500 persons living and working in what must have been the most desolate outpost in the West. Puerta Isabel was 157 river miles downstream from Fort Yuma, also a small community marooned in the middle of the Sonoran Desert, but north of the border. The Colorado Steam Navigation Company, then holding a monopoly on commercial river traffic, diked off between twelve and fifteen acres of tidal flats at Puerta Isabel and built carpenter, blacksmith, and machine shops, along with a company office, cookhouse, mess hall, storerooms, and cottages. An old steamer was hauled out onto the flats and served as a bunkhouse. One visitor to Puerta Isabel said the area was "destitute of every kind of vegetable growth, of every living thing except the fish, the wild fowl and the coyote." But within the diked area there was a constant surge of activity, with passengers and freight being transferred from schooners and oceangoing steamers, and vessels being repaired and built on a drydock.

The steamer traffic flourished from the end of the Civil War to 1877. There were new settlements upriver; and the soldiers needed to protect them from Indians, and the freight and passengers bound for the new gold and silver mines along the lower Colorado, were all carried on the shallow-draft stern-wheelers. In those boom times even the barren delta was looked upon as a potential bonanza. Charles Granville Johnson must have been dreaming of another land when he wrote in a pamphlet, sold in San Francisco for fifty cents, "These overflowed lands are of rich soil, and can easily be diked in and cultivated to almost any extent, and in the cheapest and most labor-saving manner. Here there are regular seasons of rain. The climate is of a warm temperature, very pleasant and productive. The nights are cool at all times of the year; blankets are, at night, necessary for comfort." Johnson was writing of a region with less than three inches of average annual rainfall and temperatures so hot it was painful to handle metal dining utensils.

But the boom was short-lived, and steamboat traffic on the river, which reached as high as the mouth of the Virgin River 590 miles above the gulf, was drastically reduced in 1877 when the Southern Pacific railroad tracks reached Yuma. Puerta Isabel was abandoned in 1878. Navigation of the lower Colorado had been, at best, a tenuous business that was at the mercy of a wildly fluctuating river. That peripatetic traveler and federal government agent, J. Ross Browne, wrote after visiting Yuma in 1864, "As a navigable stream it possesses some advantages during the dry season; boats can seldom sink in it; and for the matter of channels it has an unusual variety. The main channel shifts so often that the most skillful pilot always knows where it is not to be found by pursuing the course of his last trip." Godfrey Sykes, who was later to make the most comprehensive survey of the delta, visited the abandoned port in 1891 and found a group of Papago Indians tearing zinc plates off the hulls of the beached steamers and barges to make into bullets for an ancient smooth-bore, flintlock musket. The Indians were the only humans Sykes and his companion encountered on their trip through the delta. One of Sykes's sons, Glenton, visited the shipyard in 1975 and found it a "strange and fearful country."

When Gary and I arrived there, it spoke of abandonment and failure that could still be tangibly felt in that most hostile environment. There were hemp lines, coiled and rotted in place. Two large mounds, not unlike elongated burial plots, etched on the silt the contours of rotted ships' hulls. Piercing the mounds were iron bolts and pieces of machinery. Other odds and ends lay scattered about the drydock, whose general outline was still visible. I saw where the treasure hunters on trail bikes had dug, only to find common Tabasco bottles mixed in with clam shells. Gary strung a poncho between two upright posts and we waited out the midday sun under that primitive shelter before starting back.

Not only has Shipyard Slough been silted in during the last hundred years, perhaps because of the absence of freshwater flows from a river in periodic flood, but there have been other drastic changes in the regimen of the delta since large amounts of water were diverted from the river or stored behind dams upstream. The navigable channel used to run along the east side of the delta. Now, whatever flows remain run through a channel that has swung toward the west. Sykes traced the abandonment of the river's former channel back to 1909, the year the first dam on the main stem of the Colorado was completed. Laguna Dam, just above Yuma, cut off what little was left of commercial river traffic. Levees built to protect the agricultural fields in Mexico and the Imperial Valley from inundation helped to stabilize the flow of the river in its new channel.

With less water and the absence of spring and summer floods—dams upriver now caught and regulated the runoff from the mountains—drastic changes came about in the wildlife and vegetation of the delta. At one time the area was biologically the richest region in the Southwest. The naturalist Aldo Leopold explored it by canoe in 1922 and sensed "it had lain forgotten" since the first Spanish explorers had landed. Leopold described its "lovely groves" and "awesome jungles" where there was "a welter of fowl and fish." He hoped to see one of the jaguars that were occasionally spotted there, and remarked upon the "troops of egrets" resembling a snowstorm. Other early visitors commented upon the lushness and abundance of the region. Photographs and written accounts show that although some of those characteristics of "green lagoons," as

Puerta Isabel, a shipyard used to repair the steamers that plied the lower river in the late nineteenth century.

Leopold referred to them, remain along the watercourses, they are greatly diminished.

The delta has more than its fair share of strange tales, some of them revolving around the wildlife at the mouth of the river and in the equally remote gulf. Although Pattie, the peripatetic trapper, spoke of shooting "an animal not unlike the African leopard," it was probably one of the stray jaguars that sometimes wandered up from central Mexico. But I am not able to find a rational explanation for what a Jesuit priest witnessed. Father Johann Jakob Baegert, who lived in Baja California for seventeen years during the mid-eighteenth century, said there were large alligators living in the Colorado River "capable of devouring a full-grown man." The priest, who related this tale once he got back to his native Germany in a generally negative book about his experiences, said he had seen several of these creatures, which he described as a cross between a lizard and a turtle. Alexander S. Taylor, a federal court clerk and honorary member of the California Academy of Science, described strange goings-on in the gulf in what was otherwise a rather straightforward, mid-nineteenth-century account of the West's mineral resources. There were immense swordfish "known to attack vessels and leave their shafts in its timbers." Gigantic, eight-armed octopuses or squid, "the size of a 74-gun frigate," were known to have subdued sperm whales. The dugong, or sea cow, was described as a mammal "which almost answers to the sailor-myths of the mermaid." No greater than three feet long or ten pounds in weight, this mini-mermaid possessed a head "similar to a dog or a sea otter, skin smooth and without scales or hair above the navel" and breasts "similar to those of a man or woman." There were also "ferocious" sharks and "terrible" manta rays, known also as devil fish, inhabiting the remote night-blue waters of the gulf. Giant whirlpools and Indians who practiced cannibalism were also part of the gulf's lore.

There are also accounts of buried treasure. The proprietor of Robinson's Hotel at the mouth of the river, a steamboat captain in his working hours and later overseer of the shipyard, lived in that remote spot so he could spend his spare time hunting for the gold thought to have been lost by a French nobleman when he fled the Mexican state of Sonora after fomenting a revolution in that northern province. The count's vessel got into the wrong channel and sank. Most of the party were killed by Indians, so the tale goes. "Old Rob," as the steamboat captain was known, never found the treasure, and the place so depressed his young bride that he had to retire to Clear Lake near San Francisco, where he operated a pleasure boat on those tame waters.

The count was not the only soldier of fortune whose attempt to seize a part of Mexico floundered in the muddy waters of the lower Colorado. William Walker, an enterprising San Francisco journalist, seized La Paz near the southern tip of Baja California with a small force and declared the peninsula an independent republic. He marched north in 1854 and was on his way to Sonora to establish that territory as a slave state when he lost most of his supplies and some of his men crossing the river. He escaped across the border and surrendered to the U.S. Army. Walker was subsequently acquitted of any wrongdoing in a San Francisco trial. For a long time Americans regarded the Colorado River Delta as their fiefdom, an attitude that was to influence nationalistic feelings in Mexico when problems with the quality of river water arose in the 1960s.

The delta, of course, is a much larger area than just the most obvious end of the river where the reliefless salt flats stretch south from the last cultivated fields in the Mexican states of Sonora and Baja California. All that sand and silt and clay washed out of the canyons, mesas, and mountains of the West for millions of years have partially filled a trough 100 miles wide and 800 miles long, extending from San Gorgonio Pass north of Palm Springs to the mouth of the Gulf of California. Deltaic sediments 20,000 feet deep in places were deposited in the trough by ancestral streams of the Colorado River system and interwoven with marine deposits from periodic intrusions of the gulf and inland seas. These deposits date back ten to fourteen million years.

This seems like a lot of material to have washed down a river, even a river referred to in its predam state as having the greatest silt load in the world, until one imagines the Grand Canyon picked up and placed within this trough. During one well-documented flood on the Bill Williams Fork, a minor tributary of the lower Colorado, ten million cubic yards of silt were deposited on the delta in a few days. The floods of 1891 were so great on the lower river that silt and water once again flowed from the Colorado to the Salton Sea. Godfrey Sykes journeyed by boat partway, and an expedition sponsored by the *San Francisco Examiner* floated from the river to the inland sea. It took the *Examiner* party five days to cross what Sykes termed the "semi-navigable desert," later to become known as the Imperial Valley.

The Salton Sea basin, whose lowest depth is 273.5 feet below sea level, has periodically filled since ancient times with Colorado River water. The name given by anthropologists to the body of water during those periods is Lake Cahuilla. The shoreline of the ancient lake can still be traced at a point 42 feet above sea level. Most probably the lake, extending to just north of Indio with an outlet to the gulf, last filled between 1000 and 1500 A.D. The accidental filling of the Salton Sea between 1905 and 1907 only reached a height of 196 feet below sea level; it had another 238 feet to go to reach the maximum height of the ancient lake. The oral traditions of the Cahuilla Indians recorded by early white visitors spoke of an abundance of fish and waterfowl in and about the lake whose level decreased slowly because of evaporative losses but rose rapidly during periodic inundations. At least once the Indians were forced to flee into the surrounding mountains because of the rapidly advancing waters, thus spawning a tale reminiscent of Noah and the Ark.

Here as elsewhere in the arid West, the Indians practiced irrigation. In the Coachella Valley near present-day Indio, water from an artesian well was held behind adobe walls for release through a half-mile-long ditch to the fields. On the winter day I walked along the edge of the Salton Sea hundreds of small fish lay belly-up in ditches just inland from the shore. The fish, victims of cold water, were *Tilapia.* In the waters of the sea that day a red tide, caused by a dense concentration of algae, had stifled other fish and cast a putrid smell over the area. The Salton water level, advanc-

Deserted recreational subdivision, inundated by rising agricultural waste waters emptying into the Salton Sea from the Imperial Irrigation District.

ing once again because of increasing agricultural flows from the Imperial Valley, had inundated recreation facilities and residential areas that now lay deserted. It seemed truly to be a dead sea.

Biblical overtones were also present the day I went to take a look at Laguna Salada. It was Sunday in Mexico, and as I climbed a ridge along the northeastern edge of the interior basin I could see a cross outlined against the blue sky. It was a comfortable January day and the air had been washed clean of dust by a recent rain, a rarity in that region. Ahead of me, blocking the path, stood an elderly woman clothed in a white robe whose hood lay against her back. She scurried up the trail as I approached and joined about a dozen similarly robed people clustered around the cross overlooking the 500-square-mile basin. I did not feel I was particularly welcome, so I stood apart and watched the service that seemed to go back into time. I heard snatches of songs and chants and every once in a while arms would be flung toward the sky. The wind blew gently around us on that winter day.

Like the mouth of the Colorado, from which it is separated by an offshoot of the Peninsular Range, Laguna Salada has its own tale of buried treasure. When it was connected to the sea, an early Spanish explorer is supposed to have sailed into the interior basin with a valuable cargo of pearls gathered from the lower gulf. The vessel went aground and was abandoned to the drifting sands. Like the Salton Sea basin, Laguna Salada has a history of periodic inundations. It filled, at least partially, six times between 1884 and 1928, and in 1923 a tidal range of six inches was recorded. Water was pouring into the basin when I visited it, transported by a canal dug by the Mexicans to keep the fields in the lower delta from being inundated and to provide the means for recreational development in that isolated basin. The unlined canal carried the water west around the southern tip of the Sierra del Mayor along the same gradient that was once utilized by natural overflows.

I like to think of the small stream that flows down the Cañon Guadalupe from the Sierra de Juárez and disappears into the sands on the west side of the basin as the last tributary of the Colorado River system. Photographer David Cavagnaro climbed up the streambed and wrote in *Baja California:*

A bit higher in the canyon, there was a good vantage point from which to look back down. In the middle distance, loose rock and soil that desert storms had washed down the canyon in huge quantities fanned out into the Laguna Salada; in the background the Sierra de los Cucapas rose bleakly from the flat desert floor. But here by the bubbling stream, the scene was anything but desolate; it teemed with life.

Not for long, however. As the climb continued, the stillness became palpable. The stream had disappeared—to flow underground. The high late-morning sun suddenly felt particularly hot. Another backward look down the canyon drove home the uncomfortable fact that the vast desert is all-encompassing. Without water, no plant leaves or palm fronds rustled in the soft breezes, no frogs called, no doves cooed.

It was into these two interior basins, both at one time being arms of the gulf, and into the flat plain that the sand and silt poured for millions of years. During the period of his survey, from 1891 to 1935, Sykes estimated that the Colorado River dumped more than 6.5 billion tons of silt onto its 3,325-square mile delta. The deposition rate during those forty-four years varied from a few tons a day to more than a thousand tons per second.

Sykes, as a person, was as fascinating as his comprehensive study of the delta. As with most others who spend some time on or near the river, the Colorado and particularly its delta became a passionate obsession with him. As a youth in England, Sykes read a nineteenth-century romantic novel, *The Headless Horseman,* which contained vivid descriptions of the Southwest. It was enough to inspire him to set out and follow the sun to the American West. First a cowboy in Texas, Sykes went on to a career in engineering and eventually wound up at the Tucson Desert Laboratory of the Carnegie Institute of Washington. Before joining the scientific community, Sykes built a 22-foot, shallow-draft sailboat and floated down the Colorado from Needles, California, to the gulf. There he experienced the tidal bore, visited Puerta Isabel, and eventually destroyed the boat in an accidental fire; he and his companion were forced to trek back to civilization by the same excruciating route taken later by the Chinese and Arthur Walbridge North.

Sykes, who claimed credit for introducing the English custom of afternoon tea and being "at home" to frontier Flagstaff, was not deterred. He repeated the trip a number of times with his wife and two small children, taking copious notes of what he observed in that unknown region. In his river travels, Sykes came across Indians on tule and willow rafts and a trapper who serenaded coyotes with mournful tunes played on his clarinet. There were a few Mexican cowboys below Yuma, but little else. The Englishman was on hand when the river broke through and flooded the Imperial Valley in 1905. He stopped for lunch in the valley town of El Centro, which "consisted of surveyors' pegs, magnificent distances, and the rose-tinted hopes of its proprietors." The farmers offered to trade their underwater lots for Sykes's wagon and team, but the Englishman would have none of it. Sykes's involvement with the river spanned its formative years, from its free-flowing state to nearly precise control.

Sunday religious service, on hill above Laguna Salada, which resembles the Dead Sea area.

Before he died in 1948, he poured his own concrete coffin with a glass lid to take advantage of the Arizona sun and wrote of his feelings about the river's change:

> Of course, as an Engineer, I fully appreciate the magnificent structures that have brought the lower Colorado under control, and the breadth of their conceptions and planning that made their erection possible, but I confess that I have much the same sympathy for my old friend, the sometimes wayward, but always interesting, and still unconquered and untrammeled river of the last and preceding centuries, that I have for a bird in a cage, or an animal in a zoo.

I steered the kayak toward the east bank of the Colorado and relaxed as that dim smudge of firmness gave me a sense of orientation. The fog cleared, and shortly after 9:00 A.M. Gary and I stopped to stretch our legs. We waded the few feet from the kayak to the shore with great caution, having had the unpleasant experience of beaching the craft on Isla Gore in the gulf and hopping out only to sink to our thighs in the primordial ooze of the tidal flats. This time the footing was firm. The landing had been used by fishermen and duck hunters. Rusting cans, old campfires, and expended plastic shotgun shells were lying about. We took turns using the one pair of binoculars. Neither of us was a birder, but we were learning quickly, with such a profusion of winged life around us.

A few more miles on the river, around a bend to the west, and suddenly there was no mistaking the fact we had arrived at our first destination, the start of the canal to Laguna Salada which is the last of many diversions on the river system. There was a mound of dirt that only an earth-moving machine could have produced in that reliefless terrain. The water passed into the canal through a headgate, flowed over some rocks that dissipated its energy, then shot west in an unnervingly straight line, something we were not used to after the sweeping meanders of the river above this point. Shorebirds walked across the river below the diversion. It was obvious we could not proceed in that direction.

Gary plotted our position. He estimated the canal ran 255 degrees to the west. "Seventy-nine degrees, water temperature," he announced, and then the last water sample was taken. I thought of the thirty-two-degree water temperature at the headwaters of the Green River. About a hundred yards below the last diversion, I threw a stick into the water and watched it float back upstream to where the bulk of the water was being drawn off at the headgate. The end of the river was now all too apparent. The mystery was gone, and I felt deflated. I walked downstream on cracked mudflats beside the remnant of a river for about a half mile. It became shallower and wider, a sure sign of quick death in that evaporation-prone climate.

I suddenly got the urge to walk across the Colorado River. It was something I hadn't had the chance to do since jumping across the freshets at the beginnings of the four tributaries, and not too many can boast of having walked across the main stem of the Colorado. In that setting one could imagine he was Moses, and perhaps the prophet did cross the Red Sea at low tide, while those who followed were confronted by a tidal bore. I withdrew hastily after sinking to my knees in the ooze before getting even one-third of the way across.

There seemed little else to do but walk back to the kayak. I had gone far enough on my quest for the end of the river. I did not want to see something once grand suffer the ultimate despoliation. The river had acquired the fragility and character of a human life for me after all the

The last diversion of the river, a canal that transports water into Laguna Salada.

time spent on or about it, and I felt I would be a voyeur if I followed it the remaining short distance to the bitter end.

We portaged the kayak around the headgate and paddled to the west on the strangest leg of our journey. To travel down that canal through a completely barren desert with the canal banks, ubiquitous tamarisk, and horizon merging like railroad tracks is like being one of those floating images in a Salvador Dali painting. We had not planned to end our trip this way, but it appeared to be a quicker route to the parked car than retracing our way back on the meandering river. It also appealed to my sense of the bizarre, which had been considerably heightened during the trip. Gary muttered, "It is absolutely surrealistic."

The journey by water ended at the San Felipe Highway bridge. It was a Saturday, and Mexican families were clustered on the banks hoping to catch shrimp in nets which, when thrown, fell in a graceful arc upon the

An abandoned attempt to carve out a recreational subdivision, complete with Venice-like canals, at Yomuri. Cranes guided trek.

frothy water. I hiked the few miles back across the desert to the car parked at the fishing camp used by gringos during duck season, A scene of almost absolute devastation marked the landscape at Yomuri. The land surrounding the primitive fishing camp had been stripped bare of vegetation and graded; fingerlike canals had been dug from the river. Whatever had been envisioned—perhaps a Venice of the delta—had failed. One more boom gone bust. The two tall cranes abandoned by the sides of the uncompleted canals guided me across the desert as night fell. And so it ended.

S E L E C T E D   B I B L I O G R A P H Y

I N D E X

SELECTED BIBLIOGRAPHY

The listing that follows is a highly selective and abbreviated one, and is intended to do no more than suggest the breadth of material on the subject and what aspects of it might be worth pursuing. (Just how abbreviated this selection is may be gauged from the fact that one bibliography dealing only with the lower river basin contains 1,425 entries, and even so is not complete. I know of no equivalent bibliography for the upper basin.) The entries are arranged by chapter, and represent in each case the more important sources for that chapter, excluding specialized or highly technical documents. I have omitted, for the most part, publications that are difficult to obtain, and sources of information where the pertinent material is scattered through voluminous testimony. In not a few instances, a given work listed under one chapter was quite helpful in one or more other chapters as well, but I have sought to place each entry where it seemed the most relevant.

CHAPTER ONE: WATERSHED

ALEXANDER, THOMAS G. "The Powell Irrigation Survey and the People of the Mountain West." *Journal of the West,* January, 1968.

BEALE, CALVIN L. *Internal Migration in the United States Since 1970.* Statement to the House Select Committee on Population, February 8, 1978.

BERKMAN, RICHARD L., and VISCUSI, W. K. *Damming the West.* New York: Grossman Publishers, 1973.

BILLINGTON, RAY ALLEN. *Westward Expansion.* New York: Macmillan Publishing Co., 1974.

CARLSON, MARTIN E. "William E. Smythe, Irrigation Crusader." *Journal of the West,* January, 1968.

Colorado River Basin Water Problems: How to Reduce Their Impact. General Accounting Office, Washington, D.C., May 4, 1979.

The Colorado River Region and John Wesley Powell. U.S. Geological Survey Paper 669. Washington, D.C., 1969.

Committee on Water of the National Research Council. *Water and Choice in the Colorado Basin.* National Academy of Sciences Publication 1689. Washington, D.C., 1968.

DARRAH, WILLIAM CULP. *Powell of the Colorado.* Princeton: Princeton University Press, 1951.

Department of the Interior. *The Colorado River: A Natural Menace Becomes a Natural Resource.* Washington, D.C., March, 1946.

Department of the Interior. *Westside Study Report on Critical Water Problems Facing the Eleven Western States.* Washington, D.C., 1975.

DIDION, JOAN. "Holy Water." *Esquire,* December, 1977.

FARQUHAR, FRANCIS P. *The Books of the Colorado River and the Grand Canyon.* Los Angeles: Glen Dawson, 1953.

FRADKIN, PHILIP L. "Down the Colorado: From Lively Stream to Brackish Pools." *Los Angeles Times,* August 26, 1973.

FREEMAN, LEWIS R. *The Colorado River Yesterday, Today and Tomorrow.* New York: Dodd, Mead and Co., 1923.

GIBSON, MCGUIRE. "The Breakdown of Ancient Desert Civilizations." Paper presented to the Conference on Alternative Strategies for Desert Development and Management, Sacramento, Calif., May, 1977.

HAURY, EMIL W. *The Hohokam.* Tucson: University of Arizona Press, 1976.

HEWETT, EDGAR L. *The Chaco Canyon and Its Monuments.* Albuquerque: University of New Mexico Press, 1936.

HIIBNER, CALVIN W. "Urbanization in the Colorado River Basin." Paper presented at the Colorado River Basin Environmental Conference, Salt Lake City, October, 1973.

JACOBSEN, THORKILD, and ADAMS, ROBERT M. "Salt and Silt in Ancient Mesopotamian Agriculture." *Science,* November 21, 1958.

LA RUE, E. C. *Colorado River and Its Utilization.* U.S. Geological Survey Water-Supply Paper 395. Washington, D.C., 1916.

LEOPOLD, LUNA B. *Water: A Primer.* San Francisco: W. H. Freeman and Co., 1974.

LICHTENSTEIN, GRACE. "The Battle over the Mighty Colorado." *The New York Times Magazine,* July 31, 1977.

Lower Colorado Region Comprehensive Framework Study. Main report and eighteen appendixes. Lower Colorado Region State-Federal Interagency Group, June, 1971.

MERK, FREDERICK. *History of the Westward Movement.* New York: Alfred A. Knopf, 1978.

MORRISON, PETER A. *Current Demographic Change in Regions of the United States.* Rand Paper Series. Santa Monica, Calif., 1977.

Office of the White House Press Secretary. Statement on Water Projects, with background material on the Central Arizona Project, April 18, 1977.

POWELL, JOHN WESLEY. *Report on the Lands of the Arid Region of the United States.* Edited by Wallace Stegner. Cambridge: The Belknap Press of Harvard University Press, 1962.

POWER, THOMAS M. *An Economic Analysis of the Central Arizona Project.* Phoenix: CAP Publications, 1978.

ROBINSON, MICHAEL C. *Water for the West: The Bureau of Reclamation, 1902–1977.* Chicago: Public Works Historical Society, 1979.

'75 Water Assessment, Lower Colorado Region, Summary Report, U.S. Water Resources Council, Washington, D.C., December, 1977.

'75 Water Assessment, Upper Colorado Region, Summary Report, U.S. Water Resources Council, Washington, D.C., July, 1977.

SIBLEY, GEORGE. "The Desert Empire." *Harper's,* October, 1977.

SMYTHE, WILLIAM E. *The Conquest of Arid America.* New York: Harper and Brothers, 1900.

STAMP, L. D. *A History of Land Use in Arid Zones.* UNESCO, Paris, 1961.

STEGNER, WALLACE. *Beyond the Hundredth Meridian.* Boston: Houghton Mifflin Co., 1953.

TURNER, FREDERICK JACKSON. *The Frontier in American Society.* New York: Holt, Rinehart and Winston, 1920.

UDALL, CONGRESSMAN MORRIS K. Statement to Departmental Water Projects Review Team. *Congressional Record,* March 21, 1977.

Upper Colorado Region Comprehensive Framework Study. Main report and eighteen appendixes. Upper Colorado Region State-Federal Interagency Committee, June, 1971.

VIVIAN, R. GWINN. *Prehistoric Water Conservation in Chaco Canyon.* Tucson: Arizona State Museum, 1972.

WARNE, WILLIAM E. *The Bureau of Reclamation.* New York: Praeger Brothers, 1973.

CHAPTER TWO: BEGINNINGS

ARPS, LOUISA WARD, and KINGERY, ELINOR EPPICH. *High Country Names.* Denver: Colorado Mountain Club, 1966.

BOWDEN, CHARLES. *Killing the Hidden Waters.* Austin: University of Texas Press, 1977.

CALVIN, ROSS. *River of the Sun.* Albuquerque: University of New Mexico Press, 1946.

———. *Sky Determines.* Albuquerque: University of New Mexico Press, 1948.

CORLE, EDWIN. *The Gila.* Lincoln: University of Nebraska Press, 1964.

CRAMPTON, C. GREGORY. "The Discovery of the Green River." *Utah Historical Quarterly,* October, 1952.

Final Environmental Statement, Project Skywater. Bureau of Reclamation, Washington, D.C., 1977.

GEBHARDT, DENNIS. *A Backpacking Guide to the Weminuche Wilderness.* Durango, Colo.: Basin Reproduction and Printing Co., 1976.

HASTINGS, JAMES RODNEY, and TURNER, RAYMOND M. *The Changing Mile.* Tucson: University of Arizona Press, 1965.

INGRAM, HELEN M. *Patterns of Politics in Water Resource Development: A Case Study of New Mexico's Role in the Colorado River Basin Bill.* Division of Government Research, University of New Mexico, 1969.

LEOPOLD, ALDO. *A Sand County Almanac.* New York: Sierra Club and Ballantine Books, 1970.

MICHENER, JAMES A. *Centennial.* New York: Random House, 1974.

MITCHELL, FINIS. *Wind River Trails.* Salt Lake City: Wasatch Publishers, 1975.

NICHOLS, JOHN. *The Milagro Beanfield War.* New York: Ballantine Books, 1976.

WALKER, BRYCE S. *The Great Divide.* New York: Time-Life Books, 1973.
ZWINGER, ANN. *Run, River, Run.* New York: Harper and Row, 1975.

CHAPTER THREE: HIGH COUNTRY

Alternative Plans for Water Resource Developments, Green River Basin, Wyoming. Bureau of Reclamation, Salt Lake City, May, 1972.
ATHEARN, FREDERIC J. *An Isolated Empire.* Cultural Resource Series, No. 2. Bureau of Land Management, Denver, 1976.
ATHEARN, ROBERT G. *The Coloradans.* Albuquerque: University of New Mexico Press, 1976.
BOLTON, HERBERT E. *Coronado: Knight of Pueblos and Plain.* Albuquerque: University of New Mexico Press, 1949.
BROWN, F. LEE, and others. "Some Remarks on Energy Related Water Issues in the Upper Colorado River Basin." *Natural Resources Journal,* October, 1977.
CLAWSON, MARION. *The Western Range Livestock Industry.* New York: McGraw-Hill Book Co., 1950.
Colorado River System Consumptive Uses and Losses Report 1971–1975. Bureau of Reclamation, Washington, D.C., 1977.
Colorado State Water Plan, Phases I and II. Colorado Water Conservation Board and the Bureau of Reclamation, Denver, 1974.
The Colorado Water Study, Directions for the Future. Office of the Executive Director, Colorado Department of Natural Resources, Denver, October, 1978.
DeVOTO, BERNARD. *The Easy Chair.* Boston: Houghton Mifflin Co., 1955.
FRADKIN, PHILIP L. "Craig, Colorado: Population Unknown." *Audubon,* July, 1977.
———. "Energy Search: Atomic Drilling Plan Spreads Shock Waves." *Los Angeles Times,* February 4, 1973.
———. "King of the Mountain." *West,* January 28, 1968.
———. "Lost Dutchman Breeds Violence." *Los Angeles Times,* May 31, 1970.
———. "The Eating of the West." *Audubon,* January, 1979.
GATES, PAUL W. *History of Public Land Law Development.* Public Land Law Review Commission, Washington, D.C., November, 1968.
GOWANS, FRED R., and CAMPBELL, EUGENE E. *Fort Bridger, Island in the Wilderness.* Provo, Utah: Brigham Young University Press, 1975.
HYMAN, SIDNEY. *The Aspen Idea.* Norman: University of Oklahoma Press, 1975.
LARSON, T. A. *History of Wyoming.* Lincoln: University of Nebraska Press, 1978.
McCARTHY, G. MICHAEL. *Hour of Trial.* Norman: University of Oklahoma Press, 1977.
Modernization of 1872 Mining Law Needed to Encourage Domestic Mineral Production, Protect the Environment, and Improve Public Land Management. General Accounting Office, Washington, D.C., July 25, 1974.
MORGAN, DALE L. *Jedediah Smith and the Opening of the West.* Lincoln: University of Nebraska Press, 1953.
One Third of the Nation's Land. Report to the President and the Congress by the Public Land Law Review Commission, Washington, D.C., 1970.
O'REAR, JOHN and FRANKIE. *The Aspen Story.* New York: A. S. Barnes and Co., 1966.
PURDY, WILLIAM M. "Green River: Main Stem of the Colorado." *Utah Historical Quarterly,* July, 1960.

Revision of the Mining Law of 1872. Report of the Senate Committee on Energy and Natural Resources, April, 1977.

Serving Two Masters: A Common Cause Study of Conflicts of Interest in the Executive Branch. Common Cause, Washington, D.C., 1976.

STEGNER, WALLACE. *The Uneasy Chair.* Garden City, N.Y.: Doubleday and Co., 1974.

TWAIN, MARK. *Roughing It.* New York: New American Library, 1962.

VOIGT, WILLIAM. *Public Grazing Lands.* New Brunswick: Rutgers University Press, 1976.

Water and Related Land Resources, Yampa River Basin, Colorado and Wyoming. Colorado Water Conservation Board and Department of Agriculture, Denver, 1969.

WEATHERFORD, GARY D., and JACOBY, GORDON C. "Impact of Energy Development on the Law of the Colorado River." *Natural Resources Journal,* January, 1975.

The Wyoming Framework Water Plan. Wyoming State Engineer's Office, Cheyenne, 1973.

YOUNG, OTIS E. *Western Mining.* Norman: University of Oklahoma Press, 1970.

CHAPTER FOUR: SEPARATE NATIONS

ABBEY, EDWARD. *Slickrock.* San Francisco: Sierra Club Books, 1971.

ALEXANDER, THOMAS G. "An Investment in Progress: Utah's First Federal Reclamation Project." *Utah Historical Quarterly,* Summer, 1971.

American Indian Policy Review Commission. *Final Report.* Washington, D.C., May 17, 1977.

ARRINGTON, LEONARD J., and BITTON, DAVIS. *The Mormon Experience.* New York: Alfred A. Knopf, 1979.

BERKHOFER, ROBERT F. *The White Man's Indian.* New York: Alfred A. Knopf, 1978.

CALLAWAY, D. G., and others. *The Effects of Power Production and Strip Mining on Local Navajo Population.* Lake Powell Research Project Bulletin No. 22. Los Angeles, 1976.

CATHER, WILLA. *Death Comes for the Archbishop.* New York: Random House, 1971.

CHAMBERS, REID PAYTON, and PRICE, MONROE E. "Regulatory Sovereignty: Secretarial Discretion and the Leasing of Indian Lands." *Stanford Law Review,* May, 1974.

Final Environmental Impact Statement, Proposed Kaiparowits Project. Bureau of Land Management, Washington, D.C., 1976.

Final Environmental Statement, Bonneville Unit, Central Utah Project. Bureau of Reclamation, Washington, D.C., 1973.

GREGORY, HERBERT E., and MOORE, RAYMOND C. *The Kaiparowits Region.* U.S. Geological Survey Professional Paper 164. Washington, D.C., 1931.

Issues Paper: The Central Utah Project, Part I. Citizens for a Responsible CUP, Salt Lake City.

JACKSON, RICHARD H. *Righteousness and Environmental Change: The Mormons and the Environment.* Charles Redd Monographs in Western History, no. 5. Provo, Utah: Brigham Young University, 1975.

KLUCKHORN, CLYDE, and LEIGHTON, DOROTHEA. *The Navaho.* New York: Natural History Library, 1962.

MacMeekin, Daniel H. "The Navajo Tribe's Water Rights in the Colorado River Basin." Unpublished paper of Navajo legal services lawyer, April, 1971.

Native Americans and Energy Developments. Anthropology Resource Center. Cambridge, Mass., 1978.

Nelson, Lowry. *The Mormon Village.* Salt Lake City: University of Utah Press, 1952.

Nelson, Michael C. *The Winters Doctrine: Seventy Years of Application of "Reserved" Water Rights to Indian Reservations.* Arid Lands Resource Information Paper No. 9. Tucson: University of Arizona Office of Arid Land Studies, 1977.

O'Dea, Thomas F. *The Mormons.* Chicago: University of Chicago Press, 1975.

O'Neill, Floyd A., and MacKay, Kathryn. *A History of the Uintah-Ouray Ute Lands.* American West Center. Salt Lake City: University of Utah Press. n.d.

Price, Monroe E., and Weatherford, Gary D. "Indian Water Rights in Theory and Practice: Navajo Experience in the Colorado River Basin." *Law and Contemporary Problems,* Winter, 1976.

Reserved Water Rights for Federal Indian Reservations: A Growing Controversy in Need of Resolution. General Accounting Office, Washington, D.C., November 16, 1978.

Smith, Henry Nash. *Virgin Land.* Cambridge: Harvard University Press, 1950.

A Study of Administrative Conflicts of Interest in the Protection of Indian Natural Resources. Prepared for the Senate Subcommittee on Administrative Practice and Procedure, Washington, D.C., 1971.

Toward Economic Development for Native American Communities. A compendium of papers submitted to the Joint Subcommittee on Economy in Government, Washington, D.C., 1969.

Underhill, Ruth M. *The Navajos.* Norman: University of Oklahoma Press, 1956.

U.S. Supreme Court. *Arizona* v. *California.* Reporter's Transcript, pages 2433–2498, July 6, 1956.

CHAPTER FIVE: CANYONLANDS

Austin, Mary. *The Land of Little Rain.* New York: Ballantine Books, 1971.

Collins, Robert O., and Nash, Roderick. *The Big Drops.* San Francisco: Sierra Club Books, 1978.

Crampton, C. Gregory. *Land of Living Rock.* New York: Alfred A. Knopf, 1972.

———. *Standing Up Country.* New York: Alfred A. Knopf, 1973.

Draft Environmental Statement, Proposed Colorado River Management Plan, and various technical reports. National Park Service, Grand Canyon, 1977.

Dutton, Clarence E. *Tertiary History of the Grand Canyon District.* Salt Lake City: Peregrine Smith, 1977.

Garrett, W. E. "Are We Loving It to Death?" *National Geographic,* July, 1978.

Goldwater, Barry M. *A Journey Down the Green and Colorado Rivers.* Phoenix: H. Walker Publishing Co., 1940.

Hughes, J. Donald. *In the House of Stone and Light.* Grand Canyon Natural History Assn., 1978.

Jacoby, Gordon C., and others. "Law, Hydrology, and Surface Water Supply in the Upper Colorado River Basin." *American Resources Bulletin,* October, 1976.

Kolb, Ellsworth L. *Through the Grand Canyon from Wyoming to Mexico.* New York: Macmillan Co., 1928.

Lake Powell: Jewel of the Colorado. Bureau of Reclamation, Washington, D.C., 1965.

LEOPOLD, LUNA B. *The Rapids and the Pools—Grand Canyon.* U.S. Geological Survey Professional Paper 669. Washington, D.C., 1969.

LEYDET, FRANÇOIS. *Time and the River Flowing.* San Francisco: Sierra Club Books, 1964.

MCPHEE, JOHN. *Encounters with the Archdruid.* New York: Farrar, Straus & Giroux, 1971.

MANN, DEAN E., and others. *Legal-Political History of Water Resource Development in the Upper Colorado River Basin.* Lake Powell Research Project Bulletin No. 4. Los Angeles, September, 1974.

NASH, RODERICK. *Wilderness and the American Mind.* New Haven: Yale University Press, 1967.

———, ed. *Grand Canyon of the Living Colorado.* New York: Sierra Club and Ballantine Books, 1970.

PORTER, ELIOT. *The Place No One Knew.* San Francisco: Sierra Club Books, 1963.

POWELL, JOHN WESLEY. *The Exploration of the Colorado River and Its Canyons.* New York: Dover Publications, 1961.

REILLY, P. T. "How Deadly Is Big Red?" *Utah Historical Quarterly,* Spring, 1969.

ROOSEVELT, THEODORE. *A Book-lover's Holidays in the Open.* New York: Charles Scribner and Sons, 1919.

STANTON, ROBERT BREWSTER. *Down the Colorado.* Norman: University of Oklahoma Press, 1965.

STRATTON, OWEN, and SIROTKIN, PHILIP. *The Echo Park Controversy.* The Inter-University Case Program No. 46. University of Alabama Press, 1959.

A Survey of the Recreational Resources of the Colorado River Basin. National Park Service, Washington, D.C., 1950.

WALLACE, ROBERT. *The Grand Canyon.* New York: Time-Life Books, 1973.

WATKINS, T. H., ed. *The Grand Colorado.* Palo Alto, Calif.: American West Publishing Co., 1969.

CHAPTER SIX: DESERTS

Arizona State Water Plan, Phases I and II. Arizona Water Commission, Phoenix, 1975 and 1977.

BARR, JAMES L., and PINGRY, DAVID E. "The Central Arizona Project: An Inquiry into Its Potential Impacts." *Arizona Review,* April, 1977.

BROWNE, J. ROSS. *Adventures in the Apache Country.* New York: Harper and Brothers Publishers, 1871.

BROWNELL, HERBERT, and EATON, SAMUEL D. "The Colorado River Salinity Problem with Mexico." *American Journal of International Law,* June, 1975.

CABRERA, LUIS. "Use of the Waters of the Colorado River in Mexico: Pertinent Technical Commentaries." *Natural Resources Journal,* January, 1975.

Carl T. Hayden, Memorial Addresses and Tributes. Senate Document No. 92-68. Washington, D.C., 1972.

CASEY, HUGH E. *Salinity Problems in Arid Lands Irrigation.* Tucson: University of Arizona Office of Arid Lands Studies, 1972.

ELY, NORTHCUTT. *Light on the Mexican Water Treaty from the Ratification Proceedings in Mexico.* A report to the Colorado River Water Users' Assn., Washington, D.C., 1946.

FAULK, ODIE B. *Arizona: A Short History.* Norman: University of Oklahoma Press, 1970.

Final Environmental Statement, Central Arizona Project. Bureau of Reclamation, Washington, D.C., 1972. Also later statements for specific units.

Final Environmental Statement, Colorado Basin Salinity Control Project, Title I. Bureau of Reclamation, Washington, D.C., 1975.

FURNISH, DALE B., and LADMAN, JERRY R. "The Colorado River Salinity Agreement of 1973 and the Mexicali Valley." *Natural Resources Journal,* January, 1975.

HOFSTADTER, DAN. *Mexico 1946–1973.* New York: Facts on File, 1974.

HOLBURT, MYRON B., and VALENTINE, VERNON E. "Present and Future Salinity of Colorado River." *Journal of Hydraulic Division, Proceedings of the American Society of Civil Engineers,* March, 1972.

HUNDLEY, NORRIS. *Dividing the Waters.* Berkeley: University of California Press, 1966.

———. "The Colorado Water Dispute." *Foreign Affairs,* April, 1964.

IVES, JOSEPH C. *Colorado River of the West.* Washington, D.C.: Government Printing Office, 1861.

JOHNSON, K. F. *Mexican Democracy: A Critical View.* Boston: Houghton Mifflin Co., 1971.

JOHNSON, RICH. *The Central Arizona Project.* Tucson: University of Arizona Press, 1977.

KAHRL, WILLIAM L., ed. *The California Water Atlas.* The Office of Planning and Research, Sacramento, 1978.

LINGENFELTER, RICHARD E. *Steamboats on the Colorado River, 1852–1916.* Tucson: University of Arizona Press, 1978.

MANN, DEAN E. "Politics in the United States and the Salinity Problem of the Colorado River." *Natural Resources Journal,* January, 1975.

MARTIN, WILLIAM E. "Economic Magnitudes and Economic Alternatives in Lower Basin Use of Colorado River Water." *Natural Resources Journal,* January, 1975.

MEREDITH, H. L. "Reclamation in the Salt River Valley, 1902–1917." *Journal of the West,* January, 1968.

MEYERS, CHARLES J., and NOBLE, RICHARD L. "The Colorado River: The Treaty with Mexico." *Stanford Law Review,* January, 1967.

NADEAU, REMI A. *The Water Seekers.* Salt Lake City: Peregrine Smith, 1974.

Need for Controlling Salinity of the Colorado River. Colorado River Board of California, Los Angeles, 1970.

Options for Resolution of the United States–Mexico Colorado River Salinity Problem. Federal Interagency Task Force, Washington, D.C., 1972.

OYARZABEL-TAMARGO, FRANCISCO, and YOUNG, ROBERT A. "International External Diseconomies: The Colorado River Salinity Problem in Mexico." *Natural Resources Journal,* January, 1978.

STANLEY, MILDRED. *The Salton Sea Yesterday and Today.* Los Angeles: Triumph Press, 1966.

TAYLOR, PAUL S. "Public Policy and the Shaping of Rural Society." *South Dakota Law Review,* Summer, 1975.

———. "Water, Land, and Environment in Imperial Valley: Law Caught in the Winds of Politics." *Natural Resources Journal,* January, 1973.

WATERS, FRANK. *The Colorado.* New York: Holt, Rinehart and Winston, 1946.

WRIGHT, HAROLD BELL. *The Winning of Barbara Worth.* New York: A. L. Burt Co., 1911.

YATES, RICHARD, and MARSHALL, MARY. *The Lower Colorado River: A Bibliography.* Yuma: Arizona Western College Press, 1974.

CHAPTER SEVEN: ENDS

BAEGERT, JOHANN JAKOB. *Observations in Lower California.* Berkeley: University of California Press, 1952.

BROWNE, J. ROSS. *Resources of the Pacific Slope.* New York: D. Appleton & Co., 1869.

FLINT, TIMOTHY. *The Personal Narrative of James O. Pattie.* Philadelphia: J. B. Lippincott Co., 1962.

GERHARD, PETER, and GULICK, HOWARD E. *Lower California Guidebook.* Glendale, Calif.: Arthur H. Clark Co., 1970.

JOHNSON, CHARLES GRANVILLE. *History of the Territory of Arizona and the Great Colorado of the Pacific.* San Francisco: Vincent Ryan and Co., 1868.

JOHNSON, WILLIAM WEBBER. *Baja California.* New York: Time-Life Books, 1972.

KNIFFEN, FRED B. *The Natural Landscape of the Colorado River.* Berkeley: University of California Publications in Geography, 1932.

NORTH, ARTHUR WALBRIDGE. *Camp and Camino in Lower California.* New York: Baker and Taylor Co., 1910.

SYKES, GLENTON G. "Five Walked Out." *Journal of Arizona History,* Summer, 1976.

SYKES, GODFREY. *The Colorado Delta.* American Geographical Society Special Publication No. 19. Washington, D.C., 1937.

———. *A Westerly Trend.* Tuscon: Arizona Pioneer Historical Society, 1944.